7/12 - do not withdraw
- classic

WITHDRAWN

A Social History of Engineering

TECHNOLOGY TODAY AND TOMORROW

Edited by P. F. R. Venables, Ph.D., B.Sc., F.R.I.C.

Principal of the College of Advanced Technology, Birmingham

*

A SOCIAL HISTORY OF ENGINEERING

W. H. G. Armytage, M.A., *Professor of Education in the University of Sheffield*

A HISTORY OF ELECTRICAL ENGINEERING

P. Dunsheath, C.B.E., D.Sc., M.A.

A HISTORY OF MECHANICAL ENGINEERING

A. F. Burstall, D.Sc., Ph.D., *Professor of Mechanical Engineering and Director of the Stephenson Engineering Laboratories at King's College, University of Durham*

THE PROFESSIONAL TECHNOLOGIST

P.F.R. Venables, Ph.D., B.Sc., F.R.I.C.

A
SOCIAL HISTORY
OF
ENGINEERING

W. H. G. ARMYTAGE

*Professor of Education
in the University
of Sheffield*

PITMAN PUBLISHING CORPORATION
NEW YORK

Printed in Great Britain

'I heard, then, that . . . in Egypt, was one of the ancient gods of that country, the one whose sacred bird is called the ibis, and the name of the god himself was Theuth. He it was who invented numbers and arithmetic and geometry and astronomy, also draughts and dice, and, most important of all, letters. Now the king of all Egypt at that time was the god Thamus. . . . To him came Theuth to show his inventions, saying that they ought to be imparted to the other Egyptians. But Thamus asked what use there was in each, and as Theuth enumerated their uses, expressed praise or blame, according as he approved or disapproved. The story goes that Thamus said many things to Theuth in praise or blame of the various arts, which it would take too long to repeat; but when they came to the letters, "This invention, O king," said Theuth, "will make the Egyptians wiser and will improve their memories; for it is an elixir of memory and wisdom that I have discovered." But Thamus replied, "Most ingenious Theuth, one man has the ability to beget arts, but the ability to judge of their usefulness or harmfulness to their users belongs to another; and now you, who are the father of letters, have been led by your affection to ascribe to them a power the opposite of that which they really possess. For this invention will produce forgetfulness in the minds of those who learn to use it, because they will not practise their memory. Their trust in writing, produced by external characters which are no part of themselves, will discourage the use of their own memory within them. You have invented an elixir not of memory, but of reminding; and you offer your pupils the appearance of wisdom, not true wisdom, for they will read many things without instruction and will therefore seem to know many things, when they are for the most part ignorant and hard to get along with, since they are not wise, but only appear wise."'

Plato *Phaedrus* 275 (translated H. N. Fowler).

'Shall we say, for example, that Science and Art are indebted principally to the founders of Schools and Universities? Did not Science originate rather, and gain advancement, in the obscure closets of the Roger Bacons, Keplers, Newtons; in the workshops of the Fausts and the Watts; wherever, and in what guise soever Nature, from the first times downwards, had sent a gifted spirit upon the earth?'

Thomas Carlyle, 'Signs of the Times'
Edinburgh Review xlix (1829), p. 449

'Is it their luck, or is it in the chambers of their brain,—it is their commercial advantage, that whatever light appears in better method or happy invention, breaks out *in their race*. They are a family to which a destiny attaches, and the Banshee has sworn that a male heir shall never be wanting.'

R. W. Emerson, *English Traits* (Boston, 1856), p. 97

Preface

This book has three aims: to chart technological developments with especial reference to Britain, to indicate how they have affected and been affected by social life at certain stages and to offer some clues as to the origins of innovations and institutions. So that readers can answer other, perhaps more relevant, questions selected bibliographical references are given for each chapter. These amplify statements and facts in the text and indicate (as far as possible) sources of further information. The professional institutions mentioned in Appendix B are most kind in lending books or giving information.

There is also need to abstain now, more than ever, from the retailing of packaged history and predigested fact. This, it is hoped, is more of a sketch map of some determinants of contemporary life. Contemplating the activity of others in this field, I am only too conscious of being amongst the epigoni, following up and making small parcels of their work. Dedications are often missed, but I would like to offer this to members of the Newcomen Society and especially to the late Engineer-Captain E. C. Smith, R.N., who, with his fellow enthusiasts, have done so much for the history of engineering. The frequency with which the abbreviation *T.N.S.* occurs in the references indicates my debt to their *Transactions*. Other abbreviations include:

A. of S.	= *Annals of Science*
Econ. H.R.	= *Economic History Review*
J.H.I.	= *Journal of the History of Ideas*
J.T.H.	= *Journal of Transport History*
N. & R.R.S.	= *Notes and Records of the Royal Society*
Proc. Roy. Soc.	= *Proceedings of the Royal Society*

Unless otherwise stated, books listed in the chapter bibliographies are published in London.

Preface

To the general editor of the series, Dr. P. F. R. Venables, I am grateful for much advice and help; to my colleagues in the Faculties of Science and Engineering who so willingly open up discussions on some of the themes in this book I am continuously obliged; and to the group of typists, Miss Hilary Crowson, Miss Barbara Maydew, Mrs. Jean Kitson and Miss Winifred Syner who grappled with the original manuscript I am very much indebted. This should have been a better reward for all their efforts.

Contents

11

Illustrations

13

Illustrations

14

Illustrations

15

Illustrations

16

CHAPTER 1

From Stone to Metal

At the end of the second ice age, 370,000 years ago, the mortal remains of a recognizable human were fortunately preserved for us in a cave near Pekin. The flakes of stone and burned bones which surrounded him show that fire and tools had been discovered. Fire enabled him to live in cold countries, prepare food, frighten animals away and lighten the gloom of his caves. In time he could use it to make pottery and stabilize his life. Its symbolic importance can be seen in the legend of Prometheus, who saved the human race from destruction and gave them fire stolen from heaven. Tools were no less important. As *homo faber*, man the maker, his first tools were of stone. Thus the multi-millennial stages of his existence before recorded history are called paleolithic, mesolithic and neolithic.

It is only in our own time, with the acquisition of recently discovered techniques, like radioactive carbon dating, aerial photography and pollen analysis, that the hidden tracts of man's history are being fitfully and intermittently illuminated. Recent discoveries in China should warn us against assuming that man's evolution is any other but a global matter. But since this book is basically concerned with Britain, revealing evidence from other lands must be omitted unless it is relevant to our story.

Consecutive phases of development in these early times are deceptive and confusing. Cultures described as paleolithic, mesolithic and neolithic flourished at different times and in different places. So for convenience it is perhaps best to begin our story in 3000 B.C.

THE URBAN REVOLUTION

By 3000 B.C. the communities cultivating the alluvial plains of the Nile, the Tigris-Euphrates and the Indus valleys were producing so much foodstuff that they could use labourers to dig canals, employ artisans to manufacture tools, and support merchants who bought other produce. They used sails to propel boats, oxen to draw ploughs, wheels to bear carriages and metals to fabricate tools and facilitate arduous manual operations. This conglomerate of skills enabled them to foregather in towns in Sumeria, Egypt and India.

The Sumerians who long before 3000 B.C. had drained marshes along the lower Euphrates near the Persian Gulf, irrigated land by canals and used animals to draw wheeled carts, foregathered in Ur, Kish, Erech and Nippur. Though conquered by Sargon (2637–2582 B.C.) their culture was strong enough to mould his Semitic followers. The great founder of Babylon, Hammurabi (1728–1686 B.C.), became the master of Mesopotamia and it is from his reign that most of our information comes, through the script of wedge-shaped signs which was developed from pictographs on clay. Sumerian mathematics was based on a sexagesimal system due to the fact that they thought there were 360 days in the year. We follow them in dividing a circle into 360 degrees. From their brick towers or ziggurat (the best known being the Ziggurat of Ur, excavated by Sir Leonard Woolley, and not unconnected with the Tower of Babel mentioned in *Genesis* ii, 1–9) priests could and did make astronomical observations.

The Egyptians built even more impressive pyramids and obelisks in the Nile Valley. Beginning with King Zozer of the Third Dynasty (*c.* 3000 B.C.) whose 200-foot step-pyramid at Saggara near the old capital, Memphis was followed a century later by Khufu (*c.* 2900 B.C.) or Cheops of the Fourth Dynasty, who built one some 480 feet high. For this, the largest building up to that time, relays of 100,000 men were conscripted every three months by Cheops who was so obsessed that he sold his daughter's virtue for money to finish it. His daughter built a small pyramid of the stones given her by her lovers. His brother Chephren built another.

Fourteen centuries later, in the Eighteenth and Nineteenth Dynasties (1580–1205 B.C.) obelisks were built from granite quarried at Aswan, just below the first cataract. In the quarry there lies an obelisk (rejected because of imperfections) some 137 feet long, weighing over a thousand tons. So stupendous was the task of building these that the engineer architects responsible for them were often given tombs in the necropolis and statues in the temple. We know the names of two of them, Senmut and Beknekhonsu. Senmut built the great temple of Deis-al-Bahari and several obelisks for Queen Hatshepsut (1495–1475 B.C.) and Beknekhonsu built, a century later, the obelisk

18

that was taken from Luxor by the French naval engineer, J. B. A. Lebas, and reconstructed in Paris in A.D. 1836. Another obelisk from Heliopolis was re-erected on the London embankment in 1878, and a third in New York in 1881.

The Egyptians probably opened up copper mines in the Sinai Peninsula as early as the third millennium B.C., whilst their quarries at Turra near Cairo show that they gave up open mining in favour of shafts. Shafts were also used for wells: Joseph's well at Cairo goes 300 feet into the earth. The beginning of civilization among peoples coincided with mining: a science which gradually evolved and spread. These mines were uniformly worked by slaves, obtained by war. Sometimes they worked with their bare hands, to judge by the clay which is found in ancient mines bearing the impress of thousands of fingers. More often they used hammers and wedges, bones and horns. They tunnelled and constructed galleries, often so narrow that children had to be used to squeeze through them. For blasting they used fire and vinegar: a technique which survived to Hannibal's time. The Egyptians also had windlasses to draw up the water in the mines.

The metals so obtained, gold, silver, lead, iron, and later copper and tin were used in various ways. Legend, in the person of Herodotus, had it that the Ethiopian kings bound their prisoners with gold chains. The Argonauts, who in about 1350 B.C. raided Colchis in Georgia, were looking for gold. Silver was used for handles on shields.

From before 1000 B.C. sixteen papyri on Egyptian mathematics are extant: Alimose the scribe, who compiled one of the most important of them in 700 B.C., began his work with the words 'Rules for enquiring into nature, and for knowing all that exists', and discusses fractions, division and easy equations with unknown quantity.

The papyrus on which it was written is itself a technological advance, for it is the root of our modern paper. In addition to this, glass making was practised with materials from the Caucasus, fabrics were woven, cisterns dug, mining practised and metals smelted with the aid of blowpipes and bellows.

Rivers are liable to flood and crops must be planted. To plan these things in advance a calendar is needed, and in Egypt one such exists— an official calendar of 365 days and a lunar calendar of 364. (In Mesopotamia they clung to the lunar calendar for all purposes.) To co-ordinate the two, careful observation of the moon and the sun were necessary and by 747 B.C. the Assyrians had managed to draw up tables. This in turn stimulated the study of mathematics, and the construction of time pieces like the water clock and the sundial.

From Stone to Metal

Whilst the riverine cultures were well advanced in the Bronze Age, Britain underwent a neolithic revolution. The neolithic culture took root in Britain on the Downs and uplands from Sussex to East Devon and is perhaps best typified by Windmill Hill at Avebury: a stable camp. Picks and rakers made of antlers and shovels from the shoulder blades of oxen enabled them to mine flints with success. These flint-miners of Norfolk, Sussex and Wessex were able to barter their products. Gordon Childe has called them 'the founders of British industrialism'.

About 2000 B.C., other small bands of colonists landed on the western coasts and left their mark in the form of large chambered tombs. They were followed about 1750 B.C. by a Megalithic religion which found expression in collective tombs—long pear-shaped barrows which are often 300 feet long and at Maiden Castle extend for a third of a mile.

These were in turn invaded about 1750 B.C. by the Beaker folk who imposed upon Great Britain and Ireland a remarkable uniformity of religion and culture. They brought the use of metals with them and beads of Egyptian design. Their great monuments Avebury and Stonehenge indicate this. Stonehenge is actually built of spotted dolerite brought from the Prescilly mountains in South Wales, a distance of 145 miles, even if brought overland.

This sarsen stone structure at Stonehenge is the major structural feat up to this time in Britain. The fact that the stones were brought from Wales is a further indication of the trade routes that existed between these Beaker folk. The skilful dressing of these stones, their erection, the mortising of the lintels and their raising have excited engineers and archaeologists and the reader is recommended to the masterly summary by R. J. C. Atkinson for the best account of their speculations and findings.

Not the least of the intriguing phenomena is the carving of what is virtually a Mycenaean dagger, which proves an interesting link with another culture where architectural progress was outstanding and deserves a brief mention.

MYCENAE

Mycenae was a famous Greek city, the golden town of the Cyclops, mythologically connected with the labours of Hercules and his taskmaster, Eurystheus, associated with Perseus and the House of Acresius and known, through Homer, as the stronghold of Agamemnon. Thanks to the excavations of Schliemann (in A.D. 1877) and subsequently of Tsountas and Wall, we know that the citadel was first occupied in the early Bronze Age, and it was growing in power at the time Stonehenge was being built. The Princes of Mycenae developed the copper mines, built a new palace, enlarged the citadel and exhaled an influence over the Greek mainland to Egypt.

They built an underground cistern outside the town with secret access from within so that the inhabitants could have a secure supply of water. The heyday of Mycenae, from 1400–1100, saw the construction of the galleries on a cantilever system with elaborate counterweights. The Treasury of Atreus shows evidence of having been planned beforehand with great care. Of its construction A. J. B. Wace remarked, 'The unknown master of the Bronze Age who designed and built the Treasury of Atreus deserves to rank with the great architects of the world.' Certainly the engineering skill which enabled these Mycenaeans in their Aegean fortress to control flood waters and divert streams which threatened their town leaves us with little doubt of the experimental skill that existed towards the end of the second millennium B.C.

THE RISE OF THE SMITH

The next major event, about 1400 B.C., was the production of iron in a cheap economical manner by the Hittites in Asia Minor. They could write, and brought with them new wares and new gods from beyond the Black sea. In Anatolia they had copper, silver and lead, and built cities with stout masonry walls. Their power was at its height about 1350 B.C., controlling Syria and threatening Egypt. By 1200 B.C. iron technologies spread through the Euphrates valley, especially among the Assyrians who, beginning with Ashurnasipal II (884–859 B.C.), forged the empire of Ninevah. Under Ashur-bani-pal (668–625 B.C.) this reached its finest flower. This empire was destroyed in turn by the Chaldaeans, whose most notable ruler was Nebuchadnezzar (605–551 B.C.). In turn the Chaldaeans fell under Persian rule for two centuries (536–332 B.C.).

The march of man was punctuated by conquests, not only of his fellows, but of nature. The rotary hand mill, the use of horses and chariots, the chain of pots and the pulley are all characteristic of this Asian culture.

21

From Stone to Metal

THE PHOENICIAN ALPHABET

Meanwhile along the eastern shores of the Mediterranean a subject race of the Egyptians achieved emancipation. The downfall of Mycenae (when the Dorians overran the Peloponnese and islands as far as Crete) enabled this emancipated race, the Phoenicians, to establish a trading network that soon straddled the Mediterranean with factories.

Centred at Tyre (*Ezekiel* 27: 13–25), their 'factories' or 'colonies' competed with the Greeks and later with the Romans. Carthage, their first settlement on the African coast, was established early in the first millennium B.C., and they settled on both the east and west coasts of Spain. From Egypt, Mesopotamia and Arabia they obtained goods which they sold. They also, by their ubiquity, established a consonantal alphabet, an indispensable corollary of trade.

Their greatest service was in the transmission of technological ideas. The half-legendary Cadmos, son of a Phoenician king, is said to have brought the art of mining to the Greeks, being the first to work the gold and silver mines of Pangaion in Macedonia. Thasos, another Phoenician prince, worked gold mines in an island of the North Aegean named after him. Cyprus became in Sarton's words 'the metallurgical centre of the Mediterranean' where such engineers and builders as Eupalinus (*c.* 600 B.C.) made their names.

Before the Persians embarked on their conquests there was, in the Aegean coast of Asia Minor, the trading colony of Miletos in Ionia where Thales advised the Ionian cities to federate against the Persian threat. Thales (*c.* 624–548/5 B.C.) was a Phoenician by ancestry who had assimilated Egyptian learning. Sarton sees in him 'the first man in any country to conceive the need for geometric proposition'. Thales was the founder of magnetism, for he knew of the properties of the lodestone. He held that water is the original substance. His friend Anaximander (610–545 B.C.), some fifteen years his junior, went on to write the first treatise on natural philosophy in the history of mankind, and also tried to define the primary substance which he called *apeiron* (or the undetermined). Anaximenes, who followed him, thought of the stars as situated on a rolling sphere.

CHAPTER 2

The Greeks

SAMOS

The first known civil engineer in history was Eupalinus of Megara, who built an aqueduct at Samos. According to Herodotus, this was one of the three greatest works to be seen in any Greek land. The aqueduct was a double-mouthed channel pierced for a hundred and fifty fathoms through the base of a high hill, seven furlongs long, eight feet high and eight feet wide. Other Samians whose ingenuity excited Herodotus were Rhoicos, who built the greatest temple Herodotus had seen, and Theodoros, who built another at Ephesus for the worshippers of Artemis. Samos sheltered Herodotus, the historian, after he had been expelled from Halicarnassus. Herodotus had visited Egypt and Tyre, and became the great historian of antiquity.

Samian engineers were employed by others. Darius I, King of Persia (521–485 B.C.), employed Mandrocles of Samos to bridge the Bosphorus for him with boats when he marched against the Scythians and made him a handsome gift as a reward. The Persians themselves acquired similar skills and Darius' successor, Xerxes (485–465 B.C.), employed two Persians, Artachaies and Bubares, to cut a 2,500 yard canal across the Athos peninsula from north to south to save his fleet. When Artachaies died soon after, he was buried with great ceremony, as befitted a great engineer.

THE FIRST TOWN PLANNER

The epitome of Greek civilisation, especially in the years 480–431 B.C., was Athens, head of the Ionian League in its life and death struggle with the Persians. As with Venice, two thousand years later, the focal point of its effort was its harbour—the Peiraeus, geometrically planned before 466 B.C. by Hippodamus of Miletus.

23

Hippodamus was so capable that in 433 B.C. when Pericles of Athens colonised Thurii on the Gulf of Tarentum in Lucania he sent Hippodamus with Herodotus, the historian, and Lysias, the orator, to lay out the town. Hippodamus has been vividly described by Aristotle (*Politics* II 8, 1267B) as 'a strange man, whose fondness for distinction led him into a general eccentricity of life, which made some think him affected (for he would wear flowing hair and expensive ornaments; but these were worn on a cheap but warm garment both in winter and summer); he, besides aspiring to be an adept in the knowledge of nature, was the first person not a statesman who made enquiries about the best form of government'. Perhaps that is why he was exiled.

ATHENS: CRUCIBLE OF IDEAS

Here, in Athens, Miletan and Pythagorean scientific thought was systematized and clarified, though the thinkers were not always popular: Anaxagoras, who taught that the sun was a mass of red hot metal larger than the Peloponnese, had to flee for his life. Democritus of Abdera (born c. 460 B.C.), the great universal debunker and atomic theorist, and the first Greek scientist known to have visited Babylon in person, found, when he came to Athens, that no one knew him. And, tantalizingly enough, we know just enough about him today to wish we knew far more, for his intuitive grasp of atomic theory is astonishing. Even an Athenian like Euctemon felt that his native city was so uncongenial that he left to become a colonist in Amphipolis. Meton (c. 430 B.C.), the astronomer who had an enlightened scheme of calendar reform based on the 19-year cycle, was the subject of Aristophanes' wit in his plays, the *Clouds* (423 B.C.) and the *Birds* (414 B.C.). Aristophanes also parodied Socrates in the *Clouds* because Socrates had shown a keen interest in astronomy and physics. But in old age, however, Socrates challenged the experimental point of view and started a conflict between science and philosophy by declaring it was impious to investigate what God had not intended us to know.

Socrates had in turn a pupil, Plato, who was even more vehemently opposed to practical testing of hypotheses by mechanical devices. To Plato it destroyed all the good of geometry. Practical manipulation was vulgar, banausic, and fit for slaves. One might have heard more of Eudoxus (c. 409–356 B.C.) and Archytas, Plato's Pythagorean friends, had he not discouraged them from putting some of their theories about the propagation of sound to the test by building apparatus.

When Plato (429–347 B.C.) established his Academy outside Athens in 385 B.C., he placed over it the inscription: 'Let no man destitute of geometry enter my doors.' To encourage reasoning, he forbade his adult pupils to use anything more than a ruler and compasses. In the *Philebus* (55 D) Socrates says, 'if arithmetic, mensuration and weighing be taken away from any art, that which remains would not be much'. Unfortunately Plato and his followers tended to turn their backs on the Pythagorean school and regard pure numbers and pure geometry as the only worthy subjects of study. 'I cannot think', says Socrates, this time in the *Republic* (529 A), 'of any study as making the mind look upwards except one which has to do with unseen reality.'

Aristotle (384–322 B.C.), whose opinion of Hippodamus we have quoted, gave us the classification of the sciences in which we habitually work. Aristotle joined Plato's Academy in 367 B.C., at the age of 18. Invited by Philip of Macedon, in 349 B.C., to undertake the education of his 13-year-old son Alexander, he returned to Athens 13 years later to found the Lyceum. Here a library was collected and research was organized, and the first history of science was compiled.

ALEXANDER AND THE DIFFUSION OF THE GREEK SPIRIT

Aristotle's pupil, Alexander the Great, went on to build a great army in which he included Aristotle's nephew, Callisthenes, and surveyors like Baeton, Diognetus, Philonides and Amynites. With this army Alexander conquered Asia Minor and marched triumphantly to India. As the missionary of Greek culture by founding (or refounding) a number of cities, 70 in all (of which 25 are verifiable), he won the reputation of being the greatest city-builder of all time. Some stemmed from existing towns like Alexandrettei, Troy and Smyrna; others were new cities like Alexandria in Egypt and Chodjend. Not all of them were on the sea, nor did they all have a Greek population. Through these cities Alexander hoped to fuse Europe and Asia, for through them Greek culture was to be (and was) diffused through Asia Minor, and even to India. More than anyone else, Alexander made Greek science and the Greek language universal in his day and age.

Apart from Plato's Academy and Aristotle's Lyceum, the Greeks had no institutions where scientific thought could be collected or collated. Aristocrats by temper, the Greeks discouraged *hubris*, the intellectual arrogance of those who investigated heavenly mysteries. They did not wish to improve upon the technical achievements of the past. Moderation was the theme of Aristotle's teaching; moderation and a sense of value. As he said in his *Metaphysics* (980 d): 'But as more arts were invented, and some were directed to the necessities of life, others to recreation, the inventors of the latter were naturally always regarded as wiser than the inventors of the former, because their branches of knowledge did not aim at utility.'

To the Greeks life was to be understood, not changed. Nor could projects for lightening human labour find favour when they had slaves producing, in the mines of Laurion, the economic lubricant of their society: silver.

THE MUSEUM AT ALEXANDRIA

No Athenian inhibitions against practical experiments were evident in Alexandria, where, after Alexander's death, one of his generals, Ptolemy, set up as an independent king of a part of Alexander's empire. Ptolemy invited Strato of Lympsacus, head of Aristotle's Lyceum in Athens, to take over the Museum at Alexandria.

This was the first example of research being supported by the state. Learning, in the shape of libraries, had been patronized for centuries in the east, but the Museum, with half a million books, became the largest ever. Its lecture rooms, zoo, observatory, a garden and dissecting rooms, provided employment for over a hundred teachers. From here, for a century and a half, a regular intellectual dynasty was nourished. Four only need be mentioned: Euclid, Aristarchus of Samos, Eratosthenes and Archimedes. Euclid (*c.* 300 B.C.) superseded the work of earlier mathematicians by his *Elements*. Aristarchus of Samos (*c.* 310–230 B.C.) sustained the heliocentric hypothesis that the 'fixed stars and the sun remain unmoved, and that the earth revolves about the sun in the circumference of a circle, the sun lying in the middle of the orbit'. Eratosthenes of Cyrene (*c.* 275–194 B.C.) was an all-rounder who made the first scientific attempt to fix the dates of political and literary history, and calculated the circumference of the earth and the magnitude and distance of the sun and moon.

The greatest of them all, however, was Archimedes (*c.* 287–212 B.C.), who was a friend of Eratosthenes. He invented a water screw, a technique for moving great weights by a small force (and established the theory of the lever), experimented in specific gravity (and thereby outlined a science of hydrostatics). After studying in Alexandria he returned to Sicily, where King Hiero employed him as a military adviser. In this capacity, Archimedes launched a galley by means of pulleys, and constructed a battery of military machines to defend Syracuse. He died at the hand of a Roman soldier, and hoped that on his tomb there would be carved a sphere in a cylinder with the formula for the volume by which the cylinder exceeded the sphere.

Conic sections, the basis of all geometry up to the time of Descartes in the seventeenth century, were first systematized by Apollonius of Perga (*c.* 240 B.C.) who worked on the subject as left him by Menaechmus (a pupil of Plato), Euclid and Archimedes. His work survives.

The end of the Alexandrine period threw up Hipparchus (died *c.* 125 B.C.) who abandoned the heliocentric theory in favour of a geocentric theory. As a compensation for this retrogressive step he outlined the procession of the equinoxes and mapped the position of nearly a thousand stars. It was he who, in the course of his astronomical observations, drew up a table of chords of arcs in a circle subtending angles of different sizes. This, in the hands of others, was to become trigonometry.

EARLY TECHNOLOGISTS

Three vague and shadowy figures tried to apply their scientific knowledge. They were toy-makers rather than technologists and were Alexandrians rather than Athenians. Ctesibius of Alexandria, who is generally thought to have lived at the beginning of the second century B.C., compressed air to work a gun, built a water clock and a water organ and sprayed fire through a hose. Philon of Byzantium followed in his footsteps and was the author of an encyclopaedia of applied mechanics. Hero of Alexandria, probably active in the first century B.C., constructed pneumatic devices like fountain syphons, water organs and ingenious toys of all kinds, including a much discussed steam turbine called an aeolipile. His work was known and translated in medieval Sicily and has excited the interest of historians of the steam engine for many years. He gave an account of five simple machines which became the basis of early technology: the wedge, the pulley, the lever, the wheel and axle, and the endless screw.

But it is not for these shapes of things to come that the Greeks should be remembered. Five centuries of deductive reasoning from Thales to Hipparchus sharpened the very language of science and engineering. Though in their mathematics they used letters of the alphabet as numbers, yet they gave the name *mathema* to science. So also their capacity for *analusis* (loosening) a problem before *synthesis* (or putting it together) endowed modern mathematicians with the techniques of their trade. *Proposition* was Pythagoras' name for statement and proof of a geometric fact, as *parabole, ellipsis, hyperbole* were for relationships of figures equal, shorter, or longer than each other. As for Aristotle's work, the classifications which he set posterity lasted for nearly two millennia. Indeed, it used to be said that science itself, up to the fifteenth century, was a series of footnotes to his work.

CHAPTER 3

The Romans

The Romans had one advantage over their predecessors: a good cement. The volcanic ash of Vesuvius, mixed with their lime, gave them pozzuolana. This enabled them to pioneer new constructional techniques like the barrel vault. They also seem to have made efforts to train their engineers. Crassus, Julius Caesar's early friend, is said to have had a technical school for slaves, and so had Trajan, Severus, Constantine, Julian and Justinian.

THE FLAT CAMPAGNA

One of the earliest Roman engineers was the blind Appius Claudius who in the days of the Republic was in charge of public works. He built the Via Appia as far as Capua, and paved it. Described by Statius as 'the queen of roads', it was laid on a substructure of heavy stones, cemented with lime mortar and surfaced with gravel. It later was extended to Brindisium. This gave access to the rich country of the Campagna, the wide plain which surrounded Rome. This plain was first spanned by the same man with the Aqua Appia (312 B.C.) which brought pure water from the Anio and the hills south of the town, and was the first of fourteen systems supplying water to the capital. It was eleven miles long and, like previous aqueducts at Samos and Pergamon, was mainly underground.

The second century B.C. witnessed an extraordinary outburst of such buildings. This manipulation of bold masonry arches in them was successfully extrapolated to bridges. The Pons Aemilius was arched in 142 B.C., two years after the Aqua Marcia (third of the capital's aqueducts) was constructed with 18-foot arches. Arched bridges like the Pons Mulvius (109 B.C.) and the Pons Fabricus (62 B.C.) lengthened in span from 60 to 80 feet. Sewers, like the Cloaca Maxima, harbours (like Ostia), temples, baths, triumphal arches and

29

even private houses, all show the Roman arch in its multiple exploitations.

The flat Campagna also encouraged the building of amphitheatres. Sulla's Colony in 80 B.C. at Pompeii had an amphitheatre, and a second was built at Pozzuoli. The first to be built in Rome was of wood in 46 B.C. They spread all over the empire in the towns which the Romans planted and planned, like Turin (26 B.C.) and Aosta (25–23 B.C.). The largest of all was the Flavian Amphitheatre, begun by Vespasian A.D. 69–79 and dedicated by Titus in A.D. 80. This, usually known as the Coliseum, had walls of mortar and could hold 45,000 people. It is their greatest example of co-ordinated support of an auditorium on a framework of corridors and staircases, based on barrel vaults.

VITRUVIUS

Vitruvius is perhaps the best known of the Roman engineer architects. He flourished under Julius Caesar and Augustus. His famous book *De Architectura* is dedicated to Augustus between the years 29 B.C. and 11 B.C. Architecture was a science to him: 'Architecti est scientia pluribus disciplinis et variis eruditionibus ornata' (The science of the architect is embellished with many disciplines and divers skills). He designed the Basilica of Fano which, as both Choisy and Revoira remark, 'reveals talent of an exceptional kind'. His work attracted great attention in the Renaissance and after. He said that the architects of Rome had to be of good birth and sound training, and from him we learn that they were also the military engineers who would construct siege works, engines of war, camps, roads, earth works, drains, market places, baths and amphitheatres. He also mentions Cossutius, who erected the temple of Olympian Zeus in Athens at the request of Antiochus Epiphanes (175–164 B.C.), and Gaius Mucius who designed the Temple of Honour and Virtue in Rome.

WATER ENGINEERING

Rome was a great consumer economy and its largest consumption was of water. Baths, thermae and grottoes, as well as household needs, were using up some 300 million gallons daily in the time of Nerva (A.D. 97). To provide this, nine aqueducts, some 250 miles in length, were built, the fifth—some 50 miles—being on stone arches. Nerva's surveyor of aqueducts, Frontinus, wrote a treatise describing them which became a classic. Two more, making eleven, were built after his time. These aqueducts which straddled the Campagna were copied in the Roman provinces. At Nîmes the Pont du Gard was 160 feet high, at Segovia and Tarragona they were 102 feet and 83 feet high respectively.

Under the Republic the censors looked after the water supply of the city. Agrippa, who built the Aqua Julia and the Aqua Virgo, trained a body of slaves to maintain the aqueducts which, on his death, in 12 B.C., became the property of the emperor Augustus. Augustus appointed a board of three senators called the *curatores aquarum* to assume responsibility for them. Thus the senate were responsible for their maintenance, as they became also for temples and public buildings. Tiberius also appointed a senatorial commission to deal with the flooding of the Tiber. Roads too were given at first to the senate but by the Flavian period they were sustained by the imperial treasury. In no way were the Romans more ingenious (to use the word in its true Latin sense) than in building aqueducts. The 100-mile aqueduct that carried water from the Jebel to Leptis Magna (the modern Homs) in Libya also carried olive oil. The oil was simply poured with the water in the aqueduct at the Jebel end, flowed down to Leptis where (since water and oil are immiscible) it was skimmed off and put into a separate container for export.

PUBLIC BUILDINGS

Trajan, who succeeded Nerva in A.D. 98, commissioned Appolodorus of Damascus (who had designed the Drobetae bridge across the Danube in the second Dacian Campaign of A.D. 104) to build for him a magnificent Forum in Rome. It was laid out like a Roman Camp and was one of the grandest architectural compositions ever realized.

Arches were the underlay of the Pantheon, a round domed structure 142 feet in diameter and 140 feet high, with an eight-columned portico at the front. The dome—the biggest yet constructed—was supported not by an uninterrupted wall but on eight piers of great strength and resistance linked by arches thrown across them. Probably the work of Valerius or of Decrianus of Ostria, it marks the culmination of Roman ingenuity, for the tremendous thrust of the dome has withstood eighteen hundred years of the depredations of man and nature.

Constructed under Hadrian in the years A.D. 120 and 124 to replace an original structure of Marcus Agrippa, it was repaired in A.D. 202 and abandoned in A.D. 399. In A.D. 608 it was consecrated as a Christian church and the Pope had twenty-eight cartloads of early martyrs' bones buried under the fabric. In A.D. 663 the gilded bronze tiles of the rotunda were stripped from the dome by the Byzantine Emperor Constans II, and a later Pope was subsequently forced to cover it with lead. By the eleventh century it had become a fortress. In A.D. 1632 another Pope stripped 200 tons of bronze beams from the portico roof so that Bernini could make a new baldaquin for St. Peter's and eighty cannons for the papal armoury. It has always excited admiration and emulation. Among the admirers was Sebastian Serlio, whose *Regole Generali de Architectura* (A.D. 1540) was a standard manual for European builders and who remarked:

'Among all the ancient buildings to be seen in Rome, I am of the opinion that the Pantheon (for one piece of work alone) is the fairest, wholest, and best to be understood; and is so much the more wonderful than the rest because it hath so many members, which are all so correspondent one to the other.'

The emulators included Palladio, Bernini, Flitcroft, Decimus Burton and the builder of the Jefferson monument in the U.S.A. The Pantheon embodied Roman civilization as the Parthenon had embodied the Greek.

ROMAN BRITAIN

Active as they were above ground, the Romans seemed singularly lethargic below it. To many Romans, the classification animal, vegetable or mineral would have little meaning, since many of them believed that metals grew like vegetables. Iron was so scarce and valued that engagement rings were often made from it, even for the wealthiest girls. Nor did they like mining operations, and preferred to use criminals to undertake them, which did not make for improvements.

In no part of the Roman Empire was there a more active mining industry than in Britain. According to Tacitus, one of their historians, it was why they invaded Britain in A.D. 43. R. G. Collingwood, after examining all the archaeological evidence, concludes that coal was extensively mined.

The three legions and their auxiliary troops stationed in Britain stimulated other native industries. Potteries (as at Holt in Denbighshire), smithies (as at Templeborough, near Sheffield), a gold mine (at Dolaucothy), lead mines (in Derbyshire), tin (in Cornwall), fulling centres (at Darenth in Kent and Chedworth in Gloucester), uniforms (at Winchester); all these developed over three centuries. Southern England also exported grain to Gaul.

An early mechanical harvester of the second century A.D. discovered in Belgium: one of the Roman granaries. It was pushed through the corn by oxen pushing from behind.

Liburna or animal-propelled paddle boat as outlined by "The Anonymous".

MANPOWER

Apart from transport difficulties (which the Romans tried to overcome by making roads), their main problem was manpower. This was ample enough when the Empire was establishing itself. But the plague of A.D. 165–6 ravaged the Empire for fifteen years and initiated a downward trend in the population. This was accentuated by rebellion, civil war, barbarian invasions and a further plague which hit the country in the time of Marcus Aurelius (A.D. 121–80).

Diocletian (A.D. 284–305), in trying to recruit 150,000 new troops to the 500,000 already in service, drew still further on the farm labouring class and probably the ensuing hardships made it impossible for the peasantry to raise families large enough to maintain the population at its existing level. For the Roman legionary served for 20 years, and his obligations were hereditary. By A.D. 367 legionaries were so scarce that the minimum height requirements were lowered by 3 inches from 5 foot 8 inches to 5 feet 5 inches. By the time of Honorius in A.D. 406 even slaves were allowed to enlist.

Man-power shortage spread to the ancillary services like the minters (*monetarii*), coin circulators (*collectarii*), armourers (*fabricenses*), silk workers (*gynaeciarii*), clothing workers (*textores*), collectors of shell fish for dyemaking (*murileguli* and *conchylileguli*), miners (*metallarii*) and transport workers (*bastagarii*).

The six western mints at Rome, Aquileia, Siscia, Lugdunum, Arles and Treves were staffed by guilds whose duties were obligatory and hereditary. Seventeen silk works and two linen factories also worked under the same compulsive discipline. The *murileguli* had even more onerous obligations: if they failed to produce the required quota of shell fish their property was confiscated, and if their daughters bore sons these sons were compelled to join the guild of shell collectors.

The *fabricenses* or armament workers were branded on the arm so that if they deserted their factories they could be quickly identified and brought back. Miners, especially of gold, had to be bribed to undertake service and preferred to abscond to distant parts rather than to undertake such work. The manpower shortage was not remedied by ordinances and laws, so the government had to take over the superintendence of enterprises formerly private. Building suffered so much that Constantine I ordered the colleges of *dendrofori* (woodsmen) and *centonarii* (makers of mats) to unite with the *fabri* (building trades) to make up their numbers.

THE RISE OF INGENIUM

Ingenium, as an accepted term, was being used by Tertullian (an early father of the Christian Church) to describe a warlike machine that was the product of invention as early as A.D. 200. In pacific operations, like milling, water power was being employed in spite of the opposition of the millers' guilds. From the fourth century B.C. there has been a marked development of the type of mill with a horizontal shaft, driven by an undershot waterwheel. Since it was familiar to Vitruvius in *De Architectura* (x. v. 2), it is known as Vitruvian mill, though Vitruvius described it as a machine rarely used in his day.

There were water mills and saws for cutting marble near Treves about A.D. 370 and they seem to have become standard from the fourth century onwards. They were to be found in fifth-century Athens as well as Byzantium, where overshot variants began to develop. Called *hydraulae* by Vitruvius, these mills were later called *aquariae molae* by Palladius in the fourth century and recommended as a replacement for animal and slave labour. When they became common the name denoting the motive power dropped and the word used was *mola, molina* or *molendinum,* from which comes our modern word mill.

'THE ANONYMOUS' AND LABOUR-SAVING

To combat this manpower shortage, an anonymous author who lived in the second half of the fourth century A.D. put forward a variety of suggestions for mechanizing the army. He had seen the Danube, was interested in his country's war against the Persians and, in his concern at the shortage of labour, drafted a memorandum to the emperor, probably hoping to secure employment as a technical adviser. Written probably between A.D. 366 and 375 and entitled *De Rebus Bellicis,* this memorandum was, in the opinion of Professor Thompson, 'probably intercepted by a civil servant and pigeon-holed'. In them, the Anonymous stressed the utility (*utilitas*) of his proposals, which included the reform of the mint, a curbing of imperial expenditure and several labour-saving devices.

He can claim to be the first man to canvass the potentialities of propelling a ship without either oars or sails by means of a paddle-wheel, for one of his proposals was for a *Liburna* or war ship in which oxen walk round a rotating capstan to drive six paddle-wheels. Other labour-saving devices like a scythed chariot, a portable bridge and various kinds of ballistic engines are also explained. The ballistic engines are of two types: the *ballista quadrirotis*, or a mobile field army assault weapon, and the *ballista fulminalis*, for static use. The latter was powerful enough to send its arrow-like missiles across the Danube. The propulsive force of this artillery was not the torsion of gut or rope or human hair (hitherto the standard technique among the Greeks and Romans) but a bent bow of iron or steel (*arcus ferreus*).

These steel bow-like *ballistae* were later to be employed by the great Belisarius in his defence of Rome in A.D. 536–8. The *de rebus bellicis* is apparently the oldest extant work to mention this invention, and its author may at least be credited with an attempt to introduce a machine that was remarkably successful when used one hundred and sixty or seventy years later.

THE FALL OF THE ROMAN EMPIRE IN THE WEST

In the same decade as this ambitious technological memorandum was composed and ignored, two-thirds of the Roman army of the East, its ablest generals, and the Emperor Valens himself were destroyed by the Visigoths at Adrianople in A.D. 378. This battle marks a watershed between the ancient and medieval worlds, for the Roman foot soldiers were beaten by Gothic cavalry. The Goths were henceforward accepted as self-governing allies by Theodosius (A.D. 378–95) and, at his death, the Roman Empire was permanently divided into Eastern (centred at Constantinople) and Western (centred at Ravenna) parts. By A.D. 410 Visigoths were marching through the streets of Rome. As the Roman Empire of the West split up into the German kingdoms, that of the East became more and more of an oriental monarchy, fed economically, technologically and artistically from Syria, Egypt and Anatolia, and they in turn fed from cultures farther East.

Byzantine engineering was defensive and religious. The great wall of Constantinople was built. Half a century later the Emperor Zeno commissioned Theodore, King of the Ostrogoths, to reconquer Italy. Cities (especially their walls) were restored, aqueducts, baths, amphitheatres and roads were rebuilt. The chief architect, Aloisius, was instructed by Theodoric 'to translate my projects into accomplished realities . . . which shall be the admiration of new generations of men. It will be your duty to direct the mason, the sculptor, the painter, the worker in stone, in bronze, in plaster, in mosaic. What they know not you will teach them. The difficulties which they find in their work you will solve for them. What various knowledge you must possess, therefore, thus to instruct artificers of many sorts. If, however, you can direct their work to a good and satisfactory end, their success will be your eulogy and will form the most abundant and flattering reward you could desire.'

At Constantinople he employed two Asian architects, Anthemius of Tralles and Isodore of Miletus, to build the Cathedral of St. Sophia. In less than five years, A.D. 532–7, this vast brick rectangle was crowned by an enormous dome resting on four great arches. It housed splendid mosaics—which the Turks later covered over with whitewash.

Justinian (A.D. 527–65) strengthened the protective walls of towns, repaired their aqueducts and replenished their reservoirs, baths and fountains. Tripolitania, Leptis Magna and Sabratha (founded by Severus) were improved and many new cities were established. Procopius, the private secretary and legal adviser to Belisarius described in *The Buildings* their vast underground cisterns, aqueducts and public buildings. Justinian tried to rehabilitate his authority in the west by employing Belisarius with heavy Asian cavalry (*Cataphracti*) armed with bows.

When besieged in Rome by the Goths, who in A.D. 536 cut the water which drove mills on the Janiculum Hills, Belisarius transferred them to the Tiber where they were slung on a bridge of boats.

There is a story that Justinian was the first to attempt to cultivate raw silk in Europe in A.D. 530. Someone smuggled silkworm eggs from Khotan in a hollow stick. But the main source of raw silk, as of a number of other things, remained in the Far East till the twelfth century.

CHAPTER 4

The Legacies of the East

THE CHINESE

It is to the Chinese, according to Dr. Needham, that we must look for the conversion of rotary into reciprocating motion, as effected by the crank. They also used a trip hammer worked by a mill. They were also aware of the property of pivoted lodestones, to point in the direction of the Great Bear. By the sixth century they realized that this pointing property was possessed by pieces of iron stroked by the lodestone. Here we have what seems to be the origin of the compass. They are also credited with the discovery of the sternpost rudder, an indispensable adjunct of a bamboo raft or its descendant —the junk.

Bamboos in the Chinese economy had a multiple function: sawn up in individual lengths and filled with a nitrous mixture obtained from salt pans or manured ground, they made fire crackers; hence yet another Chinese legacy to the West—gunpowder. The first reference to this is in A.D. 850 and by A.D. 1100 they were giving clear descriptions of bombs and grenades, and by A.D. 1044 the first recorded prescription for it.

The Chinese had been making paper from vegetable fibres in the first century B.C., and printing on it arose consequentially. Thanks to the Arabs paper arrived in Europe in the twelfth century, and the art of printing with movable metal types arrived later.

Perhaps their most significant contribution was the horse collar. This opened to the West a new source of power, for horses are more rapid and more efficient than oxen though needing greater care. Neither the Greeks nor the Romans could really master them: hooves split and harnesses tended either to strangle or dissipate the power of a team. Hence where oxen could not serve, slaves did. With a collar

37

(which reached the West in the ninth century A.D.) the horse could breathe freely and throw his whole body into the work. In tandem harness several horses could draw great weights. Nailed horseshoes prevented slipping and extended the period of usefulness. As Lynn White remarked: 'Taken together these three inventions suddenly gave Europe a new supply of non-human animal power, at no expense or labour. They did for the eleventh and twelfth century what the steam-engine did for the nineteenth.'

THE INDIANS

India was an intellectual entrepôt as far back as the time of Alexander the Great; who, on his arrival at Taxila in the spring of 326 B.C., encountered an institution which, with its school of philosophy and medicine, was virtually a university. It had been virtually a university for three hundred years before he arrived, and was to continue so for twice that time after he was dead. As it declined, an equally if not more famous institution arose at Nalanda, which impressed the Chinese savant Huien Tsiang, and attracted Japanese and Korean students as well. Indian schools of medicine established such a reputation that Haroun-al-Rashid the celebrated Caliph of Baghdad, not only sent men to India to study but invited Indian doctors to his capital, where he fostered the translation of Indian medical works into Arabic.

Copious clinical accounts of the ways in which the Christian West was suckled on Arabian science have indicated that much early technology (and even some of Greek science) was refracted to the Arabs from India *via* Ujjain, Marw, Bactria and Sogdiana.

The magnificent temples of Elephanta and Ellora were fine expressions of the Hindu renaissance. Also a virtual revolution in mathematics was initiated by the introduction of zero. From the fifth century A.D. with scientists like the two Aryabhatas and Uirahamihira, to the seventh, with Brahmagupta, Hindu mathematicians had succeeded in the perfection of a number system that was to reach the West *via* Syria. The first mention of the zero in the west was by a Monophysite bishop in Syria called Severus Sebockt. The Indians also developed a method of dealing with unknown quantities, which takes its name from the title of the Moslem al-Khwarizmi's encyclopaedia *Hisab al-Habr w-al-Mugabala:* algebra. *Gebr* is the operative word describing the restoration of something broken. Thus algebra is the Spanish and Portuguese for a bone-setter. Algebra still retains its original function of completing an expression, since al-Kwarizimi outlined it as a device with which to calculate inheritances.

Three major artefacts have their roots in the far east. The windmill seems to have been inspired by the Tibetan prayer wheel, the steam engine by the air guns used in Malayan jungles and the concept of perpetual motion to have been clearly formulated by the Indian astronomer and mathematician Bhāskāra in his Siddhānta Siromani (c. 1150). These and other ideas continued to be carried to Europe by slaves brought in Genoese ships to Italy right up to the fifteenth century.

THE ARABS

Traffic is an Arabian word and it was by traffic with the Hindu and Chinese world that their contributions to European science and technology were obtained.

The Arabs succeeded the Persians (who first used sand bags for defence) in a wave of conquest on the Mediterranean littoral that began in A.D. 641–2 with the capture of Alexandria and ended with the wresting of the whole of North Africa from the Byzantine Empire.

With the Persian and Byzantine provinces they had conquered they took over Hellenistic science. There was even a mosque built in Constantinople under Leo III (A.D. 717–41). Arab pirates in Crete (which they had captured in the eighth century A.D. following Cyprus in the seventh) made life very insecure for Byzantines. Incidentally, the Arabs stimulated the Byzantine Emperors to build a navy. The Byzantine ships used 'Greek Fire', said to be an invention of Callicinus of Heliopolis in the seventh century. With its help the Arabs were prevented from adding Constantinople to their large conquests, though they assaulted it in A.D. 670–7 and again in A.D. 718.

In Arabian capital towns like Baghdad, and to a lesser degree Jundishapur, the Caliphs subsidized translations of Greek scientific works. A bureau of translation, Dar el Hikhma, was founded by al-Mamun where not only Greek works were arabized but Persian and Indian ones too. Utility, not culture, seems to have animated this salvage operation, hence most of the Greek texts transmitted to the West before the fifteenth century were mainly concerned with mathematics. For Arab merchants in their computations used Hindu numerals, and Arab officials in their surveys developed trigonometry. Arab harems (and the hot weather) called for the distillation of perfumes on a large scale by the Arabian kerotakis or alembic, an early still.

39

These stills represent a notable phase of chemical development. The great alchemist Jabiribn Hayyan (A.D. 721–?), a notable figure in Baghdad at the court of Haroun-al-Rashid (who presented a water clock to Charlemagne), tried to combine sulphur and mercury and obtained cinnabar. He discovered nitric acid and recommended that:

'The first essential in chemistry is that you should perform practical work and conduct experiments, for he who performs not practical work nor makes experiments will never attain to the least degree of mastery. . . . Scientists delight not in abundance of materials, they rejoice only in the excellence of their experimental methods.'

Razi (A.D. 866–?) was the first to list equipment for melting metals and to attempt a systematic classification of chemical substances that had been defined by experiment under the headings animal, vegetable, mineral and derivative. They found native aluminium sulphate (alum of Yemen) which they treated with stale urine and other materials to make alum—a 'fixer' for dyes. The production of alum was, in fact, the earliest chemical industry.

As warriors the Arabs needed swords to fight with and damask for ostentatious display. Damascus supplied both, whilst Fustat, a suburb of Cairo, provided fustian. Farther east, the Persian tafta became well known in the West as taffeta. Since eye disease in tropical countries was endemic, their physicians studied eyes and Ibn-al-Haitham's *Optical Thesaurus* (c. A.D. 1038) contains a remarkable recognition of the eye as a lens. The study of optics was to lead to other extensions of man's visual powers in the future. By A.D. 1065 the first Arab university was founded in Baghdad. One of its greatest teachers was Omar Khayyam. Just as his name has become familiar, so have other Arabian words like cheque, lamp, calibre, magazine and tare.

CORDOBA: THE TRANSMITTER TO THE NORTH WEST

At the western end of the Arabian empire, Cordoba in Spain (conquered in the eighth century), became under the Omayyad Caliphs from A.D. 928–1031 the most civilized city in Western Europe. It contained 70 libraries and 900 public baths. The Emperor Constantine Porphyrogenitus sent 140 columns to the Spanish Caliph Abd-er-Rahman III, who at that time was building Medinat ez-Zahra, his palace.

The sugar cane which the Arabs introduced to Palestine, spread to the Greek Islands, Sicily and Spain. Sugar itself is an Arab word, for the West had used honey for sweetening. Irrigation spread in Spain and Italy with the Arabian techniques of branching the life-giving waters. Here not only their desert experience but the machinery of the shadbof and the waterwheel was made available.

Arabian astronomers in Spain, like Al-Zarkali (d. A.D. 1087) wrote on the Astrolabe, and others compiled tables based on the meridian of Toledo, which was to be for some time the chief meridian of reference in Europe. With the recapture of Toledo in A.D. 1085, a number of Arabian writings fell into Christian hands, and were translated into Latin. Gerard of Cremona had a passion for such translation and before he died two years later gave Euclid's *Elements* and Ptolemy's *Almagest* to the West. He was followed by Englishmen like Adelard of Bath (who translated the first fifteen books of Euclid) and Robert of Chester (who translated al-Khwarizimi's algebra). Thus the commingling of the Mediterranean and the North West European civilizations increased to the advantage of the latter.

THE BARBARIAN ADVANCES IN THE NORTH WEST

As all these ingenious devices were developing in China, India and the Mediterranean, the barbarians who had taken over the Western Roman Empire were themselves improving the technology of the Romans. As Lynn White has written:

'Our view of history has been top-lofty. . . . In technology, at least, the Dark Ages mark a steady and uninterrupted advance over the Roman Empire. Evidence is accumulating to show that a serf in the turbulent and insecure tenth century enjoyed a standard of living considerably higher than that of a proletarian in the reign of Augustus.'

The colder latitudes of the north stimulated the making of butter instead of olive oil and the cultivation of rye, oats, spelt and hops. These foods demanded a new type of agriculture, which in the north was the basic industry. As early as the first century B.C. the northern wheeled plough was used in Britain. With its coulter, horizontal share and mould-board it greatly facilitated the cultivation of rich, heavy soils, saved labour but demanded greater power. By the eighth century there was a system of rotation of crops through the three field system. This, in the opinion of many, was really responsible for the enhanced importance of the people living on the plains of Northern Europe.

These modifications had affected even the native inhabitants of

Rome itself. The use of trousers and boots (*tracae vel tzangae*), though forbidden in Rome at the close of the fourth century, spread. So did the habit of allowing the hair to grow and of wearing fur garments (*indumenta pellium*), though attempts were made to stop such practices.

Even more startling was the 'barbarian' habit of washing with soap. The Fanti of West Africa and the Gauls of the first century A.D. apparently discovered soap independently. For although the preparation of alkaline dyes from the ashes of plants was known from early time, its preparation with tallow and vegetable oils seems to have been confined to medicinal ointments. To Pliny (*Natural History*, Book XXVIII, Cap. 51) soap was 'an invention of the Gauls for giving a reddish tint to the hair' and a propaedeutic for scrofulous sores. For the Romans used fullers earth for cleansing, and cosmetics for adornment. Not until the time of Aretaeus (*c.* A.D. 250) is soap mentioned as being used for washing clothes, and then only by the Gauls, though Aretaeus recommended it for bathing. By the seventeenth century, however, the importance of manufacture of soap was recognized by the formation of a guild of soap makers in Italy.

The people of North-West Europe also learned to convert the energy of the wind by sails. On their seaboards square-rigged ships, with masts stepped in the body of the ship, tacked their way across the channel and up and down the coast. Oarsmen here, as in the Mediterranean (where fore and aft rigs were being used), became superfluous. These two styles were to combine by the fifteenth century. The sail was also used to trap energy for the mills, especially in flat countries where the rivers were too sluggish for the Vitruvian mill. Both sailing ship and windmill extended the Northerners' command over nature rather than peoples, and a vigorous multi-national culture grew up in Europe. The wind-powered ship traversed great distances in less time and stimulated ancillary technologies of working in wood, and, as we shall see, instrumentation. As late as the nineteenth century, blocks and tackle for sailing ships were to engage the attention of engineers like the elder Brunel.

Thanks to the ship the north-western nations penetrated the Mediterranean ever more intensively. Perhaps the most dramatic of these penetrations were the Crusades to wrest the Holy Places from the Arabs. These were to have their effect on focal structures of the Christian world: its churches.

CHAPTER 5

Anglo-French Civilization, 1100-1399

CASTLE AND CATHEDRAL

One of the possible results of the Crusades was the introduction of the pointed arch to western Christendom. Easily adaptable, it keyed the new cathedrals which, during the twelfth century, were built in France and England. The Abbey Church of St. Denis (1137) was the first of these 'Gothic' structures, characteristic in that great space and height are enclosed so aesthetically and economically. To the pointed arch was added the flying buttress and vault ribs, with the result that large windows now became possible.

The stimulus to build came partly from the feudal system (castles) and partly from religion (churches, abbeys and cathedrals). Castles presented problems of mass and size: walls 21 feet thick (as at Langres) or stretching for 2½ miles (as in the defences of Verona) are known. Churches, abbeys and cathedrals presented problems of height and strength and, if the term be not misunderstood, of grandeur. They were now increasingly built of stone, and their size and splendour were accelerated by the quarries opened up and by the increasing wealth of communities around them. Of the 80 cathedrals and 500 churches built in France during the years 1170 to 1270, Chartres cathedral (built between 1195 and 1240) is perhaps the finest of its kind.

Inevitably problems of technique arose. Consultations between experts are recorded as to whether arches were possible. Piles were driven. 'Claving grains' of stone were known. Blocks and pulleys, windlasses and even inclined tramways on the scaffolding were used to move materials. Measurement and calculation involved in the works are recorded and G. P. Jones has said that 'there can be no doubt about the capacity of medieval architects to make such geometrical drawings as were required'. The choir of St. Peter's at

Beauvais rose to a height of 150 feet and only with the advent of steel has that height been exceeded. The Architect of Cambrai cathedral was Villard de Honnecourt, whose sketchbooks of things that he had noticed in his travels between 1243 and 1251 cover a number of surveying problems such as techniques of measuring distances, and structural details of vaults. He also notes a number of interesting power devices. One was a saw mill driven by water, another a machine to afford perpetual motion, and a third a device to enable an angel to keep his finger always pointing towards the sun.

The building of cathedrals and castles generated the first theoretical treatise on structural analysis by Jordanus Nemorarius (*c.* 1237). His *Elementa de Ponderis*, though elementary, contains a discussion of statics.

Since the large windows of the cathedrals needed glazing to keep out the draughts as well as to admit the light, they were built up from small pieces of glass strengthened by leaden ribs. Glass-making in England was encouraged by Norman workers like Laurentius Vitrearius, who settled at Dyer's Cross near Pickhurst, about 1226. The industry developed, especially in the Weald, and enjoyed virtually a national monopoly both of window and vessel glass until the sixteenth century when the Venetians and Lorrainers arrived.

Lenses and Instruments

These windows acquired didactic functions too. Stained glass techniques stimulated, not only work on pigments and practical chemistry, but on glass itself, especially lenses. Robert Grosseteste (1168–1253), who held that the laws of optics were the basis of all natural explanation, was the first medieval writer to discuss this subject systematically in considering the properties of mirrors and lenses. His pupil Roger Bacon (*c.* 1219–92) pointed the way to the microscope and telescope, advancing beyond his great master Grosseteste to study parabolic mirrors and foreseeing the magnifying properties of convex lenses.

Of all the instruments that were to aid man's mastery of nature, spectacles were perhaps the humblest yet most significant. It is said that these were an Italian invention of the late thirteenth century. Two Dominican friars, Alessandro della Spira (d. 1313) and Salvino degl' Armati (d. 1317) are associated with it. As Lynn White (p. 143) remarks: 'Surely no one in the bespectacled academic world will be sufficiently discourteous to doubt that this technical development does much to account for the improved standard of education and the almost feverish tempo of thought characteristic of the fourteenth and fifteenth centuries. People were able to read more and to read in their maturer years.' Other instruments multiplied. Levi Ben Gerson of Provence (1288–1344) invented an early version of the sextant called the cross shaft, whilst an *Equatorial Planetarie* (an instrument for predicting the planets) was the subject of a book by Geoffrey Chaucer.

INGENIATOR TO ENGINEER

France and England had close bonds at this time. From the time of the Norman Conquest the French language was the language of the upper classes, and even after the loss of the French posses- sions French ideas seeped through society. The very names of the early engineers betray this. Ailnolth (*fl.* 1157–90), who worked on the Tower of London and Westminster Palace, was described as an *ingeniator*. He was a versatile man, and worked as easily with lead or glass as stone or wood. John Harvey remarks 'he was a craftsman of technical training, perhaps a carpenter as were most of the engineers of the 12th and 13th centuries, men accustomed to making the great war engines which preceded the use of gunpowder'. Such war engines were, for instance, being made at Chester and Carlisle by Gerard (*fl.* 1240–56), one of the principal technical officers in the service of Henry III (1216–72).

By the time of Edward I (1272–1307) the technical officers of the Crown were faced with the formidable task of constructing a chain of forts in North Wales following on its conquest. Here James of St. George (*fl.* 1278–1308) and Richard Lenginour (*fl.* 1277–1315) were outstanding and the results of their efforts remain today in castles like Conway, Flint and Rhuddlan no less than in cathedrals like Chester.

TOWN PLANNING

Both these kings ruled parts of France where they built a number of *villes neuves* to house those made homeless by war. Cadillac, Monpazier, Villeneuve, Villefranche, La Sauve, Sauveterre: the very names evoke their purpose. Guienne, Gascony and Languedoc were the chief areas. One of Edward I's English *bastides*, Winchelsea, owed much to Itier of Angoulême, who had done similar service for him in France, and by the end of the century Edward was able to assemble his fleet in its harbour. Another, Kingston-on-Hull, replaced the old port of Ravenser which was falling into the sea. Edward I's conquest of Wales resulted in further *bastides* at Flint, Conway, Caernarvon and Beaumaris.

The medieval towns took shape either as faubourgs (or gatherings round a feudal bourg) or mercantile centres housing the merchants and craftsmen. Their main interest, unlike that of the lords surrounding them, was everyday knowledge and information. They formed guilds to safeguard both their interests and their technical knowledge. These guilds provided the technical training needed at the time.

Two Medieval Scientists: Roger Bacon and Peter the Pilgrim

Roger Bacon (1219–92) by his preoccupations indicates that there was a real interest in mechanical problems on the part of his contemporaries: problems which have only recently been solved. In his *Epistola de secretis operibus* Bacon wrote: 'Machines for navigation can be made without rowers so that the largest ships on rivers or seas will be moved by a single man in charge with greater velocity then if they were full of men. Also cars can be made so that without animals they will move with unbelievable rapidity; such we opine were the scythe-bearing chariots with which the men of old fought. Also flying machines can be constructed so that a man sits in the midst of the machine revolving some engine by which artificial wings are made to beat the air like a flying bird. Also a machine small in size for raising or lowering enormous weights, than which nothing is more useful in emergencies. For by a machine three fingers high and wide and of less size a man could free himself and his friends from all danger of prison and rise and descend. Also a machine can easily be made by which one man can draw a thousand to himself by violence against their wills, and attract other things in like manner. Also machines can be made for walking in the sea and rivers, even to the bottom without danger. For Alexander the Great employed such, that he might see the secrets of the deep, as Ethicus the astronomer tells. These machines were made in antiquity and they have certainly been made in our times, except possibly a flying machine which I have not seen nor do I know any one who has, but I know an expert who has thought out the way to make one. And such things can be made almost without limit, for instance, bridges across rivers without piers or other supports, and mechanisms, and unheard of engines.'

Another contemporary, Peter the Pilgrim, was described by Bacon as 'conversant with the casting of metals and the working of gold,

46

silver, and other metals and all minerals; he knows all about soldiering and arms and hunting; he has examined agriculture and land surveying and farming; he has further considered old wives' magic and fortune-telling and the charms of them. . . . But, as honour and rewards would hinder him from the greatness of his experimental work he scorns them.' Bacon continued, 'For the past three years he has been working at the production of a mirror that shall produce combustion at a fixed distance; a problem which the Latins have neither solved nor attempted, though books have been written on the subject.' Peter the Pilgrim went on to issue his *de Magnete* in 1269. In the second chapter of Book 1, he wrote: 'For in investigating the unknown we greatly need manual industry, without which we can accomplish nothing perfectly. Yet there are many things subject to the rule of reason which we cannot completely investigate by the hand.'

THE POWER REVOLUTION IN MEDIEVAL ENGLAND

Spreading from the continent, the increasing use of water mills at this time was so striking and spectacular that Professor Carus-Wilson, in listing over a hundred of them established before the reign of Edward III (1327–77) concludes 'the changes in technique and organization were striking enough in themselves to merit the use of the word revolution'.

From the first mention of a water mill (in 762, near Dover) to the Domesday Survey (in 1086 when 5,624 were listed south of the Trent and Severn) there had been a great extension of this source of power to grind corn. After Domesday the use of water-powered fulling mills accelerated enormously. As a result of these water-powered mills congregating around the clear, swift streams of the north and west, a great displacement of the woollen industry took place. Up till the middle of the thirteenth century Stamford, Lincoln, Louth, Beverley and York were the centres of cloth making, but by the fourteenth and fifteenth centuries it was the West Riding, the Lake District, Cornwall, Devonshire, Somerset, the Cotswolds, Wiltshire and the Kennet Valley, which became the industrial centres. This movement of industries outside the cities needed, of course, capital and here we are faced with the beginning of capitalism in industry.

Fulling mills were followed by tanning mills (1217), paint mills (1361) and saw-mills (1376). Possession of these mills became a source of legal disputes. Their economic consequences were to move the operators out into the river areas where such mills could be operated.

The ponds, races and conduits essential for their operation led to further improvements and this in turn led to further utilization of mill wheel power for draining mines and forging metals.

The mill also contributed to the development of the mechanical clock which, it is now agreed (as opposed to the water clock that goes back to ancient Egypt), emerges round about the year 1271 with Robertus Anglicus, though Usher would put it at 1335. Perhaps one of the best known was that built for the Palais de Justice by Henry De Vick in 1364–70, as the King of France ordered the hours and the quarters to be struck in all the churches of Paris according to the time given by it. The practice spread to Germany where by the fifteenth century the sub-divisions on clocks included minutes and seconds.

THE MEDIEVAL FUELS

The five English fuel centres of Domesday Book were New Forest (Hampshire), Windsor (Berkshire), Wychwood (Oxfordshire), Wimborne (Dorset) and Gravelinges (Wiltshire). In time these were supplemented by others like the Weald (Kent), Ashdown (Sussex), Sherwood (Nottingham), Whittlewood and Salcey (Northamptonshire), Waltham, Hainault and Epping (Essex) and the Forest of Dean (Gloucestershire). These forests not only supplied the main fuel for burning, but provided the heat for lime-burning, smithing, salt-boiling and glass-making. But even more significant were their rôle in chemical engineering: tanning used the bark, and glass-making, soap-boiling and cloth-making used the alkali from woodash.

The medieval chemist used what materials he had to the limit. Salt was used in the preparation of leather, the 'rubbing of chimneys', soldering of pipes and in making distillations from wine. Above all, it preserved and flavoured meat, fish, butter and cheese. Medicinally, it cured toothache, indigestion and lassitude. Salt works were established at Lüneburg, the Low Countries and in England. English salt works or *salinae* numbered 1,195 in 1086 and they tended to concentrate on the east coast, especially in Lincolnshire. So, taken by and large, timber was in great demand to evaporate it.

As the maritime trade of the northern powers developed, tar and pitch derived from the wood were used to make the vessels more seaworthy. And, as iron began to develop, with water-worked bellows to provide it in quantity, Europe began to face a fuel famine. The wholesale plunder of the forests of Europe intensified and England, which imported timber, soon began to feel the need to regulate its own

Side view of the Pantheon showing the arch construction. (*From I Vestigi dell' Antichita di Roma* (1575) by Etienne Du Perac (1535–1604), who kept alive the Roman tradition by his etchings and became a notable employee of the French court.)

Westminster Palace A.D. 1097   Lindisfarne Priory,
                                Durham A.D. 1093–1099

Malmesbury Abbey, Wiltshire   Kirkstall Abbey, Yorkshire
        A.D. 1115–1139   A.D. 1152–1182

Gothic architecture was based on arches, as stone is strong in compression and weak in tension. Contrast this with modern buildings
which are based on girders strong in tension and relatively weak in
compression.

(From *An Attempt to Discriminate Styles of Architecture in England
from the Conquest to the Reformation* (1881) by Thomas Rickman
(1776–1841): a book which had a considerable influence in promoting
the study of Gothic Architecture in England. First published in 1812,
it was re-issued many times and revised after his death by J. H. Parker.)

natural resources of timber and to exploit other fuels. As a result coal began to be increasingly used in England by dyers and brewers who boiled liquids in quantity, and limeburners and smiths who needed high temperatures, and so the coal of Northumberland, South Yorkshire and other places was increasingly used. The use of iron was to intensify this mining.

CANNON MAKING

The increasing hunger for metals (which had to be mined) was aggravated by the wars of Edward III in France. These were the first in which cannon were used. Indeed one of the earliest illustrations (as opposed to mention) of a cannon is that of a cannon in a manuscript by Walter de Millemete dedicated to Edward III in 1327, the year of his accession. Edward III used them in the siege of Calais in 1345. Froissart, the chronicler of the time, says that 400 of them were used to capture St. Malo in 1378.

One of the seventy-three guns which William Woodward made in five years at the Tower of London weighed 7 cwt. It had a large central barrel for the discharge of stones and ten surrounding barrels for discharging lead bullets of moderate size. Randolph Hatton, the Keeper of the Privy Wardrobe from 1382 to 1396 to Edward III's grandson and successor—Richard II, was so active in placing orders for them that 'his purchases indicate that cannon founding was becoming a considerable industry in England'.

GUNPOWDER

These new 'engineers' made not only the guns for garrisons at home and abroad but the gunpowder too. They used willow charcoal, sulphur and saltpetre. Though the formula was Chinese, the methods of obtaining the raw material led to the development of chemistry. The charcoal industry was itself stimulated by the demand for guns. The sulphur industry stimulated distillation. Saltpetre needed leaching from nitrogenous soils from farms. As the need for mass manufacture developed the leaching was done by damp rollers driven by water or animal power.

The utilization of gunpowder had profound social effects. It not only helped to render both the catapult and the medieval castle obsolescent, so further weakening the feudal system: it stimulated other sciences—chemistry, metallurgy, mechanical engineering, surveying.

It intensified the hunger for alum. Attempts to explain combustion were to lead to the discovery of oxygen and the explosions in the cylindrical barrel were to lead ultimately to the steam and internal combustion engine. Finally, the very scope of engineering was itself changed. From being a builder of castles and a maker of catapults the engineer became an experimentalist.

The trials of these early gun founders were literal, for both in 1365 and 1377 there are references to their products 'weak, broken, noisome, used up and broken and wasted in trials and assays'. From such assays resulted the *ribaudequin*, a multi-barrelled mobile weapon and a small barrel carried by a man and mounted on wood. As a result the fortifications and design of towns altered considerably—defence in depth supplanting the simple keep or castle tower.

The officer who looked after the firearms in the Privy Wardrobe was known in Edward III's time as the *Artilator* or *Ingeniator*. William Byker appears first as 'engineer of the King's war slings' and then as *artillator domini regis in turre Londoniarum* when he was making cannon. T. F. Tout says that the officers who looked after firearms elsewhere than the Tower of London were 'engineers, not accounting officers or clerks'.

Edward III certainly used miners from the Forest of Dean and smiths from the City of London in his campaigns in France, for at Calais in 1346 they mustered 314 out of the total of 31,294 men in his forces. This is one of the earliest, if not the earliest record of engineers and artillery used in a siege.

THE CRAFT GUILDS

As the social embodiments of a handicraft age the guilds had multiple social functions. From the opening decades of the thirteenth century they provided technical training for the apprentices, protected workmen against sickness, competition and undercutting, set standards of craftsmanship enforced by searchers empowered to burn unsatisfactory products, acted as friendly societies and labour exchanges. They forced, by these regulations, industry outside the medieval towns. In addition, they played a very important rôle in medieval social life by secularizing drama; bringing it out of the church and into their own guild halls. They were able to do this in that they were usually pledged to the communal performance of religious duties. Whether one looks at them as survivals of the Roman collegia or derivatives from German sacrificial carousals it is

50

certain that their incorporation as Christian brotherhoods led them to put on plays which take their name from the guilds themselves: mystery plays.

The famous 'cycles' of Coventry, Chester, Townley and York were presented by the craft guilds in those towns. For instance, in York the armourers were responsible for presenting Adam and Eve being ejected from paradise; the shipwrights for Noah's Ark; and the fishermen for The Flood. These performances were given on moveable stages and formed the first secular popular entertainment of the country.

The purse-proud oligarchy which developed in these guilds expressed itself by the fifteenth century in the large companies, claiming the right, denied to the noble retinues, to wear a livery. In London such companies were henceforth known as Livery Companies. These Livery Companies were often the result of internal struggles between the merchants (who sold) and the journeymen (who made) the product, and as often the result of a struggle between trading guilds as opposed to craft guilds. With the decline in handicraft the guilds' real economic rôle declined and their decline was further accelerated by the State when it confiscated their religious possessions in 1547 and began to grant patents of monopoly to private individuals to work inventions.

CHAPTER 6

Italian Ingenuity

B y the fifteenth century the five States of the Italian peninsula, Naples, The Papacy, Florence, Milan and Venice, sustained a remarkable kindling of the human spirit which French and German writers of the nineteenth century used to call the Renaissance. Though such a term is now out of favour, it is nevertheless valid when applied to the impetus given to the visual arts by Giotto and to literature by Petrarch and Boccaccio. The essence was *virtu*, which was itself adopted and adapted as the activity of a *virtuoso*.

FLORENCE

Of these five States, Florence was perhaps the most outstanding at the beginning of the fifteenth and Venice at the beginning of the seventeenth century. Both nourished what might be called sub-renaissances within their borders.

Florence became the first industrial town in the world by manufacturing her own cloth. The *Torre di arte lana* was built in 1308 and indicates that the Florentines had probably discovered the technique of producing the sheen, richness and bright colours of the Moslem manufacturers. Certainly Florentine dyers formed a special guild, the *ufficiali della tinta*, and soon its operations were extended to dyeing imported cloth for re-export. The dye factory was built in 1309 and worked on indigo (imported from Asia) and alum (from both Asia and Castile). The specialization involved led to capitalist organization developing more obviously in medieval Florence than anywhere else. There was a yet older industry—that of silk, worked since 1187. After the devastation of Lucca in 1314 it too began to grow.

The *drappi* of the city, especially the cloth of gold and damask, were much sought after and during the fourteenth century there were eighty-three store and work shops. Slowly the *Arte della Seta* surpassed the *Arte della lana* and by the middle of the fifteenth century the Florentines began to cultivate their own mulberry trees to feed the silk-worms.

This industrial growth led to Florence becoming a great banking centre. The Medici looked after the Papal accounts between 1414–76 and 1484–92 and, after establishing branches in Antwerp, Barcelona, Bruges and London, royal accounts as well. King Edward IV of England (1461–83) borrowed money from them to provide arms. And with this developed academies. The Accademia Pontamaria (founded in 1433) and the Accademia Platonica (founded in 1442) were to be harbingers of others in the following century.

THE PALLADIAN STYLE

The most visible Italian influence on England and America was exerted by Andrea Palladio (1518–80) whose churches, convents and palaces still attract admiration. At thirty-one, Palladio won a competition to renovate the medieval Basilica at Vicenza. So began his career. Venice invited him to design the Church of San Giorgio Maggiore and innumerable merchant princes competed for his help in planning their country palaces. Working in the idiom of classical Rome, Palladio translated what had hitherto been considered a style suitable only for public buildings into private palaces. He found himself so busy that he built a drawing office in a luxurious barge on the Brenta (shaded by a yellow and black awning) to enable him to move from place to place supervising their construction. One of his showpieces was La Rotunda, a palladian villa outside Vicenza, where the dregs of wine vats were used to stain the roof tiles a rich purple red. This became, in the centuries after his death, a mecca for English and American visitors. Inigo Jones, Sir Christopher Wren and Lord Burlington were three Englishmen who, in turn, converted eighteenth-century country houses and churches into palladian semi-palaces. Jefferson used the inspirational design of the Italian master in building his own house, Monticello, and for the first non-sectarian university in the United States, the University of Virginia. The hall-marks of the Palladian style (which looks so out of place in a dark industrial town) are wide windows looking out on to water and statuary.

Palladio's first patron was the republic of Venice. First recognized by Byzantium as an independent community it later captured material and spiritual influence and obtained 'a half and a quarter of the Roman Empire'. The plunder of the East was augmented by its own industry. Metals and glass were made for most of the civilized world, with power obtained from undershot waterwheels.

THE NEW COSMOLOGY

As elegant as any of Palladio's material and earth-bound structures in Venice were the cosmological theories of a former student at the Venetian university of Padua: Nicolaus Copernicus (1473–1543) the founder of modern astronomy.

Nicolaus Copernicus graduated in 1499 at Padua and left to serve as a canon in an east German cathedral. He worked out the geometrical laws governing the motions of the planets in order to predict their behaviour in the future. His friends, who first published his work at Nürnberg in 1543, tried to avoid ecclesiastical hostility by declaring that it was a technique for computation of festivals. But the gigantic nature of his theory could not lay hidden once his book *De revolutionibus orbium coelestium* became public property. For it demolished the geocentric theories of the universe and substituted the heliocentric theory—that the earth rotates on its axis and revolves about the sun. This concentricity of orbits was, as he showed, shared not only by Venus and Mercury (which are nearer the sun) but by Mars, Jupiter, Saturn and the Fixed Stars (which are farther away). As he put it, 'the Sun rules the family of Planets as they circle round him'. The importance of Copernican astronomy is immense. Copernicus (and Venice) stimulated Galileo to publish in 1610 his *Siderius Nuntius* which enjoyed a wide circulation.

THE RISE OF A SCIENTIFIC LANGUAGE

But between Copernicus and Galileo took place the third great contribution of Italy at this time: the refinement and sophistication of a new mathematics. Obviously it did not spring spontaneously from Italian soil, for many factors had been at work beforehand. Though Moslem numbers had been popularized in Italy as early as the thirteenth century by Leonardo Fibonacci, the compulsive coincidence of the capture of Constantinople in 1453 and the opening up of the new world twenty-four years later kindled a new interest. Displaced scholars came to the West. Navigational techniques developed apace and their popularization by printing and paper followed. 'He who scorns the very great certainty of mathematics', wrote Leonardo da Vinci, 'is feeding his mind on confusion and will never be able to silence the sophistical teachings that lead only to an eternal battle of words'.

Leonardo da Vinci (1452–1519) was military engineer to the Duke of Milan for eighteen years: essential to any Italian state at the end of the fifteenth century. Such another was Biringuccio who also held for his pains the monopoly of producing saltpetre in the Duchy of Siena. A third, Simon Stevin (1548–1640) was military engineer to the Prince of Orange. All of them were involved in fortifying, building, analysing and casting.

When Leonardo began to serve Ludovico Sforza of Milan his mind teemed with projects for portable bridges, hydraulic machines, cannons, mines and covered chariots. He went on to serve Cesar Borgia and the King of France. During his thirty-six years as a consultant he always noted the endless technical and scientific ideas that came to him as he went about his business.

These jottings, comprising some five thousand pages, are scattered in museums in many continents. They were written from right to left and this did not make them very easy to read. Breech-loading and steam cannon, rifled firearms, tanks and submarines; looms, link chains, gears of various kinds, screw cutting machines, cranes and presses; helicopters and parachutes crowd the many pages of his sketches. Seeing these sketches and realizing his debt to Euclid's *Elements*, Aristotle's *Mechanics*, the writings of Hero, Archimedes and the anonymous author of the *De rebus bellicis* as well as Jordanus and others of the Middle Ages, is an ideal object lesson in the evolution of ideas.

The Leonardan tradition, partly self-advertisement, partly brilliant forays outside the contemporary engineering periphery, was eagerly taken up. His manuscripts at Vaprio near Milan were rifled by lesser men whose concrete achievements were greater. Geronimo Cardano (1501–76) was one. Like Leonardo, the illegitimate son of a lawyer but unlike Leonardo, a mathematician, his books *De subtilitate rerum* (1550) and *De rerum varietate* (1557) were printed at Nurnberg and Basel and widely circulated in Europe where he was recognized as a most distinguished mathematician. Another Leonardan, Bernard Palissy (*c.* 1510–89) was a self-taught potter, whose experience in discovering the right clays for his works in the Tuileries led him to become an authority on French natural resources. A man of independent mind and judgment, he was protected by the Court and for ten years (1575–85) he gave the results of his findings to a selected audience in Paris, thus becoming the first public lecturer on natural science in Europe. These lectures, published as a dialogue between Theory and Practice, were well attended.

MATHEMATICAL CONTESTS AND THE DEVELOPMENT OF ALGEBRA

There was a fashion in Italy at this time for professors of mathematics at the universities to take part in public mathematical debates and offer problems on such subjects as cubic equations. These were almost exclusively academic contests. The most successful problem solver was Niccolo Tartaglio (1506–59) who became the algebra champion of Italy. It was from him that Cardano obtained the secret of solving cubic equations. And Cardano in turn taught Ludovico Ferrari (1522–65) who, with Rafael Bombinelli, was one of the greatest algebraists of the day. The greatest was probably Francois Vieta (1540–1603) a French amateur mathematician who pioneered in the use of decimals, introduced the custom of using consonants for known quantities and vowels for unknown. It was Vieta's secretary, Nathaniel Torporley, who helped Thomas Hariot (1560–1621) the English mathematician.

THE BY-PRODUCTS OF COMMERCE AND WAR

Moslem numerals, increasingly employed for navigation and commerce, were now supplemented by warehouse signs like $+$ and $-$ (indicating over or under weight) and later by signs like $\times$, $\div$, $\therefore$ and $=$. The first appearance of the $=$ sign in England occurs in a book by Robert Recorde aptly entitled *The Whetstone of Witte* (1557) which was also the first book on algebra ever published in England. A second branch of mathematics, trigonometry, began to emerge; the name first appears with Pitiscus. Before that the word *hypotenuse* had been coined by Rheticus in 1551 and tangent and secant by Thomas Fincke in 1583.

Bellicose city states sustained a continuous interest in the trajectory of cannon balls. Expressed in terms of algebra this involved graphic representation and in turn, coupled to the increasing interest in cosmology, to a completely new kind of geometry where conic sections could be explained in algebraic terms. For this Vieta and Descartes were to be mainly responsible.

Tartaglio was the first to publish a treatise on ballistics (which he dedicated to Henry VIII of England) and the making of artillery which remained authoritative works until the end of the next century.

The decimals of Vieta were to be further developed by Simon Stevin, whose works on statics and applied mathematics stimulated Galileo.

GALILEO AND THE VENETIAN ARSENAL

Galileo (1564–1643) was Professor of Physics and Military Engineering at the University of Padua, which belonged to Venice. Here, at the first lay or 'civic' university in the West, Galileo began to lecture at the age of twenty-nine. Previously, he had taught at Pisa but he was forced to leave after publicly proving, by dropping three bodies of varying weights from the Leaning Tower, that Aristotle was wrong in his assertion that the velocity of falling bodies varies with their weight. Galileo was the first to construct and use a telescope scientifically. With it, he discovered that Jupiter had four satellites which he named the 'Medicean stars'.

He was much impressed by the arsenal of Venice where the Venetians built their ships, made their rope and built their fleet. To Dante it had seemed a hell, large and noisy, like the fifth abyss of the Inferno where Simonists and statesmen in public offices lay in a bath of pitch. But to Galileo its bustle and activity was a constant stimulus.

'The constant activity which you Venetians display in your famous arsenal', he wrote, 'suggests to the studious mind a large field for investigation, especially that part of the work which involves mechanics; for in this department all types of instruments and machines are constantly being constructed by many artisans, among whom there must be some who, partly by their own observations, have become highly expert and clever in explanation.'

Galileo now felt he could marry the new mathematics of Cardano, Vieta and Stevin with the practices of the arsenal. The result was his *Dialogue Concerning Two New Sciences* (from which the observations on the Venetian Arsenal in the preceding section were taken). This, as its title indicates, outlines the sciences of statics (or forces in equilibrium) and dynamics (or forces out of equilibrium). His discussions with shipwrights had borne fruit.

The third phase of his work is perhaps the best known. Having established to his satisfaction these two 'new' sciences, he published in 1632 another Dialogue—this time on the virtues of the Copernican as opposed to the Ptolemaic cosmology. He was tried, convicted of heresy and forced to recant in the following year. But his work went on with the financial help of the Medicis and the Grand Duke of Tuscany.

THE DISSEMINATION OF NEW KNOWLEDGE

One of Galileo's friends and pupils, Benedetto Castelli, utilizing Leonardo's drawings, compiled a work on the measurement of the flow of rivers and canals. Entitled *Delli Misuri dell' Acque Correnti* (1628) it is but one of many works on power mechanisms to which mathematicians were applying themselves. Others had gone before like Jacques Besson, a professor at Orleans, who had described machine tools, pumping planes and a screw-cutting lathe in his *Théâtre des Instrumens Mathematiques et Méchaniques* (1579) and Ramelli whose *Le Diverse et Artificiose Machine* (1588) described water-wheels with reversing devices and chain drives. A third popularizer of 'machinery' was Vittorio Zonca (1568–1602) whose *Novo Teatro di Machine et Edificii* (1621) was concerned with the application of power to multiple industrial purposes and furnished sketches to illustrate his points.

Two of Galileo's disciples, Viviani and Torricelli, were leading members of the Accademia del Cimento in 1657. Toricelli, Galileo's secretary, put his knowledge to good use and gave us the principle of the barometer. This was virtually sustained by the Medicis who were keen students of science and since 1647 had been running a private group for scientific experiment and discussion. The Accademia, though it only lasted for ten years, nevertheless did important work which was published in 1667. An English version by Richard Waller appeared in 1684 as *Essayes of Natural Experiments made in the Academie del Cimento*.

Waller's translation of the accounts of these Italian experiments indicates a real debt which Europe owes to the Italians of this period; for it was in Italy that the national academy of science seems to have been born. The first was that in Naples, founded in 1560 as the Academy of the Secrets of Nature. Giambattista Porta (1538–1615), the first to recognize the heating effect of light rays, was its first president. A second, of which Galileo was a member, bore the apt name the Academy of Lynxes and took shape in 1603. These two precursors of the Accademia del Cimento were also the institutional forerunners of the national academies of France and England. They indicated, by their very existence, that science was an activity that needed sharing, criticism and demonstration. And the direction in which science was being taken by the Italians is well illustrated by the title of the fourth academy—founded at Rome in 1677: the Accademia Fisico-matematica. These four academies of science (as opposed to the Florentine academies mentioned earlier in the chapter) had, as a condition of membership, the making of some discovery in natural science. As such a condition spread, there arose a professional feeling amongst inquisitors of nature that was to eliminate the aura of magic and charlatanry.

CHAPTER 7

German Miners and Metallurgists, 1450-1650

THE QUEST BENEATH THE EARTH

The embellishment of cathedrals, the building of towns, the colonization of the uplands and the increasing need for money, all needed metals. Lead for roofs, iron for implements and ploughshares, and silver for the money payments which were increasingly dissolving the service obligations of Feudal Christendom, were the three in greatest demand. But, with the advent of cannon, copper and iron were even more so.

Beginning in about 1450 there is evident a marked advance in mining technology. In 1451, for instance, Johannsen Funcken began to exploit the discovery that silver could be separated from the copper ores with which it was found with the help of lead. As J. U. Nef has said, 'no other invention had so stimulating an effect upon the development of the mining and metallurgical industries of Central Europe on the eve of the Reformation'. Secondly, John de Castro discovered, in 1461, at Tolfa in Italy rich alunite deposits. As Castro was responsible for the revenue of the papal chamber, he secured the backing of the Pope to exploit these rich deposits to emancipate Europe from its dependence on the Moors. Within two years, four mines were employing 8,000 men. The papal monopoly was reinforced by grim sanctions, as the encyclical of Julius II, issued in 1506, indicates. Thirdly, new seams of ore like cinnabar began to be worked at Almaden in Spain and Idria in Carniola. Fourthly and concomitantly with all these, came vast improvements in mining engineering: longer adits for draining, ingenious arrangements of chains of buckets worked by water and horse power and the organization, especially in Central Europe, of Saigerhütten. These were virtually metallurgical factories and needed a large outlay of capital, and were capable of dealing with great quantities of ore brought from great distances.

Naturally since the Saigerhütten were concerned with the processing of silver their natural owners were bankers like the Fuggers, who built at Hohenkirchen in Thuringia and Villach in Carinthia two of these metallurgical factories.

AUGSBURG

The financiers of this busy activity lived in Augsburg. There, the Fuggers were patrons of the painter Hans Holbein the elder and of the architect Elias Holl. They were bankers to both Maximilian I and the Emperor Charles V and re-invested their money in mines in the Tyrol, Hungary, Saxony, Bohemia and Spain. Their colleagues, the Welsers, took Venezuela as security from Charles V as a pledge and married their daughter to Ferdinand of Austria.

The Nevada of Europe was Bohemia where mining camps like Joachimsthal, Iglau and Deutschbrod grew into towns. Clustered round this quadrilateral massif other mining centres nucleated: the Harz Mountains, Mansfeld (where Martin Luther's father worked as a miner), Slovakia (where copper and silver and lead were extracted), and Styria, Carinthia and Carniola (where calamine and quicksilver were mined). Much of this too was controlled by the Fuggers.

So great did the industry become that by 1525 100,000 men were employed in German mining and smelting operations. This does not include German engineers and miners sent abroad by the Fuggers and their fellow financiers, the Welsers. More silver came from Germany than all the other countries together. The mining of copper (used for cannon founding) developed rapidly and the Fuggers extended their interests to Spain where silver and mercury were also mined.

THE SPREAD OF TECHNOLOGICAL SKILLS

The main event separating what we call the Renaissance from the Middle Ages was the invention of printing. Its importance was twofold: typographical and pictorial. Here too German technology was decisive as when Johannes Gutenberg of Mainz perfected the process of printing from movable type. His Gutenberg Bible, issued in 1454, was the first book to be so printed.

An Englishman, in 1471, William Caxton (1422–91), living in Cologne, helped to set up the type for the *De Proprietatibus Rerum*, an encyclopaedia compiled by Bartholomaeus Anglicus, an English monk. One of Caxton's fellow residents in Cologne, Colard Mansion, set up a press at Bruges, and published Caxton's own *Recuyell of the Historyes of Troy*. Caxton returned to England in 1476 to set up a press of his own which, after his death, was taken over by Wynkyn de Worde, who was probably a German. Worde moved the press to Fleet Street in 1500 and in the next thirty-five years he printed over seven hundred works. Worde introduced *italic* type into England. (There were other German printers who came to England: Theodoric Rood went from Cologne to Oxford, Reymer Wolfe from Strasburg to St. Pauls.)

Worde ordered his paper from John Tate of Hertford who is generally regarded to have been the first paper maker in England. But in 1588 Johannes Spielmann, another German, set up his great mill at Dartford which employed over six hundred men. Thomas Churchyard wrote a poem about it and King James I knighted Spielmann. Paper mills were introduced into Scotland by another German, Peter Groot Heare.

The Rise of Mining Engineering

Thanks to the printing press, books on mining began to appear. The *Bergbüchlein* and the *Probierbüchlein*, printed significantly enough at Augsburg in 1505, were followed by many others. By 1540 Biringuccio, sponsored by the Pope, the Emperor and the Senate of Venice, published the first comprehensive text book on metallurgy, in which he rejected alchemy and superstition, stressed exact measurements and impressed the Pope. Both English and French translations were commissioned. It would have been even more popular but for the appearance in 1556 of the *De Re Metallica* of George Bauer, better known as Agricola. 'With regard to the veins, tools, vessels, sluices, machines and furnaces,' he wrote, 'I have not only described them, but have also hired illustrators to delineate their forms, lest the descriptions which are conveyed by words should either not be understood by men of our own times, or should cause difficulty to posterity. ... I have omitted all these things which I have myself not seen, or have not read or heard of from persons upon whom I can rely. That which I have neither seen, nor carefully considered after reading or hearing of, I have not written about.' Agricola was a town physician of Joachimsthal and spent twenty-five years collecting his material. His twelve books cover not only every phase of the mining industry but of contemporary metallurgy too, listing no less than sixty minerals. For our purposes, it is the sixth book with its chain pumps, piston pumps and bellows which is most interesting. For Agricola was conversant with the flywheel, horse whins and windmill-driven ventilation fans, no less than with suction pumps in series. It was a George Bauer (not necessarily the same person) who was recommended to Cardinal Wolsey in 1528 as a skilful judge of metals and minerals.

Completing the trinity of sixteenth-century metallurgists was Lazarus Ercker whose *Treatise on Ores and Assaying* (1574) was authoritative. He was Chief Superintendent of Mines in the Holy Roman Empire and the Kingdom of Bohemia, and was in contact with Spanish mining in Mexico and Peru.

The Migration of Mining Engineers

In addition to English translations or plagiarisms from their illustrated works, German mining engineers themselves carried the new techniques abroad. From Augsburg Joachim Höchstetter came with six other German industrialists to England to develop mines. German metallurgists were used by Sir Thomas Gresham to restore the English currency. In 1564 German capitalists were active in exploiting English copper mines. Daniel Höchstetter came over from Augsburg, German miners followed to settle at Keswick and in 1568 the Mines Royal Company was incorporated. Their main centre of operations was at Newlands, near Keswick. Here a rich lode of copper was struck and Daniel Höchstetter took up permanent residence in 1571. He invented an engine for draining mines and even considered setting up a company to manufacture it. Höchstetter's work in Devon and Cornwall was less successful but his sons continued at Keswick till the early seventeenth century.

An English professor of chemical engineering has recently listed 136 Germans at Keswick and remarked that when the venture ended, 'it was beyond the capacity of Elizabethan England even to appreciate the degree of skill and knowledge attained by these German craftsmen'.

A second German venture, the Mineral and Battery Works, began in the same year as the Mines Royal. This discovered zinc in Somerset and in 1568 a brass foundry was set up near Tintern Abbey to utilize it.

Gerard Malynes imported German miners to work lead mines in Yorkshire and silver mines in Durham. So did Prince Rupert, who hoped to see them teach the use of gunpowder in mining operations at Ecton.

OTHER HYDRAULIC ENGINEERS

From pumping water out of mines to supplying towns is a natural transition. Towns need water, German towns especially, since they were growing. Here too Augsburg provided Hans Felber with the opportunity to build machines that gave a plentiful supply. Geronimo Cardano passed through the city in 1550 and described the principle on which they worked: a series of multiple Archimedean screws turning in layered troughs.

German engineers also built a water supply for Toledo in Spain operating through pistons driven by water wheels. These delivered water at such a pressure that all the pipes fractured. Further attempts to supply water by an ingenious arrangement of levered troughs decanting water into each other likewise failed so the inhabitants had to revert to using their donkeys.

THE REACTION ON CHEMISTRY

The practical needs of mining, together with the opportunities it afforded for detailed observation, had from early times contributed to chemistry. It was now intensified by metallurgy and by the increasing need for saltpetre. Paracelsus (1493–1534) a pupil at the mining school established by the Fuggers, virtually founded a new school of chemists who, in their endeavours to capture the 'spirit' of liquids, stimulated further distillations. One of his disciples, Andreas Libavius (? 1540–1616) of Coburg, wrote the first real chemical text book: *Alchemia* (1597). Another, J. B. van Helmont (from the Low Countries) gave us the term 'gas', which he took from the Greek chaos, and attempted to classify the various kinds. A third, J. R. Glauber (1604–68) of Karlstadt, issued *A Description of the Art of Distillation* in 1648. Though chiefly known for his famous 'salts' made from salt and alum, he also wrote a book called *The Prosperity of Germany* in which he pointed out the industrial and commercial advantages of the study of chemistry, whereby raw materials could be processed at home.

THE AIR-PUMP

But in the realm of engineering perhaps the most exciting idea came from the German town of Magdeburg, which Otto von Guericke (1602–86) was asked to fortify for the Protestants against the Imperial troops. Unfortunately it was captured and Guericke es-

The waterworks at London Bridge, which supplied the City of London with water.

Sir Hugh Middleton (1560?–1631), constructor of the New River.

caped only with difficulty. He worked for the Swedes on the fortifications of Erfurt, but returned soon afterwards. As burgomaster, he tried his best to reconstruct the town on a plan, but money was too short. He began to work on an air pump. At first he tried to exhaust a well caulked wooden barrel, but the air rushed in through the pores of the wood. Then he tried to exhaust a copper sphere, but it collapsed. In 1654 he conducted an experiment before the Imperial Diet at Ratisbon by exhausting two hollow bronze hemispheres through a stop-cock in one and harnessing a team of eight horses to each. By illustrating that they could not be pulled apart unless the stop-cock was open, he established the power of a vacuum. This public demonstration was supplemented by an account of Guericke's work by Kasper Schott in *Mechanica-Hydraulico-Pneumatica* (1657) which, in turn, influenced Robert Boyle of England to undertake further experiments on his own. For Guericke had developed the air pump: the instrument that had exhausted the Magdeburg sphere. By so doing he was clearing the way for an even more momentous advance that was to spring from mining enterprises in England: the atmospheric engines of Savery and Newcomen.

For the Harz silver mines G. W. Leibnitz (1646–1716) designed in 1679 a windmill pump, operated not by shafts and transmissions, but by pipes filled with compressed air. But he could not obtain pipes large and airtight enough for his purpose and minor mine officials opposed his ideas. So, like other engineers of his day, he turned to designing cascades and ornamental fountains for the Hanoverian court. His contributions to engineering however were more epochal and far reaching than this, for whilst he was working on them, he published in 1684, in a periodical which he had founded two years earlier, a paper called 'A new method to find Maxima and Minima'. This gave to the world the symbols dx, dy, dy/dx and the elementary rules of the differential calculus. Four years later he put forward the principles of the integral calculus, in which the sign of integration ∫ first appears.

Not the least of his interests was in the integration of Europe itself: in 1708 he submitted a plan to the Emperor of Russia for a world council.

GERMAN INGENUITY

Other products of German ingenuity were the watch (developed by Peter Henlein at Nurnberg and manufactured *c.* 1500) and the Saxony wheel (made by a citizen of Brunswick in 1533). The first was to be of great moment in navigation, the second was to serve as a model for Arkwright. Sugar was manufactured in England by Kaspar Terlin in 1597 and his countryman, Martin Begger, applied for a patent.

Well might the seventeenth-century historian, P. Heyleyn, write in his *Cosmographie* (1652), 'The greatest efficiency of this people lieth in the *Mechanical* part of learning as being eminent for many *Mathematical* experiments, strange water-works, *Medicinal* extractions, *Chymistry*, the Art of Printing, and inventions of like noble nature, to the no lesse benefit than admiration of the World.'

CHAPTER 8

The Struggle with Water in Britain 1540-1740

THE DEMANDS OF THE NAVY

The initiative in the creation of a professional British navy was taken by Henry VIII. By putting heavy muzzle loaders of the type made by Hans Poppenruyter of Mechlin into his ships, he revolutionized naval design. Since it was impossible to house them in the fore or after 'castle' of the ship, they were taken below decks and fired through ports. The ship in whose sides the first gun-ports were cut was the *Mary Rose*, in 1513. The traditional inventor of the 'broadside' was James Baker. Soon Henry VIII had set up gun-foundries of his own and his successors had to limit the export of guns to foreign parts.

Printing and publishing had broken down the secrecy and mystery (as it was called) of many arts, and now that of shipbuilding was popularized. The first publication on shipbuilding—Bayfius' *De Re Navali* (1536) was the precursor of many Elizabethan tracts based not only on the shipwrights' art but also on the experience gained from voyages. William Bourne pleaded 'progress of the arts and sciences' as an excuse for publishing his *Revises* in 1578. Dedicated to Lord Howard of Effingham, Bourne's book first mentions the ship's log and line (No. 21), the night telegraph (No. 75) and the telescope (No. 110). Three years later Robert Norman, who discovered the dip of the magnetic needle and outlined it in the *New Attractive* (1581), urged that 'men that search out the secrets of their arts and professions and publish the same to the behoof and use of others must not be condemned'.

The shipwrights' workshop and dockyard in England, like the arsenal at Venice, reinvigorated the speculations, and re-oriented the interests, of the 'pure' scientist. A typical example of this can be seen in the *De Magnete* (1600) of William Gilbert, the first account

67

of magnetism and electricity. It represents the climax of two generations of active technological work. Francis Bacon, his contemporary, remarked in his *Advancement of Learning* that 'Gilbertus our countryman, hath made a philosophy out of the observations of a lodestone'. Bacon was less than just, for Gilbert was also intensely interested in mining, and not only corrected Agricola but gave proofs that he knew a great deal of the smelting of iron, having himself forged apparatus in his laboratory. One quarter of Gilbert's epoch-making work was, it is true, taken up by problems of navigation and nautical instruments since they had sparked his theory of electricity and magnetism. Gilbert generously acknowledged his debts to the Elizabethan mariners, mathematicians and shipwrights who had played their part in annihilating the Armada in 1588. So Elizabethan sea captains, though they did not then know it, played a vital rôle in the development of British science.

Simon Stevin (1548–1640), the technical adviser to Maurice of Nassau in Holland, offers a similar case history of a 'pure' scientist stimulated by contact with artisans. His invention of decimal fractions and enunciation of the parallelogram of forces give him a unique place in the history of technology, and to this day one of the leading mathematical journals in his native land bears his name.

So Gabriel Harvey, the son of a ropemaker, might well rhapsodize: 'what profounde Mathematician, like Digges, Hariot or Dee, esteemeth not the pregnant Mechanician? Let every man in his degree enjoy his due: and let the brave enginer, fine Daedalist, skilful Neptunist, marvelous Vulcanist, and every Mercurial occupationer, that is every Master of his craft, and every Doctour of his mystery, be respected according to the uttermost extent of his publique service, or private industry'.

Frobisher's lieutenant, George Best, was very warm in his praise of 'the continuall practise and the exercising of good wittes', that had brought forth printing, the compass and navigational arts, and expressed the hope that others would be found.

They were. A number of ideas were put forward by non-naval men. Sir Hugh Platt (1552–1611) devised an oily composition to prevent rusting of ironwork, provided Sir Francis Drake with pitch to caulk his ship and gave much time to preserving foods by drying. Cornelius Drebbel (1572–1633) built a row-boat submarine which he is said to have demonstrated on the Thames and provided Richard Hawkins with a device for distilling water that was used in a voyage round South America. By 1600 when the East India Company was chartered, its needs and operations led to a great increase of shipbuilding, especially at Deptford. Shipwrighting, like millwrighting, became so important that King James I chartered a shipwrights' guild. The guild was given the supervision of all shipbuilding in England. It survived for a century. The master in 1612, Phineas Pett (who left us his autobiography), built the *Sovereign of the Seas* and used a scale model in his building. Pett's son, Peter, was followed by Sir Antony Deane (1638?–1721), who was master shipwright at Harwich at the age of 26 and commissioner of the navy in 1675. He built yachts for Louis XIV of France. Samuel Pepys, who worked with him, recorded that he was the first naval architect who could foretell the draught of water required for launching by working out the weight of material built into a ship and her volume. The level of empiric skill amongst these shipwrights was such that John Evelyn, watching the launching of the *Charles* at Deptford, remarked of its builder (not Deane) that he was 'a plain honest carpenter, master builder of this dock but one who can give very little account of his art by discourse, and is hardly capable of reading, yet of great ability in his calling'.

THE TIMBER FAMINE

The expansion of shipbuilding, as with that of the metallurgical industries (iron, copper and lead) and the concomitant development of glass making, salt boiling and gunpowder, precipitated a virtual timber famine. The forests of Sussex and the Weald of Kent sustained the activity of 7,000 smelters who, amongst other things, enjoyed the complete monopoly of the gun-making trade until 1743. The Weald was excepted from an Act of Parliament passed in 1558 prohibiting iron-works from using any trees growing within fourteen miles of the coast or a navigable river.

The furnaces near London and the south-eastern counties consumed timber so voraciously that further Acts were passed in 1581 and 1585 to restrain them. Domestic users were turning to 'Sea Coale', so called because it was imported from Newcastle. To use it without discomfort, houses began to be built with proper chimneys and flues. Attempts were made to render coal less offensive and in 1590 John Thornburgh (1551–1641), a chaplain to Queen Elizabeth, obtained a patent for removing the sulphur. By 1620 a long series of patents for coking coal, so that it might be used by brewers and maltsters, had been granted. Sir Robert Mansell (1573–1656), the admiral, obtained patents to exploit the use of coal for glass making by using covered pots and more lead flux. This resulted in a brilliant flint glass which virtually revolutionized the industry.

The reciprocal process of social and industrial change continued. Edmund Howes, a contemporary, declared that:

'within man's memory, it was held impossible to have any want of wood in England. But . . . such hath been the great expence of timber for navigation, with infinite increase of building of houses, with the great expence of wood to make household furniture, casks, and other vessels not to be numbered, and of carts, wagons and coaches, besides the extreme waste of wood in making iron, burning of brick and tiles, that at this present, through the great consuming of wood as aforesaid, and the neglect of planting of woods, there is so great a scarcity of wood throughout the whole kingdom that not only the City of London, all haven-towns and in very many parts within the land, the inhabitants in general are constrained to make their fires of sea-coal or pit-coal, even in the chambers of honourable personages, and through necessity which is the mother of all arts, they have late years devised the making of iron, the making of all sorts of glass and burning of bricks with sea-coal or pit-coal'.

It is little wonder that between 1563 and 1658 Newcastle's exports of coal grew from 32,951 to 529,032 tons a year. The 'devising' to which Edmund Howes referred was ironically true in the case of smelting iron by coal, for there were many patents registered for this process, including two by Dud Dudley of Worcestershire in 1621 and 1638. By 1642, England and Sweden were producing as much iron as the rest of Western Europe (including Russia) put together. In other words, two countries, with a combined population of 8,000,000, were producing as much iron as the rest of Europe, with a population of 60,000,000. Yet real success in the smelting of iron from coal had to wait till the time of Abraham Darby I (1677–1717) of the Coalbrookdale Works in Shropshire.

THE BY-PROBLEMS OF MINING

The more coal, iron, copper and lead were mined the deeper existing mines had to go and the more difficult it became to drain them. Burchard Kranich (1515–1578), one of the many 'Hie Almain', or German, engineers who came over in the reign of Queen Elizabeth I, was granted a patent in 1563 for 'draining of mines or conveying of water from any place whatsoever from low to high'. Kranich was the first person to use water power for crushing ore in Cornwall, an area which was to be increasingly exploited for the next three centuries for its tin and copper.

Drainage became progressively more costly as well as more difficult. By 1634 when Hannibal Vivian obtained a patent, drainage soaked up 75 per cent of the costs of mining. 'The tinners', it was said, 'have been much discouraged . . . because of the great charge in drawing of water out of the tin mines'. Further patents followed in 1682 until, in 1714 and 1724, John Coster (probably) and Francis Scobell (certainly) began to use the large water wheel.

And so the drainage engineers began to feel their way to using other power. Though steam was used as a toy by Hero in his *Aeolipile* and by Gerbert in his steam organ it was not until Cardano's time that the intelligent use of steam for creating a vacuum was discussed. Leonardo thought Archimedes had invented a steam gun. Baptista Porta in 1601 described a machine for raising a column of water by condensing steam to create a vacuum into which water flowed. Solomon de Caus had a similar idea.

From being an Italian toy, to use steam became an English obsession. David Ramsay was one of the first to obtain a patent (from Charles I in 1630) 'to raise water from low pitts by fire . . . to raise water from low places and mynes, and coal pitts, by a new waie never yet in use'. The second Marquis of Worcester, in an elaboration of Solomon de Caus' idea, actually erected a machine at Vauxhall in London which raised water to a height of 40 feet. In 1663 he obtained from Parliament rights for 99 years for his 'water commanding engine'. But his attempt to form a company to work it failed. Sir Samuel Morland (1625–96), the Master Mechanic to Charles II (1681), known for his calculating machine which Pepys described as 'very pretty but not very useful', had an assistant and successor, Isaac Thompson (*fl.* 1672–95), who was, in 1695, making pumps for draining pounds and mines as well as Morland engines. His house in Great Russell Street, opposite where the British Museum now stands, was decorated with the sign of an engine.

THE SOCIAL CONSEQUENCES OF THE PUMP

The pump was a symbol of salvation in seventeenth-century England, for it was not only the answer to flooded mines but it enabled water to be brought to the towns and removed from potential agricultural land.

London, as usual, records the earliest experiments when, in 1581,

Peter Maurice threw a jet of water from the Thames over St. Magnus's steeple and was granted a 500 years' lease of two arches of London Bridge to set up a water wheel pump to supply the water to Leadenhall and Old Fish Street. To allay the suspicions of the Water Bearers (a brotherhood dating from 1496 which had hitherto borne the burden of this trade) the Lord Mayor assured them that they would have more work, not less. Maurice and his family kept their privileges till 1701 when they were sold for £30,000.

Another 'forcier', as it was called, was made by Bevis Bulmer near Broken Wharf to supply Westcheap. He was lent part of Leadenhall to assemble his 'engine'. Henry Shaw also obtained a 500 years' lease to bring water from Smithfield. An Italian, Gianibelli, proposed in 1591 to erect a waterworks at Tyburn. But by 1600 few citizens had water laid to their houses. Even Lord Burghley had his water supply cut when it ran short.

As London grew the need for more water became acute and the corporation were authorized by Acts of Parliament to bring in water from Hertfordshire to augment supplies. At this point there appeared Hugh Middleton, a goldsmith and a friend of Raleigh. Middleton offered to undertake the task of bringing in water for a 'New River' and finish it in four years. He began in 1609 but was hampered by the opposition of landowners through whose property the 'river' was to pass. It needed the help of King James I (who agreed to pay half the cost of the work for half the profits) to finish the work in 1613. The 'New River' drew water from springs near Ware and, in the course of its sinuous 38-mile course, discharged the water into a reservoir at Islington called The New River Head. The actual engineering work was undertaken by Edward Wright (1559–1615). The injunctions to the citizens struck a hygienic note:

'no earth rubbish soyle gravell stones dogges catts or anie cattle carrion or anie unwholesome or uncleane thing nor shall wash nor clense anie clothes wooll or other thing in the said river . . . nor shall make or convey anie sincke, ditch tanhowse dying howse or siege into the said river or to have anie fall into the same'.

Middleton later leased some mines near Plynlimmon which had been flooded by water, drained them and made a large profit. He also employed Dutch workmen to drain land on the Isle of Wight in 1621.

A plan for another 'New River' from Rickmansworth to St. Giles was put forward by Sir Edward Forde in 1640. The civil war prevented this, especially since Forde was a Royalist. Fifteen years later, having made his peace with Cromwell (a not too difficult matter since his wife was the sister of Ireton, Cromwell's son-in-law), he set up a waterworks near where Somerset House is now. It was noticed by the engraver, Hollar, and the diarist, Antony Wood.

THE DRAINING OF FENS

If there was too little water in London, there was too much elsewhere. The same technique that forced Thames water into pipes could eject it from the waterlogged Fens of Norfolk, Huntingdon, Cambridge, Northamptonshire and Lincolnshire.

The same Peter Maurice who set up the water works in London in 1581 (to which we have just referred) made, six years earlier, a request to Queen Elizabeth for the sole right of making and employing certain hydraulic engines for draining of the Fens. In his application it was claimed that he had:

'to his extreme charge and with like payne travell and industrie endevord to make divers engins and instruments by motion whereof running streames and springes may be drawen farr higher than the naturall levils or couse and also dead waters very likely to be drayned from the depths into other passages'.

Two years later there was another application by Sir Thomas Golding but by 1580 Maurice was granted letters patent 'to draine by certain engines and devises never knowen or used before, which being put in practice are like to prove verie commodious and beneficiall unto the Realme'.

The Fens remained as a perpetual challenge and Maurice was followed by such men as Dr. John Dee, the noted Elizabethan alchemist and geographer, Latreille—a Frenchman, Humphrey Bradley, Master of the Dykes to Henry IV of France, Guillaume Mostart (who perfected an engine 'as was never seen in the kingdom before') and Captain Thomas Lovell who had 'invention perfected overseas' for the use of which he obtained a 21-year patent in 1597.

Such men, in addition to the inadequacy of their equipment, had to suffer from the inroads of the sea (as in 1613), local opposition to taxes and lack of sustained patronage. This latter, James I was determined to remedy. He would not suffer the Fens to 'bee abandoned to the will of the waters, to let it lye wast and unprofitable', so in 1620–1 he invited Cornelius Vermuyden (? 1590–1677) to attempt to drain them. Though Vermuyden probably owed his appointment to his brother-in-law (who had influence at the English court), he certainly justified the hopes placed in him. During his lifetime he drained

Dagenham, Windsor Great Park, the Isle of Axholme, Hatfield Chase in south Yorkshire and north Nottingham, the Great Level, Malvern Chase in Worcestershire and Dovegang Lead Mine in Derbyshire.

Vermuyden worked on purely gravitational principles, relying on the outfall to take away the water from the drowned lands by dikes. His operations at Hatfield Chase, financed entirely by Dutch capital, were unfortunate. His Dutch workmen were set upon by the 'fen slodgers' who resented the encroachment on their fishing and shooting preserves and his project succeeded in inundating land that had never before been inundated. After many recriminations he was imprisoned in 1633 but nevertheless was liberated to undertake the drainage of the Great Level of the Fens promoted by Francis, fourth Earl of Bedford, and thirteen other adventurers.

By 1653, 307,000 acres of the Great Level had been restored for food-production in one of the great land-reclamation projects of all time. The shrinkage and desiccation of the land after it was drained created fresh problems which poor administrative arrangements, whereby individual commissioners of sewers looked after individual portions of the scheme as a whole, did little to alleviate. No general pattern of pumping stations was introduced. Cromwell, himself from the Fen country of Huntingdon, appreciated Vermuyden's skill and sent him to Holland to negotiate a treaty for perpetual alliance between the two countries.

Eight years after Vermuyden's death a rhymster rhapsodized:

> '*I sing Floods muzled and the Ocean tam'd*
> *Luxurious Rivers govern'd and reclam'd,*
> *Water with Banks confin'd as in a Gaol*
> *Till kinder sluces let them go on Bail.*
> *Streams curb'd with Dammes like Bridles, taught t'obey*
> *And run as strait, as if they saw their way.*'

His rhapsody was justified, for another contribution of the drained land was to supply England with the oil needed for soap and light. Cole and rape seed crops were staple sources until petroleum and vegetable oils were brought into use.

MAKING RIVERS NAVIGABLE

To carry the manifold products of this industrial renaissance, rivers had to be made navigable. The mitre-gate pound lock, pioneered in Italy, and used in Germany and the Netherlands, was adapted for

English use. John Trew constructed a lateral cut beside the Exe in 1564–7 on which were the first pound locks. These, as A. W. Skempton remarks, 'represent the beginning of a new outlook on water transport in England'.

The integral nature of mine and fen draining and river navigation is illustrated by John Bulmer (*fl.* 1648–58) who undertook 'to find all sorts of mine and minerals and direct the working of them. To direct the making of Engines for the raising of waters for the Service of Cities, Towns etc. To find the levels of any Country, for the draining of Fennes and Lowlands. To direct an easy way of cleansing and deepening of Rivers and Harbours, choaked with sand and wracks etc.'

One of Vermuyden's minor interests was the draining of Dovegang Lead Mine in Derbyshire, a project which was continued by his son, Cornelius, until the latter's death in 1693. He was also concerned in rendering navigable the River Derwent, till it reached the Trent. Other rivers, including the Lea (1574), the Thames (1624, 1635), the Way (1651–3), the Warwick Avon (1639), the Aire and Calder (1699–1703), the Don (1726, 1729), the Kennet (1721–2), the Mersey and Irwell (1722–5) were similarly improved and their improvers, men like Sir Richard Weston, William Sandys, Andrew Yarranton, John Hadley, William Palmer, Thomas Stiers and John Hore, were in a very real sense preparing the way for the canal engineers.

Dr. G. M. Trevelyan wrote in *England Under the Stuarts* (p. 153), 'Engineers there were none except those who used petards for military purposes.' T. S. Willan, however, disagrees and in *River Navigation in England* 1600–1750 (p. 79), insists that in these drainage operations 'the mathematician merged into the surveyor and the surveyor into the engineer'. Sir Jonas Moore, Sir Edward Forde and Sir Richard Weston, he continues, 'must all bear the title of Engineer'. By the end of this period the title 'engineer' was commonly used of an 'improver' of a river and when locks and cuts were discussed before Committees of the House of Commons such experts as gave evidence were invariably described as engineers.

BACON AND TECHNOLOGICAL MOMENTUM

The constant examples of the cross pollination of ideas between the artisan and man of learning in the sixteenth and seventeenth century were given philosophic form and strength by Francis Bacon (1561–1626) who has been called the philosopher of industrial science.

'We need', wrote Francis Bacon in 1592, 'the happy match between the mind of man and the nature of things. And what the posterity and issue of so honorable a match may be, it is not hard to consider. Printing, a gross invention; artillery, a thing not far out of the way; the needle, a thing partly known before; what a change have these three made in the world in these times; the one in the state of learning, the other in the state of war, the third in the state of treasure, commodities and navigation.'

To Bacon's way of thinking, 'the sovereignty of man lay hidden in knowledge'. For the rest of his life, though he lived to become Lord Chancellor of England, his mind was preoccupied with the idea of institutionalizing science. Thus, in his private diary on 26 July 1608 he noted the possibility of obtaining control of one of the existing colleges for his ends: Westminster, Eton, Winchester, Trinity, St. John's—Cambridge, or Magdalene College—Oxford. Reflecting on his backers he lists Russell, a chemist; Poe and Hammond, court physicians; Harriot, the mathematician; Harriot's patron, the Earl of Northumberland; Chaloner Murray, treasurer to Prince Henry; Bishop Andrews, 'single, rich, sickly and professor to some experiments'; Sir Walter Raleigh and the Archbishop of Canterbury.

This college was to contain 'laborities and engines, vaults and furnaces, terraces for insulation'. It was also to produce inventions. 'Foundation of a college for Inventors', he wrote again, 'Two galleries with statuas for Inventors past, and spaces, or bases, for Inventors to come. And a library and an Inginary.' Its fellows were to experiment and correspond with people all over the world. It played (or preyed) upon his imagination for many years after that and emerged posthumously in 1627 as the *New Atlantis*, published by his secretary. In this, Solomon's House, or the College of Six Days Works, emerges the first blueprint for an institution devoted solely to scientific research, to Bacon 'the noblest . . . (as we think) that ever was upon the earth'. Its end was invention—'to effect all things possible'.

BACON'S DISCIPLES

Like Galileo, Bacon had friends and disciples. We have already met one of them in Cornelius Drebbel, a Dutchman (1572–1633), 'a great, singular, learned mechanic', as Boyle called him, who came to England and made a European reputation as a discoverer of a cochineal dye. Drebbel erected a *perpetuum mobile* at Eltham Palace and an automatic clavicord. His ovens and furnaces (used for bread, distilling water, chemistry and incubating eggs) were products of an experimental nature.

Bacon's particular pupil was Thomas Bushell (1594–1674) who possessed 'many secrets in discovering and extracting minerals'. His debts were more than intellectual, for Bacon frequently made good the losses sustained by Bushell in various experiments and speculations. Nine years after Bacon's death, Bushell obtained the grant of the Cardiganshire silver and lead mines originally worked by Sir Hugh Middleton. Finding them flooded, Bushell drained them by adits and built a mint at Aberystwyth for the silver. Not the least interesting of his work at this time was his *Miner's Contemplative Prayer in his Solitary Delves*. Bushell refined his silver and lead with turf and 'sea-coal chark'. At his house at Road Enstone, near Woodstock, he built a garden of fountains and wonders. Artificial thunders, lightnings, rain, hail showers, beating drums, automatic organs and mechanical birds and so delighted the Queen that she endowed it with her name—Henrietta. After heroic efforts on behalf of the King during the Civil War, Bushell turned to work the lead mines of Mendip in Somerset.

A third Baconian (and, like Bushell, a Royalist) was Edward Somerset (1601–67), the second Marquis of Worcester, who employed a German gun founder—Caspar Calthoff—to make models and study inventions. When released from prison the Marquis of Worcester set up, in 1654, Calthoff with a house at Vauxhall as a 'college for artisans'. Worcester's devotion to mechanics led him to collect material for a *Century of Inventions* (1655), in which he suggested (No. 68) a machine 'for driving up water by steam' and also (No. 84) a calculating machine.

THE ROYAL SOCIETY

Bacon's ideas were so exciting, that, during the Civil War, a number of men of science in London met to discuss 'the new philosophy'. Dispersed during the protectorate, some of them met in the room of John Wilkins (who married Cromwell's sister) at Oxford. The small group who continued to meet at Gresham College, London, formed a nucleus for the reformation of the full group after the Restoration. A meeting was held there on 28 November 1660 to discuss 'the founding of a colledge for the promoting of Physico-Mathematicall Experimentall Learning'. Charles II was interested and gave them a charter as the Royal Society in 1662. A second charter extended their privileges and a third gave them lands in Chelsea, which they never obtained. (It became, instead, the Pensioners' Hospital.)

Though unendowed, with members reluctant to pay fees, in ten years the Royal Society was considering the national problems we have already discussed, like naval architecture and navigation and the supply of timber, and publishing them in its *Philosophical Transactions*, the first scientific journal in the English-speaking world. Housed at Gresham College from 1660–1710 (with an interlude at Arundel House (1665–93)), it moved to Crane Court until 1778, then to Somerset House until 1857. From that time till the present day it has been at Burlington House, maintained by the Government though independent of its control.

The Royal Society was a real incubator of British science as this period ends. Prince Rupert communicated to them a paper dealing with the melting and preparation of lead for shot in 1666. Eleven years later a Mr. Greaves gave an account of 'trying the Force of Great Guns' at Woolwich, in butts constructed of oak and elm to simulate the sides of frigates. Iron works in the Forest of Dean were described in 1677–8 and in 1693 the Fellows considered 'The Manner of Making and Tempering Steel'.

Sprat, in his history, written to defend it against attacks, emphasized that the Society proceeded by 'action rather than discourse' and remarked, 'the genious of experimenting is so much dispersed that . . . all places and corners are busy and warm about the work'.

Science now had a forum, a clearing house of ideas and a court of truth, working on the maxim of Robert Boyle (1627–1691) that 'experimental philosophy may not only itself be advanced by an inspection into trades, but may advance them too'.

Robert Hooke (1635–1703), its curator of experiments, gave his name to the law that extension is proportional to force, invented the balance wheel and micrometer and prepared the way for the steam engine. He did as much as anyone to make the Royal Society a success. In the draft preamble to its Statutes in 1663, he wrote:

'The business of the Royal Society is: To improve the knowledge of naturall things, and all useful Arts, Manufactures, Mechanick practices, Engynes and Inventions by Experiment.'

Denis Papin (1647–1712) who assisted Hooke in his secretarial duty from 1629 to 1681, showed his 'digester'—an early steam pressure cooker—to the Fellows and, after working for a scientific society at Venice from 1681 to 1684, returned as temporary curator of experiments until 1687. Papin read over a hundred papers to the Society and showed them numerous experiments. He had been introduced to Boyle by Huygens in 1675 and in Boyle's own laboratory carried out his experiments with 'double barrelled pumps'. He could never leave England for long: at Cassell, in 1697, he invented a machine to raise water from mines by the force of fire and in about 1707 invented a boat with paddle wheels, with which he intended to come to England. In 1707 he offered to build a ship of 80 tons worked by 'a fire engine', if he could get £400: but the Fellows did not respond. In 1711 he outlined the design of bellows to create 'great and lasting blasts of wind to melt and refine ores', which H. W. Robinson has described as 'probably the first suggestion for the creation of blast furnaces such as are in use today for refining'.

John Evelyn, the diarist—whose grandfather had made a fortune manufacturing nitre out of guano—listed the trades which the Royal Society intended to investigate. He divided them into eight categories: 'useful and purely mechanic', then 'meane trades', 'servile trades', 'Rusticale', 'Femal artifices', 'Polite and More Liberal', 'Curious Exotick and very rare seacretts'. 'Enginere', with 'merchand' and 'Navigation' is in the sixth category, 'Clocks Watches Dialls and all Automata' in the seventh, whilst 'Aeolipiles', 'Inventions to fly' are in the eighth.

TECHNOLOGICAL LITERATURE

The arts, mills and engines excited Sprat's colleagues in the Royal Society and elicited from them a steadily increasing flow of literature devoted to what Joseph Moxon called *Mechanick Exercises or the Doctrine of Handworks* (1683). Moxon was very conscious both of the innovatory nature of his book and of his debt to Bacon. Two other Fellows, John Houghton and John Harris, published encyclopaedias of the technical knowledge of the day. John Harris published a *Lexicon Technicum* in 1704, before becoming secretary of the Royal Society in 1709. Ridiculed in 1716, he yet stood high as a popularizer of the scientific knowledge being incubated at that time.

One of the Fellows was Sir Godfrey Copley, a country squire of Sprotbrough, near Doncaster, himself a keen promoter of hydraulic engineering and canals, who supported the efforts of George Sorocold. George Sorocold of Derby was consulting engineer for the first Liverpool Docks in 1708 and probably for the first wet docks at Rotherhithe. He built a great waterwheel for the Derby Silk Mills (one of the birthplaces of the industry of silk spinning) which was regarded as one of the wonders of the century: Daniel Defoe, in his tour of 1727, mentions the Derby Mill as the only one of its kind in England. Sorocold was also engaged to drain mines, and for the Earl of Mar he designed a waterwheel with a crank and beam to drain his Clackmannanshire Colliery.

Sorocold's real genius showed itself in the installation of public water supplies. By using sets of pumps worked by waterwheels to force water up to cisterns, he was instrumental both in supplying Derby and Leeds, and in replacing Peter Maurice's system at London Bridge (which had been destroyed in the Great Fire of London in 1666). Macclesfield, Wirksworth, Yarmouth, Portsmouth, Norwich, King's Lynn, Deal, Bridgenorth, Islington, Bristol, Exeter and Sheffield all employed this versatile engineer.

By 1699 Thomas Savery was demonstrating to the Royal Society an engine for 'raising of water, and occasioning motion to all sorts of mill works, by the impellant force of fire, which will be of great use for draining mines, serving towns with water, and for working of all sorts of mills, when they have not the benefit of water nor constant winds'. Savery was a shrewd business man, as well as a capable engineer. He issued, in 1702, a pamphlet entitled *The Miner's Friend*, which described an improved form of his engine which was designed to raise water in the Cornish mines. Like the pumps in *De Re Metallica*, they had to be staged at various levels of deep mines, pumping up into the sumps of each other. They were very small— the boilers were only $2\frac{1}{2}$ feet in diameter—and very suitable for gentlemen's houses. One was installed at Campden House, Kensington, in 1712 which raised 3,000 gallons an hour. Another was at Syon House. Mines in both Cornwall and Staffordshire used them.

The first steam engine had arrived.

NEWCOMEN'S ENGINE

Another westcountryman, Thomas Newcomen, who used to supply iron tools to the Devonshire tin mines, also took up the drainage problem and began to work on an engine which was an advance on Savery's, in that the vacuum-creating cylinder was cooled internally rather than externally, and the air was released from the boiler by a snifting valve. Since Savery had covered the exploitation of his engine by a patent, Newcomen went into partnership with him and 'a Company of the Proprietors of the Invention for Raising Water by Fire' was formed in 1711. This company built machines which literally saved the mining industry of Cornwall, Staffordshire and New-

Denis Papin (1647–1712?), regarded as the first to apply the force of steam to raise a piston. His "steam digester" (1679) for softening bones was exhibited to the Royal Society.

Jacques Vaucanson (1709–1785), whose collection inspired Jacquard to perfect his famous loom.

castle. The first steam beam piston pumping engine that he erected in the following year near Dudley in Staffordshire raised 120 gallons of water a minute over 153 feet. It had an overall thermal efficiency of 0·5 per cent. Sometime between 1713 and 1718 Newcomen's engines were fitted with automatic valve gears.

The first calculation of its powers was made by Henry Beighton (1686?–1743) who lived near Griff Colliery, near Coventry, where a Newcomen engine was installed. As an F.R.S. and, incongruously enough, editor of the *Ladies' Diary*, he wrote, 'It were much to be wish'd, they who wrote on the Mechanical Part of the Subject, would take some little Pains to make themselves Masters of the Philosophies and Mechanical Laws of (Motion or) Nature; without which, it is Morally impossible to proportion them so as to perform the desired End of such Engines. We generally see, those who pretend to be *Engineers*, have only guess'd, and the Chance is, they sometimes succeed.' Beighton was called upon to erect a steam engine at Newcastle.

Newcomen's engines were exported and Isaac Potter, an Englishman, was invited to construct one for mine drainage at Konigsberg in 1722. Another was built in France at Passy, then outside the city of Paris, to pump water from the Seine for the French capital; and another at Liège. There was a project for building one at Toledo in Spain, also to provide water. Martin Triewald, a Swede who assisted in the construction of one at Newcastle, returned to his native country to construct one at Dannemora in 1737. He, too, became an F.R.S. and a founder of the Academy of Science in Sweden.

ITS APPLICATION OUTSIDE MINING

In 1726 the Newcomen engine was applied in London to provide water. The York Buildings Water Works, established fifty years before at the Lower End of Villiers Street (near the present site of Charing Cross Railway Station) had, like other early works (such as Bulmer's at Broken Wharf and Forde's at Somerset House), pumps worked by horses. A Savery engine, erected there in 1712, had been a failure. This company raised money to purchase estates in Scotland, forfeited by the rebellion of 1715, and on these estates they embarked on coal mining and iron working.

So the definition of an 'engineer' given by Campbell in his *London Tradesmen* (1747) has an authoritative ring.

'The Engineer makes Engines for raising of Water by Fire, either for supplying Reservoirs or draining Mines. . . . The Engineer requires a very mechanically turned head. . . . He employs Smiths of various sorts, Founders for his Brass work, Plumbers for his Lead-work, and a Class of Shoe-makers for making his Leather Pipes. He requires a large stock (at least £500, it is said elsewhere) to set up with, and a considerable Acquaintance among the Gentry. . . . He ought to have a solid, not a flighty Head, otherwise his Business will tempt him to many useless and expensive Projects. The Workmen . . . earn from Fifteen to Twenty Shillings a week; and the Fore man of a Shop, who understands the finishing of the common Engines, may earn much more.'

CHAPTER 9

The Power Revolution, 1740-1800

THE ROADS

Such a variegated, expanding economy needed better roads. War had sharpened the need. When Lancashire and Yorkshire were roused by the invasion of Prince Charles Edward, the young pretender in 1745, Jack Metcalfe, a blind carrier from Knaresborough, was one of the first to rally to the side of George II. He became a recruiting sergeant, joining a company which marched to join the Commander in Chief, General Wade the road-builder, at Newcastle. When Wade was superseded by the Duke of Cumberland as Commander in Chief, Metcalfe accompanied the army to Edinburgh and then to Falkirk, where he took part in the battle. When he was asked how, blind as he was, he dared do so, Metcalfe replied that if he possessed eyes he would never have risked them on a battlefield.

After the rebellion, Metcalfe began to run the first stage coach between York and Knaresborough. Having learnt the value of good roads in Scotland, his experiences as a stage coach driver led him to take advantage of an Act of Parliament which had been passed authorizing the construction of a turnpike road between Harrogate and Boroughbridge. He offered to make 3 miles of it and engaged men, got lodgings for them on the spot, and used to walk the 5 miles from Knaresborough with half a hundredweight of meat on his shoulder for their midday meal, arriving at six in the morning to set them at work. His talent for knowing the exact contour of the land and the nature of the soil through which his road had to pass amazed observers. One of them later described it to the Literary and Philosophical Society of Manchester, saying:

'I have several times met this man traversing the roads with the assistance only of a long staff, ascending steep and rugged heights, exploring valleys and investigating their several extents, forms, and situations, so as to answer his design in the best manner. I have met this blind prospector while engaged in making his survey. He was alone as usual, and amongst other conversation, I made some inquiries respecting this new road. It was really astonishing to hear with what accuracy he described its course and the nature of the different soils through which it was conducted.'

This was the first strand of a web which Metcalfe constructed over the West Riding. From Wakefield to Doncaster, to Huddersfield and to Halifax he plodded with his stick, and the construction gangs followed him. He linked Yorkshire and Lancashire by roads from Stockport to Mottram Langley, and Burnley to Skipton. He crossed the rocky Peak District with roads from Macclesfield to Chapel-en-le-Frith and from Whaley Bridge to Buxton.

Along these arteries, excellent in their day and age, the new industrial communities received new life. Jack Metcalfe even tried to enter industry in 1781, at the age of 64. But he found that road-making was a more suitable and profitable business and we find him still, at 72, engaged on his most difficult task to date, the construction of a road between Bury, Accrington and Blackburn, where local conditions and labour were more intractable than ever before.

THE MILLWRIGHTS

The only natural sources of rotary power at this time were the windmill and watermill. The windmill was used extensively in the Fens, where since Vermuyden's time, the reclaimed land had sunk and the beds of the outfall drainage channels had become as high as the lowest parts of the Fens. Charles Labalye (1705–1785), probably the last of the foreigners who were capable of showing native engineers what to do in this matter, reported on the Fens in 1745 and as a result the new Denver Sluice was rebuilt between 1748 and 1750. This completed, a large number of private and local Acts were passed for the better drainage by engines which worked large scoop wheels. In 1759 the Royal Society sent John Smeaton (1724–92) to Holland and Flanders to study windmills. Smeaton popularized the multi-sailed windmill and constructed a five-sailed flint mill at Leeds. Stephen Hales, F.R.S., had already ventilated prisons and hospitals by windmills in 1752.

The windmill was being brought to a degree of sophisticated efficiency. With its fan-tail (patented by E. Lee in 1745) which automatically kept the sails facing the wind, it was the staple of many an itinerant lecturer's discourse. John Ferguson (1710–76) for instance, used the millwright's craft to adorn his *Select Subjects in Mechanics* (1760) which went to eight editions in his lifetime and were still being published in the nineteenth century. A spring sail with multiple hinged shutters (patented by Andrew Meikle in 1772) increased efficiency. By 1787, thanks to Thomas Mead's patent, a centrifugal regulating governor made them virtually automatic. Mead's friend, Dr. Alderson, was reported in the *Mechanics Magazine* on 17 September 1825 saying:

'I do not give Mr. Watt any credit for his governors or centrifugal regulators of valves, as some have done. The principle was borrowed from the patents of my late friend Mead, who, long before Mr. Watt had adapted the plan to his steam-engine, had regulated the mill-sails in this neighbourhood upon that precise principle, and which contrived to be so regulated to this day.'

Nor were water wheels neglected. In 1760 Smeaton used one to drive two pistons moving in cast iron cylinders to provide blast air for a furnace, belonging to John Roebuck.

THE REVOLUTION IN MATERIALS

Windmills relied on wood, Savery's engine on iron, with lead pipes, and Newcomen's all on iron. But of iron Britain was particularly short and from 1710 to 1765 annual imports increased from 15,000 to 57,000 tons. In 1720 there were only 17,000 tons of native pig iron produced, bought from some sixty blast furnaces distributed over places as far afield as Sussex, Yorkshire and Glamorgan.

To remedy this shortage, British ironmasters bent their energies. Abraham Darby succeeded in 1709 in using coke to smelt native ore at his works at Coalbrookdale. Slowly this discovery penetrated to the blast furnaces of the iron districts as far afield as Northumberland and, by 1760, reached Scotland where John Roebuck had set up his famous Carron Ironworks. Roebuck's guns, known as carronades, were to play their part in the wars in which England was engaged, as were the cannons of John Wilkinson, whose ironworks at Broseley, Bersham and Bradley were to receive an enormous stimulus from the American War of Independence. In 1774 Wilkinson devised a way of boring cast iron cannon: a technique which he was to exploit when asked to bore cylinders for a steam engine. Other ironworkers who rose to prosperity at this time were the Walkers of Rotherham, who built a firm second only to that of Wilkinson. Many ironmasters migrated to Swansea, Merthyr Tydfil and Cyfartha or to Yorkshire. They were further stimulated by being able to produce wrought iron through a puddling process patented by Henry Cort, a dockyard contractor, in 1783 and 1784. This consisted of beating the metal in a reverbatory furnace, 'puddling it' with a bar to oxidize the sulphur, then rolling the product.

The original benefactors of the iron industry, ironically enough, refused to profit by the war trade: Abraham Darby II diverted his attention from the American War of Independence to building, in 1779, the first iron bridge in the world. His associate in this venture, Wilkinson, built the first iron boat and had an iron currency for his workers. He built iron chairs, iron vats and iron pipes for the Paris water supply. He even had an iron coffin in which he was buried. Thanks to this easily available iron, the spirit of invention had material on which to work.

ENTHUSIASM FOR THE NEWCOMEN ENGINE

The year after Newcomen's death John Allen, M.D., in *Specimina Ichnographica* (1730) suggested that a crew pumping air, or two Newcomen engines pumping jets every 5 seconds, could drive a man-of-war at 3 miles per hour. Allen's reticence in developing his ideas was due to his fear of presenting such warships to the world. He also considered propelling the boat by exploding gunpowder in a dry chamber. A more practical use of the Newcomen engine was evident in 1732 at the Coalbrookdale Works of Abraham Darby to raise water for a water wheel. Jonathan Hulls, an English clock repairer of Campden, Gloucestershire, secured a patent on 21 December 1736, and in the following year described what was virtually a paddle-propelled tugboat in his *Description and Draught of a new-invented machine for carrying Vessels or Ships . . . against Wind and Tide or in a Calm*. So the industrial application of its use for rotary motion was early foreshadowed. It was also to be used for providing the blast that enabled Abraham Darby II to turn coal into coke.

In 1733, four years after the death of Newcomen, the Fire Engine Act, giving exclusive rights to his company, expired. This led to a spread of steam engine manufacturing, further stimulated in 1741 when an Act of Parliament waived the duty of 5s. 5d. per chauldron on coals used for pumping in mines. By 1744 J. T. Desaguliers, F.R.S., (1638–1744), a popular scientific lecturer in London, published a full account of the history and operation of the steam engine in his *Experimental Philosophy* (II, p. 435). It was Desaguliers who also put into action the simple form of the reaction water turbine invented by Dr. Robert Barker in 1743, and known as 'Barker's Mill', the first advance to a proper turbine since Hero's aeolipile.

In 1753 the first Newcomen engine was exported to America where it was erected two years later.

Its Effect: A Case History

The imaginative explorations of the potentialities of this new device should not obscure its very real social effects. A case history of this can be seen in Derbyshire, where the Newcomen engine helped to clear the deeper lead mines of water. Three were at work at Winster in 1730 and soon ten were at work in the Peak. Their high fuel consumption greatly stimulated coal production. Coal was also required for a further innovation in lead smelting: the cupola, which obviated the necessity for kiln-dried wood. So began an intensive development of the Alton Seam (which outcrops near a village of that name), the thinnest and poorest of the seams which lies near deposits of fire clay (hence its other name—Ganister Coal). The mining of this fire clay was later to provide the raw material for a refractories industry. To provide food for the inhabitants, more land had to be brought under cultivation and in the Peak and north Derbyshire that involved treatment of the soil with lime. Arthur Young commented that it was customary to use 100 bushels of lime per acre round Chesterfield and as much as 350 bushels are recorded at Chatsworth and Tideswell. So, since the coalfield was flanked both on the east and on the west by limestone formations, coal was taken to quarries to burn down the stone to lime. This lime also served to supply the building trade which was flourishing and using brick instead of stone. The lead merchants built themselves mansions of brick: one of them, Isaac Wilkinson, built Tapton House, outside Chesterfield, which was later to be bought by Robert Stephenson. Bricks were also used for lining canal tunnels and locks, and must have led to a still further demand for coal—and hence for Newcomen engines.

Such case histories could be multiplied all over the country. And the deeper the seams, the more efficient the pumping machinery had to be. In the great coalfields of Tyneside, where 14 engines had been built before Savery's patent expired, there were 137 engines constructed up to the year 1778.

IMPROVEMENTS TO THE NEWCOMEN ENGINE

Such a growth in popularity and diffused knowledge of its working brought a new type of person to undertake its maintenance and improvement. John Smeaton, elected F.R.S. in 1753 at the age of 29, was soon contributing to the *Philosophical Transactions* on 'Improvements in Air-pumping', 'A New Tackle or Combination of Pullies' and 'An Account of De Moura's Improvements in Savery's Engines'. His experiments on model windmills and watermills revealed that the secret of producing uniform motion was to use cycloidal teeth for gear wheels. Smeaton made the first cast iron blowing machinery for the Carron ironworks in Scotland and his improved version of the Newcomen engine had an overall thermal efficiency of 0·8 per cent. He experimented to determine the best proportions for the atmospheric engine and in turn used cast iron for the gearing and shafting of windmills. By 1774 Smeaton's version of the Newcomen engine had an overall thermal efficiency of 1·4 per cent and in 1777 he erected one at Cronstadt Docks in Russia.

Even millwrights like James Brindley (1716–72) were tempted to go over to steam. Brindley at Leek made a flint mill for Josiah Wedgwood to grind material for the pottery industry and machinery for the manufacture of tooth and pinion wheels for a silk mill at Congleton, Cheshire. He then tried to drain mines by Newcomen's engine and patented an improvement for it in 1758.

A Newcomen engine used in the practical physics course of the University of Glasgow broke down in 1763–4. Mending it, James Watt, a 27-year-old instrument mechanic, with a good knowledge of the principles of latent heat (on which his employer Professor Black was an authority), noticed its two salient defects. The first was that after each piston stroke more fuel had to be burned to raise the temperature in the cylinder and the second was that the condensation of steam was not efficient.

Watt's reaction to these two defects was to build a condenser separate from the cylinder and then to utilize the steam (instead of the atmosphere) to work the piston. Continuous experimentation followed.

Unfortunately Watt had to earn his living and his research was put by in order to draft plans for the Caledonian Canal. Then Professor Black introduced him to John Roebuck, who needed pumps for his coal mines at Borrowstounness on the Forth. Realizing the significance of Watt's work, Roebuck suggested a contract whereby he paid

Watt's debts up to £1,000 and guaranteed his experimental expenses for a return of two-thirds of the profits when Watt's engine began to work. And so it was that in 1768 Watt's steam engine, known as Beelzebub, was set up at Kinneil House near Edinburgh. In 1769 the separate condenser, steam jacket and closed top cylinder were patented. It was still a one-cylinder beam engine and it did not work at all satisfactorily. The mines got flooded and Roebuck went bankrupt in 1773. It was at this point that Watt migrated south to Birmingham. In America the 'shot that was heard around the world' was soon to be fired. History had reached one of its great turning points.

THE CORTEX OF CHANGE: BIRMINGHAM

Eight years before Watt moved to Birmingham, there was built on a barren heath, on the bleak summit of which had previously stood a single hut, the Soho factory of Matthew Boulton. Finished in 1765, at a total cost of £9,000, it had five buildings and a reservoir from which water played on a large wheel which drove the factory. The factory made toys and the kind of hardware that was to make the products of Birmingham resound, if not shine, in every house in the land. To these Boulton proposed to add china. He copied *objets d'art*.

The year after his factory was finished he had consulted Benjamin Franklin and Dr. Erasmus Darwin on the possibility of a Newcomen engine adding motive power to his work. Roebuck, another of his friends, told him of Watt's experiments and in 1767 he offered to partner Watt in manufacturing engines 'for all the world'. Roebuck demurred and wanted Boulton to confine himself to Warwickshire, Staffordshire and Derbyshire. This Boulton refused to do. Roebuck's bankruptcy in 1773 gave Boulton his chance and he took it. The engine from Kinneil House, with Watt to superintend it, was working satisfactorily in the Soho works by November 1774 and three months later, on 23 February 1775, Boulton applied for an extension of Watt's patent (due to expire in 1783) for a period of twenty-five years. It was granted and the preamble of the extension (15 Geo. III, c. 61) recognized that '... several years and repeated proofs will be required before any considerable part of the public can be fully convinced of the utility of the invention and of their interest to adopt the same'.

On 1st June 1775 Watt and Boulton signed an agreement for twenty-five years.

THE DEVELOPMENT OF THE STEAM ENGINE

It took twelve years for Boulton and Watt to make a commercial success of their venture. At first they merely fulfilled replacement orders for Newcomen engines in collieries, water works, Cornish tin mines and in Paris. Then, as they built their own, they asked for the cost plus one-third of the value of fuel saved if one of their engines replaced a Newcomen. By 1781 Watt took out his second patent for rotary motion. This turned what looked like being a losing venture into a wildly successful one. When to this he added the throttle valve for regulating the power and actuated it by a centrifugal governor, it became difficult to cope with the demand. And, since Boulton and Watt engines replaced horses, their power was calculated in terms of horse power, a fixed unit which Watt introduced: 33,000 foot pounds per minute. It remained for the electrical age, which ultimately supplanted the steam age, to name its unit of power in his honour.

In 1777 when the first Watt low-pressure engine began operating at Wheel Busy, Chacewater, its economy in coal consumption (from a third to a quarter of the Newcomen engine) was a great advertisement and it enabled pits to be sunk far deeper than before. But the high royalties on the cost of fuel saved stimulated them to defy Watt's patent and encourage one of their own engineers to build his own engines. The litigation which followed led to great bitterness. One of Watt's workmen, William Murdock, who had the tiresome task of preventing the Cornish mineowners from taking political action to cancel the Boulton–Watt patent, was the first to utilize this new power as a means of traction, for in 1784 he built a model locomotive that ran at 8 miles an hour. He also suggested the utilization of gas, which was first used to light the Soho Works in 1798. He patented the long D slide-valve for engines and probably also introduced the eccentric. His improvements, probably suggested by watching the engines at work in the Cornish mines, were not inconsiderable.

THE STEAM ENGINE REPLACES THE WATER MILL

In 1785–8 a symbolic change occurred. A Watt rotative beam engine, fitted with a centrifugal governor acting on the steam throttle valve, was built for the Albion Flour Mills. The mill work was designed by J. Rennie (1716–1821), a pupil of Andrew Meikle.

He represents a new type of engineer, in whom theory and practice were blended. After leaving the University of Edinburgh in 1783 (during which time he had continued to work as a millwright in the vacations), he set up in business. He saw the Bridgewater Canal, the Liverpool Docks and, inevitably, the Soho Works of Boulton and Watt, where he was invited to undertake the planning of the machinery for this mill. Before he accepted, he completed his first bridge—the first on the London to Edinburgh to Glasgow turnpike road at Stevenhouse Mill, two miles west of Edinburgh. Boulton and Watt wished to bind him to them not to work on his own account but he refused.

The Albion Mills which Rennie built for Boulton and Watt were revolutionary. To begin with they marked the first application of steam power to a purpose other than to pump water out of mines. The Soho engines—two double self-acting engines of 50 horse power each drove twenty pairs of millstones, and worked hoists, cranes, fans, sifters and dressers: Rennie himself worked out the diverse uses of the motive power and in the four years it occupied him other innovations were made. One was the adoption of wrought and cast-iron wheels in place of wood. This effected a great saving of power—and led manufacturers of all kinds to consult him on machinery. The corporations of London, Edinburgh, Glasgow and Perth followed his lead in these matters and, in July 1798, he was called in to redesign a new Royal Mint. This, effected by 1810, made it possible to turn out 1,000,000 sovereigns in 8 days. Similar mints in Calcutta and Bombay were on the same scale—that at Calcutta producing 200,000 pieces of silver every 8 hours.

The Albion Mills at Blackfriars excited the admiration of Smeaton and, it should be added, the suspicion of others, for after only three years of operation they were destroyed by fire and never rebuilt.

STEAM BOATS

By 1786 Watt's engines were driving a paper mill, a corn mill, a cotton spinning mill and were supplying power to a brewery and Wilkinson's iron works. Since they were so successful on land, further attempts were made to utilize them to propel boats, especially on the rivers of North America, where circumstances at this time were particularly favourable to experiments. In 1787 James Rumsey (1743–92) was experimenting with jet propulsion (by injection and ejection of water) on the Potomac and John Fitch (1743–98) on the Delaware. Fitch actually started a freight service between Philadelphia and Bordentown but Rumsey came to England where his steam boat was tried on the Thames in 1792, just before he died.

THE SPIRIT OF INVENTION

The patenting of Watt's steam engine and the opening of Roebuck's Carron Iron Works were two events of a decade that was also marked by the patenting of Arkwright's spinning machine. This consisted of four pairs of rollers, rotating with progressive rapidity to draw cotton thread out with ever increasing fineness. It necessitated the building of the first cotton factory in 1771 on the banks of the Derwent, at Cromford; the first of many. Arkwright's mass production of thread by the rotatory power of a mill was the climax of thirty-eight years of reciprocal invention in the textile trade, beginning with Kay's flying shuttle which in turn stimulated Hargreaves' spinning jenny, enabling yarn to be produced to keep pace with the demand. After Arkwright's factories began to operate, Crompton in turn improved the fineness of the yarn and Cartwright and Horrocks its spinning.

None of these later machines, which were taken up by the woollen industry in the next century, could be satisfactorily made of wood, and so the demand for iron increased still more. Nor could they be worked satisfactorily by water (unless, as was done by the Strutts in Derbyshire, large dams were constructed as insurances against rivers running dry). So when the steam engine was applied in 1785 to the spinning industry the factory age may be said to have begun.

The steam engine, as well as the bricks and mortar in factories and canals, need coal. Hence coal production increased by 60 per cent between 1780 and the end of the century; from 6,000,000 to 10,000,000 tons. Iron was even more voraciously consumed when factory owners like Strutt began to build around cast-iron frames in order to make their factories fireproof. Strutt's warehouse at Milford, built in 1793, was the earliest fire-resistant mill. The building of such fire-proof structures as these still further stimulated the firm of Boulton and Watt, which also supplied iron beams for such purposes.

THE LAST YEARS OF THE MONOPOLY

Of all the multiple commissions in which Boulton and Watt executed, their Cornish orders were perhaps most significant for the future. There, their engines were serviced by travelling engineers, paid a fixed monthly sum by mineowners. One of them, Richard Trevithick, had a son of the same name, whose high spirits made him difficult to manage at school, so he was taken around the mines to learn. Though both Trevithicks were opposed to Watt, Richard Trevithick the younger was anxious to join the Soho firm but, it is said, was opposed by Murdock. So, at the age of 20, when his father died, Richard Trevithick began to work on his own idea of a high pressure steam engine. Much to Watt's annoyance, Trevithick chose, instead of Watt's cube-like boiler, a cylindrical one in which he housed a cylindrical firebox with a return flue, together with the cylinder itself, so that it could be kept hot. His first experiments ended in explosions—not unnaturally in view of their cast-iron construction. But their economy in size and weight could not be gainsaid, especially since they could be used for transport.

At the close of the eighteenth century, Trevithick was working on steam boats, road engines and locomotives. Murdock had the same ideas but his employers, Boulton and Watt, inhibited further experiments because of his duties as a maintenance engineer for their products. Boulton and Watt, however, could not inhibit Trevithick, especially after 1800 when Watt's patent expired. And when that happened the country was also at war with France.

ENGLISH MECHANICAL ENGINEERS IN FRANCE

Such expertise was exportable. In France, two leading entrepreneurs in the engineering and machine industry were John Holker, the Lancashire Jacobite, and Michael Alcock. Alcock told the French engineer D. C. Trudaine in 1758 that St. Etienne and La Charité could become the Birmingham of France. Trudaine was impressed and Mrs. Alcock was sent to England to recruit workers, whilst Gabriel Jars (a product of the École des Ponts et Chaussées) came over in 1764 as an industrial spy, touring the provinces. Jars was followed by Marchant de la Houlière in 1755, who invited William Wilkinson to come over from England to establish a royal cannon foundry at Indret in the Loire Valley. Wilkinson suggested Le Creusot as a centre for blast furnaces which would be linked with Indret by the Carolais Canal. J. C. Périer (who also established the first water works in Paris) gave money for the enterprise, as well he might since he was interested in steam pumps. Périer was in England in 1777 and again in 1779, visiting the Soho foundry of James Watt. Watt was very impressed by Périer's works at Chaillot and described it as 'a most magnificent and commodious manufactory for steam engines where he executes all the part(s) exceedingly well'. The French were equally appreciative of British skills. In 1786 in their *Voyage aux*

Montagnes F. and A. de la Rochefoucault-Liancourt admired, in the English factories:

'. . . their skill in working iron and the great advantage it gives them as regards the motion, lastingness, and accuracy of machinery. All driving wheels, and in fact almost all things, are made of cast iron, of such a fine and hard quality that when rubbed up it polishes just like steel. There is no doubt but that the working of iron is one of the most essential of trades and the one in which we are most deficient.'

That was not true of other engineering advances which the French had made, as we shall now see.

CHAPTER 10

French Influence and Example

The French engineer D. C. Trudaine who had been so impressed by Alcock's view that St. Etienne and La Charité could become the Birmingham of France, was responsible for founding an engineering school which in 1747 became the École des Ponts et Chaussées. Its products, in grey uniforms embroidered with gold and silver, were as solidly grounded in test techniques as the bridges of its first director, J. R. Perronet (1708–94). Perronet built many, including that of Pont-Sainte-Maxence using a waterwheel worked by the river to lift the 2,000 pound rammer of his pile driver. Perronet thought thick piles might actually harm a bridge by increasing the flow velocity of the river and thus cause cavitation. One of his pupils, E. M. Gauthey (1732–1806) became a Director of the Burgundy Waterways and later Inspecteur-Général des Ponts et Chaussées in Paris and a real pioneer of experimental methods, which he used in the building of the Church of St. Geneviève, perhaps better known as the Panthéon. His nephew, L. M. H. Navier (1785–1836), was undoubtedly the founder of modern structural analysis. Perronet also made extensive time studies on the manufacture of No. 6 common pins and set a standard of 494 per hour in 1760, thereby becoming a pioneer of production engineering.

A friend of both Perronet and Bélidor, Charles Labelye (1705–81), had come to England in 1725 where he associated with J. T. Desaguliers, F.R.S., and was appointed 'engineer' of the project to build a stone-piered bridge at Westminster to replace the wooden one then in use. Labelye's construction of wooden caissons, 80 feet by 30 feet, in which to build the piers was a novelty in its day and when opened in 1750 the bridge was the largest work of its kind executed up to that time. Labelye bequeathed to Perronet his papers and a model of the bridge.

95

French Influence and Example

THE CANAL DU MIDI AND ITS EFFECT

Frenchmen admired British steam engines, but the British both admired and emulated French canals. As early as 17 February 1670 a deputation from the Royal Society were treated to an account of the way in which two French engineers had overcome many obstacles to join the Atlantic and Mediterranean by the Canal du Midi. This was not the first of such major projects: the Plaine de la Crau had been irrigated by tapping the Durance River north of Marseilles in the previous century and the Loire had been joined to the Seine in 1642.

But the Canal du Midi surpassed these in sheer magnitude, for it rose through its 148-mile length for some 200 feet and dropped 600 feet by means of 100 locks. Tunnels (blasted by gunpowder), aqueducts, a reservoir and other devices made an impression on the English delegation. Begun in 1666 and finished in 1681 it continued to be one of the sights of the 'Grand Tour' undertaken by English noblemen of the eighteenth century.

The young Duke of Bridgewater saw it in 1753 and, after being crossed in love, he threw himself into making a similar if short project to improve his estate. He employed James Brindley to design for him a perfectly level canal which involved the construction of an aqueduct over the Irwell. It halved the price of coal in Manchester. This was the first of some 365 canals in which Brindley had some share. Other noblemen and manufacturers followed suit. Wedgwood was treasurer of the Mersey–Trent canal (he cut the first sod in 1766) which enabled him to bring Cornish china clay to his famous Etruria factory.

THE NEEDS OF THE COURT: CONSPICUOUS CONSUMPTION

In England the mines posed major engineering problems, but in France it was Versailles, seat of the Court. Courts must be comfortable and impressive and every refinement of engineering skill was applied to ensure that this was so. Water was essential, both to drink and use. Jets and fountains were deservedly popular in noblemen's gardens. So a Fleming, Lintlaer was employed to construct pumps on the Seine to supply water to the Louvre and the Tuileries. The service was extended to Versailles by a Dutchman called Rannequin who raised water 553 feet above the river by means of fourteen undershot waterwheels working, by complicated linkage rods, sixty-four

Sir William Congreve (1772–1828), standing by one of his rocket guns of the type used at the Battles of Bladensburg and Leipzig.

Thomas Telford (1757–1834), one of his famous aqueducts and bridges in the background, and plans on the table before him.

pumps at one reservoir, seventy-nine pumps at a second, and eighty-two at a third. This trail of rods and chains running up the hill led to the machine being called a 'Monument of Ignorance', since it dissipated 80 per cent of the power produced by the wheels. Further attempts to eliminate this led to twisting of the pipes. Four years later Edmé Mariotte (1620–84) in his *Traité du mouvement des eaux et des autres corps fluides* (published posthumously in 1686) gave the earliest rules for assessing the strength of water pipes. His interest in hydrostatics and hydrodynamics was appreciated in England by, amongst others, Sir Godfrey Copley. Mariotte appreciated a phenomenon usually credited to his English contemporary, Robert Boyle, that air when compressed has a 'spring'.

Courts need tapestries and bright clothes. Colbert established tapestry-making as a French public service in 1662 and imported the Dutch dyer van Kerchoven who, with his son, was responsible for the famous Gobelins scarlet, a colour which takes its name from the family which took over the Government tapestry and dye works in 1691.

Centring around the extensive use, in aristocratic France, of tapestries were the efforts of eminent chemists like Charles Dufay de Cisternay (1698–1739), Jean Hellot (1685–1766), Pierre Joseph Macquer (1718–84), Claude Louis Berthollet (1748–1822) and Chevreul (1786–1889). Two of these, Hellot and Macquer, were not only directors of the Dyeing Industries of France but also of the Royal Porcelain Factory at Sèvres. Berthollet was associated in the reform of chemical nomenclature and was responsible for the introduction of chlorine as a bleaching agent. Macquer's dictionary of chemistry was translated by James Keir, F.R.S., who published it in England in 1776. To Boulton it was 'the most compleat work ever yet published' and he sent it to his son in Germany. Keir himself applied the chemical knowledge he had gained by establishing a great chemical works at Tipton in Staffordshire, where amongst other things, he manufactured sulphuric acid.

THE ACADÉMIE ROYALE DES SCIENCES

Large, wealthy, rigidly centralized and intensely interested in trade and navigation, colonial expansion and war, France was the dominant nation of the late seventeenth and early eighteenth century. Colbert, minister to Louis XIV, capitalized on the scientific legacy of such distinguished French scientists as Gassendi (who had been holding informal meetings as early as 1620) and Mersenne (who had been the Hartlib of Europe) to found the Académie Royale des Sciences in 1666. Mariotte was one of the early members. A later one, R. A. F. de Réaumur (1683–1757) was commissioned in 1711 to examine every manufacturing process. During the years 1761–81 the Paris Academy published some twenty volumes with full descriptions of inventions up to that time.

Academicians worked on practical problems set them by the Government. They increased in number from twenty to seventy and were affiliated to some thirty-seven provincial academies of which Montpellier (1706), Bordeaux (1716) and Toulouse (1746) were perhaps the most outstanding in applying science to practical ends.

Its reputation was such that scientists from other countries came to work in Paris, like Christian Huygens of Holland (1629–95) who employed Denis Papin, just as Robert Boyle employed Hooke in England. And in the same year (1678) as Hooke had outlined an engine worked on the atmospheric principle, Huygens outlined one worked by the explosive force of gunpowder in a cylinder. Denis Papin, though he worked for Hooke, appropriated Huygen's idea, substituted steam for gunpowder and was successful in producing the first steam engine with a piston. In this, steam was injected into a cylinder, ejecting the air. When condensed, the vacuum created in the cylinder enabled the atmospheric pressure to force the piston down. This engine, he envisaged, could be used for raising water from mines, throwing bombs and propelling ships. But the way in which it was to drive the boat illustrates how far he was really thinking of steam power: for the pumping engine was to throw up water to turn a waterwheel which was to turn the paddles.

The traditionalism of the French shipwrights was broken down in 1643 by Richelieu's order that every shipbuilder had to pass an examination before a competent jury and every ship's design had to be similarly tested. Colbert went further and convened a conference of naval architects and admirals in 1680 and 1681. Three schools for the theory of construction were opened. A school of design was opened after 1739 by Duhamel, which had a number of distinguished professors and graduates like Pierre Bouguer (1698–1758), who tried out the propeller in 1746, and E. L. Camus. By 1741 the old title of *maître charpentier* had been changed to *constructeur*. Just before the French Revolution J. N. Sané (1754–1831) had built a fast sailing man-of-war which was repeatedly copied in both England and America.

The French established the principle of model testing. The Chevalier de Birda, a member of the Academy of Sciences, made tests in 1763–7 and in 1776 d'Alembert, Condorcet and the Abbé Bossut (1730–1814) made further tests by towing boats in a basin 100 feet long. They decided that friction was of no importance.

THE GÉNIE OFFICERS

Fighting neighbours from her borders, France needed above all good communications and strong border fortresses. Vauban, himself a soldier, became in 1678 Inspector General of French Fortresses and in 1703 Marshal of France. His star-shaped fortifications and defended areas were only matched by his bold plans for connecting the coastal ports to the hinterland by a network of canals. Rhine–Rhône and Alsace–Rhône were two of the axial water arteries he envisaged.

To execute his projects he suggested to the Minister of War, Louvois, the creation of a *Corps des Ingénieurs de Génie Militaire*. Vauban's *génie* officers were the first real civil engineers and their status was further enhanced by the establishment in 1716 of the *Corps des Ingénieurs des Ponts et Chaussées*—a phrase first used by Colbert. A special school was created in 1748 to train them, unique in Europe. Taught and led by such men as Gautier (1660–1737) the road and bridge builder, and B. F. de Belidor (1693–1761), the students utilized such mathematics, statics, geometry and tests as existed, and their mentors, Belidor especially, provided them with excellent texts. Belidor's *La Science des Ingénieurs* (1729) and especially his *Architecture Hydraulique* (1737–9) span the gap between Vauban and Navier, the famous engineer of the early nineteenth century. Belidor's work on pump design was itself a landmark. This school or institution taught civil as well as military engineering in the first two years of its foundation. Amongst its pupils were Monge, the famous mathematician, Lazare Carnot, the great war minister of the Revolution, and Cugnot, who designed the first steam carriage.

In casting about for means to accelerate transport of materials and intelligence, Nicholas Joseph Cugnot (1725–1804), attempted to substitute a steam tractor for horses to draw artillery. In 1769 and 1770 he developed an automobile on three wheels with a steam engine powering the single wheel in front. It was top heavy and lacked steam pressure. Its two cylinders worked by means of levers, pawls and ratchets. Its maximum speed was four miles per hour.

Others experimented on flight. At Versailles on 19 September 1763 the brothers Joseph-Michel (1740–1810) and Etienne-Jacques (1745–99) Montgolfier launched a balloon lifted by hot air. A sheep, a cock and a duck were caged in a wicker basket beneath it. The object of the experiment was to see whether or not the rarified atmosphere would have any effect on the physiology of animals not accustomed to it. The balloon remained aloft for 8 minutes and descended in a forest $1\frac{1}{2}$ miles away, with the animals unharmed, except for a wing of the cock damaged by a kick from the sheep. A month later, on 15 October 1783, J. F. Pilâtre de Rozier became the first human to take the air in a tethered balloon. After a number of such ascents, he undertook a free flight of 25 minutes across Paris.

Jean-Baptiste-Marie Meusnier (1754–93), a lieutenant in the Corps of Engineers, presented a memoir to the Academy of Sciences in Paris suggesting a ballonet which could be filled with air or emptied, to vary the height of the balloon. He went on in the following year to suggest an ellipsoid balloon propelled by three manually operated air screws. A car was to be suspended from the balloon and a portable tent to protect it was to be carried. Meusnier was later killed in battle, but not before his ballonet was made by the brothers Robert, financed by the Duke of Chartres. They made a trial flight in 1784 but, having no valve fitted, only escaped disaster by the Duke's dexterous puncture of the silk envelope with a flagstaff which they were carrying.

Another French officer, Baron Scott of the Dragoons, published a proposal for two ballonets—one in the front and one in the rear, to obtain height or to descend. His suggestion was carried out in 1804 by John Pauly, a Swiss gunsmith and engineer in Paris.

INVENTION IN FRANCE AND ENGLAND: A CONTRAST

Whereas in France engineering (mainly for war) was organized from above, in England inventions (in textiles, hardware and machinery) arose spontaneously from below. John Smeaton (1724–92), the first Englishman to describe himself as a civil engineer, was also concerned with power mechanisms or mechanical engineering. He was also typical of British engineers of this period in his desire to gather round him a group of like-minded men to discuss their work. This he did on 15 March 1771 when the Society of Civil Engineers first met at the King's Head Tavern in Holborn. It is often known as the Smeatonian Society.

Smeaton's was not the first of such groups. The drainers of fens and mines had professional associations. So did the road and canal makers. John Grundy, successor to Captain John Perry (1670–1732) as engineer to the adventurers of Deeping Fen and author of a scheme for 'restoring and making perfect the navigation of the River Witham', was a leading member of the Gentleman's Society at Spalding, one of the many local societies for the advancement of knowledge by mutual exchange of opinion.

Plans for a national Chamber of Arts to preserve and improve 'operative knowledge, the mechanical arts, inventions and manufactures' were mooted by Peter Shaw (a disciple of Bacon) in 1721–2. It should consider, he said, 'the best manner of Improving lands, employing the poor, Draining, and Working Mines'. Francis Peck had also proposed in 1738 a subscription for the encouragement of Arts and Sciences. This current of opinion finally crystallized in 1754 when a Society for the Encouragement of Arts, Manufactures and Commerce was formed.

The Society of Arts was formed in 1754 by William Shipley (1714–1803) a friend of Franklin who hoped to stimulate industry by means of prizes for good ideas. He was supported by four Fellows of the Royal Society, among them Stephen Hales, the physiologist and ventilating engineer. Its membership increased rapidly and, with the money they subscribed, premiums were given based on the recommendations of six committees. These, dealing with Agriculture, Chemistry, Polite Arts, Manufactures, Colonies and Trade would often meet for each night of a week considering the ideas sent to them. All such ideas and inventions were published, either as pamphlets or as articles in periodicals like the *Gentleman's Magazine* or in its own *Transactions* (1783–1851). Here contemporary advances in mezzotint-engraving helped to put models before the public. Especially valuable were the descriptions of machines in the repository, made available by William Bailey and his son from 1772 onwards. These machines included a number of ploughs and agricultural implements and show the influence of Arthur Young. By 1776 the Abbé Beaudeau had founded in Paris 'A Free Society for the Encouragement of Inventions which tend to perfect the application of the Arts and Trades in Imitation to that of London'. Though it only lasted for five years it succeeded in stimulating the foundation of others at Rheims (1778), Rouen (1787) and Amiens (1789).

Josiah Wedgwood had a scheme for an industrial research organization in the pottery industry, Joseph Priestley another for uniting the investigators of electricity, 'the cleanest and most elegant that the compass of philosophy exhibits'. Both schemes were independent of a third group of which both were members, called the Lunar Society of Birmingham, established some time between 1765 and 1768 by Dr. William Small. Small had lived in Virginia (where he had been a friend of Benjamin Franklin) and on his return to Birmingham gathered round him, perhaps on the analogy of Franklin's junto, a small group of men interested in science. On his death this became the Lunar Society of Birmingham—so called because its members met on the Monday nearest the full moon. These members included Joseph Priestley, the scientist, and Josiah Wedgwood the potter. Small is said to have been responsible for bringing Boulton and Watt together.

John Smeaton attended a meeting of the Lunar Society and members of the Lunar Society (like Boulton, Watt and Priestley) were members of the Smeatonian Society. Thus the Smeatonian Society enjoyed the services of men who were in correspondence with French and American scientists. Watt's son, James, spoke French and German fluently and corresponded with Argand, Lavoisier and others. So too did Priestley's and Boulton's sons.

The Smeatonian Society's first president, Thomas Yeoman, had been employed on draining the fens for some thirty years as well as upon numerous river improvements. He was also a ventilating engineer and undertook the fitting of Dr. Hales' ventilators in Hertford, Bedford, Aylesbury, and Maidstone Gaols, as well as St. George's Hospital. Its connection with the Gentleman's Society of Spalding was emphasized by the presence of John Grundy (1719–1783), also a fen engineer, son and namesake of one of the founders of that society who had died in 1753.

The influence of French science was felt in the small scientific subscription libraries which were being formed all over England in the eighteenth century. Thus the one maintained by the Derby Philosophic Society took the *Journal des Sçavans*, the *Histoire de l'académie royale des sciences* and the *Journal des sciences utiles* as well as the *Memoirs of the American academy of arts and sciences at Boston* and the *Transactions* of the Royal Societies of Edinburgh and London. The Derby Philosophical Society, established in 1784, had as its president Erasmus Darwin and amongst its members was William Strutt, F.R.S., the textile technologist, and John Whitehurst, F.R.S., the Derby watchmaker. And Strutt, describing the process of 'invention', disproves the notion that these eighteenth-century technologists were simple fellows, for he said, 'We must reason from what we know & inventing is only looking at all sides of a thing and putting it in different points of view & by long habit & a great store of ideas this becomes almost mechanical.'

The co-operation in these myriad, interlocking, protean groups between manufacturers and scientists in England enabled a lead to be built up over France in the industrial race, a lead which was to become apparent in the war that clouded the end of the eighteenth century and the beginning of the nineteenth.

THE NAVAL ASPECT OF BRITISH CIVIL ENGINEERING

Just as French civil engineering advances arose from the needs of the army, so British advances stemmed from an increasing export trade and naval exigencies. B. F. de Belidor (1698–1761) was the real inventor of the shell commonly attributed to the British Major Henry Shrapnel. Pontoon bridges, designed by F. J. Camus in 1710 and D'Herman in 1773, carried the French armies marching three abreast over such rivers as they encountered. Gun carriages designed by C. F. Berthelot (1718–1800) set a pattern for French artillery. Polygonal defences, designed by the Marquis de Montalembert, supplanted the previous defensive ideas of Vauban.

But in Britain, civil engineering projects had a more specifically dual purpose rôle. Lighthouses and harbours, it is true, serve navies, but also mercantile argosies. And it is here too that the almost transcendental ubiquity of Smeaton reveals itself again.

Smeaton's own genius was first revealed when he built a lighthouse on the Eddystone reef outside Plymouth harbour in 1759. This led him to make an exhaustive study of currents and utilize a mixture of pozzolana from Italy and Aberthaw lias. With his cement he set about building the third Eddystone lighthouse: the first had been destroyed by a gale in 1703, the second by fire in 1755. Both were wooden structures, though the second was partly lined with stone. Smeaton, after a thorough examination of the reef, built in three years a model of interlocking stone. Work began on 3 August 1756 and was finished on 16 October 1759 when the light, supplied by twenty-four candles, was exposed for the first time. It lasted for 118 years and was only removed because the reef on which it stood was being undermined. This monument to his capacity led him to be employed as a bridge engineer. Bridges at Perth, Banff and Coldstream were followed by others at Edinburgh (a reconstruction) and Hexham (unfortunately swept away in a flood).

Amongst his many commissions was one for an engine for the New River Head: an enterprise which, in the eighteenth century, enlisted the service of Robert Mylne (1734–1811) a member of a family which from the early sixteenth century had provided master masons to the Crown in Scotland. Mylne was concerned with the Gloucester and Berkeley Canal and in 1767 became the joint and, in 1770, the sole engineer of the New River project, holding the post for forty-one years till succeeded by his son, W. C. Mylne. His most recent biographer remarks that 'he was the last great architect of note who combined to any great degree the two avocations of architect and engineer'. Mylne's bridges—most notably that at Blackfriars—showed a sense of style derived from his studies in Italy. He also specialized in building the smaller country house.

One of the leading members of the Smeatonian Society, which in 1778 spent one of its meetings 'canallically, hydraulically, mathematically, philosophically, mechanically, naturally and socially', was John Rennie who spent his student days in the University of Edinburgh (where he studied Belidor's works) and in the mill workshops of Meikle. This was why he was so well equipped to undertake the building of the Albion Works.

The Albion Works had one permanent result: it anchored Rennie in London where he opened a business in Holland Street, Blackfriars, as a civil engineer. The Kennet and Avon Canal, the Rochdale Canal, the Lancaster Canal and the Royal Canal of Ireland from Dublin to the Shannon are some evidences of his activity. The East and West India Docks, Holyhead Harbour, Hull Docks, Ramsgate Harbour and naval dockyards at Sheerness and Chatham, like the Albion Mills, generated further improvements and inventions like the diving bell (used at Ramsgate), the bucket-chain dredger (used at Hull) and hollow walls.

But he never lost interest in his first love—bridge building. His Scottish ventures at Kelso and Musselburgh were innovatory in the flatness of the roadway which they offered to travellers. Because of their utility and beauty, Waterloo (1810–17), London and Southwark bridges were commissioned from him before he died.

With Rennie, the magnitude of the engineers' task became apparent. One of his largest projects was the construction of a mile-long breakwater at Plymouth, composed of well over three-and-a-half million tons of stone. He conceived it and his son completed it. This son, Sir John Rennie, was engaged on the building of Waterloo Bridge when he was only 19. Six years later, in 1819, he was sent to study the continental works of engineering and in 1821 took over the civil engineering side of his father's business.

In April 1793, Rennie, with Mylne and others, had resurrected the Smeatonian Society of Civil Engineers which had lapsed the previous year on Smeaton's death. This time, members were divided into two classes: real engineers employed as such and honorary members who were men of rank and fortune. It was the first engineering society to be formed in the world and and for over a century its treasurers were members of the Mylne and Rennie families. And its roots may be detected in one of the toasts, 'success to waterworks, public or private that contribute to the comfort or the happiness of mankind'.

This striking emergence of the British civil engineer was coterminous with the acceleration of other developments during the Revolutionary wars. France, embattled against a continent in the last decade of the eighteenth century was for nearly a generation to be in a state of intermittent but intensive warfare with Britain. As the next chapter shows, this war was one of the crucibles of modern technology.

CHAPTER 11

The Franco-British Wars, 1793-1815

THE INCREASING IMPORTANCE OF CHEMISTRY

A close association of scientists with the French government stemmed from the relentless pressure of events during the Revolution. Antoine Lavoisier (1743–94) was not only a member of the constituent assembly but secretary of the committee on weights and measures that secured the adoption of the metric system. His subsequent execution was not because of his discoveries but because he had been associated with the tax-farming system and it in no way indicated the hostility of successive French governments to science. On the contrary, J. A. Chaptal (1756–1832), author of the standard chemical textbook of his day, was recalled from the mountains where he had fled by the Committee of Public Safety. 'Chemistry', said the committee, 'is a profession from which the republic should obtain the most powerful aids for its defence.' And Chaptal responded by establishing the famous powder factory at Grenelle which in eleven months made 22,000,000 pounds of gunpowder. Chaptal subsequently became Minister of the Interior under Napoleon, establishing a central pharmacy for the economic distribution of chemicals, importing machinery for spinning and weaving, and encouraging the growth of sugar beet.

Cut off from supplies of colonial cane sugar during the Napoleonic wars the French took up with eagerness the manufacture of sugar from beet. Originally outlined by Marggraff in a paper to the Berlin Academy in 1747 the process was pioneered by Marggraff's pupil, Achard, in a factory at Cunern, where beets were sliced, the liquid squeezed and boiled, treated with lime and inspissated. The French factory at Passy, built by Chaptal in 1808, followed hard on the heels of Cunern and was the first of many. When Delessert in turn developed the modern manufacture of beet sugar, Napoleon personally decorated him in the Passy factory. Sugar refining in England owed much to E. C. Howard (1774–1816), consultant to the West Indies Association of Merchants and Planters, who devised a technique of refining by means of fixed coils, heated by steam, immersed in liquid. Howard also discovered fulminate of mercury as a detonator for explosives in 1799.

Another shortage was of alkali. An appeal to French chemists 'to render vain the efforts and hatred of despots' elicited several suggested processes, the most successful being that of the apothecary Leblanc (1742–1806) who showed that it could be produced from salt, sulphuric acid, chalk and coal: all materials in good supply. The salt was treated with sulphuric acid to produce sodium sulphate; this, when heated with chalk and coal in a reverberatory furnace and lixiviated with water, produced sodium carbonate on evaporation. This could be used as washing soda or, when treated with lime, used as caustic soda for soap and candles. It was so successful that before the end of the war it had been introduced into England and in 1823 the first heavy chemical works for its production was established near Liverpool by James Muspratt (1793–1885).

The need for preserving food led the French government to offer a prize in 1795 for the most practical method of preserving food for the army and navy. In 1809 this was won by François Appert (1750–1841), a Parisian confectioner, for his use of heat-sterilized glass bottles. His discovery, published in French in 1810, was translated and published in English in the following year as *The Art of Preserving all kinds of Animal and Vegetable Substances for Several Years* (1811). It is interesting to note that the bacteriological importance of the discovery was quite unappreciated. Adopted in England by Bryan Donkin (1768–1855), a millwright who in 1803 had begun to make paper milling machinery at Bermondsey, it was further modified by substituting tins for the glass bottles. Donkin was, incidentally, a friend both of Henry Maudslay (with whom he took out a patent for a combination of spur gears and an epicycloid train) and of Joseph Bramah (to whom he sold his patent of a steel pen). His enthusiastic forays into food preservation culminated in the presentation of samples to the Prince Regent in 1813 and the establishment of a business known as Donkin, Hall and Gamble. From Britain the discovery spread to America, where by 1820 William Underwood and Thomas Kensett were canning food in Boston and New York.

Napoleon's blockade in 1805 and rupture with Russia created a shortage of tallow and increased, amongst other things, the price of candles. As a result there was a great increase of substitute illuminants, like coal gas. In France and England the coking of coal for smelting iron was becoming a major industry, especially for armaments, and the gas so produced was easily tapped. In 1799 Phillipe Lebon, a teacher of mechanical engineering at the *École des Ponts et Chaussées* had succeeded in patenting a fishtail burner to use with gas produced from wood, oil or tar. Another Frenchman, Aimé Argand, had the idea of burning oil by means of a wick, shielded by a glass chimney with air introduced through the centre of the wick holder. Boulton bought the right to use his patent in England, but a series of law suits followed. Boulton's Cornish agent, William Murdock, produced gas from coal and illuminated the Soho Factory in 1798. The celebration of the Peace of Amiens in 1802 offered a chance for further display of the new illuminant and the habit spread. Samuel Clegg (1781–1861), Murdock's pupil, joined the Gas Light and Coke Company which was founded, after some trouble, in 1812 by a German emigré, Winsor.

Clegg laid down the techniques of modern gas engineering, whereby gas was scrubbed, measured and stored. He even had to light the gas lamps of Westminster Bridge himself. From Westminster Bridge the light spread. The Royal Society, worried about the safety of the gas containers, sent a deputation to examine them. Clegg's technique of reassuring them was to drive a hole in the side of one with a pickaxe and light the gas that came through. Clegg's successful demonstrations led to the rapid spread of gas illumination to Boston (in 1822), New York (in 1823), Hanover (in 1825) and Berlin (in 1826). Clegg's partner, F. C. Accum (1769–1838), was a consultant chemist and by joining the board of the London and Westminster Chartered Gas Light and Coke Company set a precedent as the first chemical director of a public concern. His *Practical Treatise on Gas Light* (1815) was a trade textbook. Another associate, Alexander Garden (after whom the gardenia is named), discovered naphthalene in the residues of the distillate of coal.

THE GARNERING OF TALENT IN FRANCE

To a country embattled against a continent one of the most precious assets is human ability. Lazare Carnot (1973–1823) the organizer of victory for the French armies in the early stages of the war had been a pupil of the brilliant geometer Gaspard Monge (1746–1818) whose lectures on fortifications at the Mézières military academy he had attended. These lectures did more than teach Carnot: from them sprang Monge's development of descriptive geometry and his pupils' own development of analytic geometry.

In 1794 Monge forged an instrument which would garner technological talent for France: the École des Travaux Publiques, modified a year later to become the École Polytechnique. Monge himself was the director and round him clustered other teachers like Fourcroy, Vauquelin, Berthollet and Guyon de Nourveau, chemists as notable and distinguished as Chaptal himself, who also lectured there. Admission was by ability to profit from a rigorous three-year course, of which the first two were in pure science. The *Journal Polytechnique* became one of the leading mathematical publications of the world. So did the text books written for the students by Monge, Prony, Lacroix and Haüy. Here for the first time engineers were being trained in fundamental science by means of laboratories and experiments.

The enthusiastic polytechnicians were to serve not only France but all Europe as well. Two of them bear mention. One, Charles Dupin (1784–1873) assisted in the construction of the Arsenal at Antwerp, surveyed the ports of Holland and was afterwards sent to Genoa and Corfu. His papers appeared from time to time in the *Recueil des Savants Étrangers* and he founded the maritime museum at Toulon. He wanted to defend his friend Carnot, the great French engineer, who had served Napoleon but Carnot preferred to go into exile. Dupin made a visit of inspection to England in 1816, an account of which was published as *Voyages dans la Grande-Bretagne* (1820–4): a popular work on both sides of the channel. As a professor at the Conservatoire des Arts et Métiers he exerted an influence in moral and political as well as engineering affairs. Another was Victor Poncelet (1759–1867) who carried forward Monge's interest in geometry even in a Russian prison. There his meditations were so fruitful that on his release he was to become recognized, not only as the founder of projective geometry, but a contributor to the theory of the water turbine.

The spirit of the Polytechnique breathes through the work of one of its most famous pupils who entered in 1814: Auguste Comte (1798–1857). Comte's system of 'positive' philosophy, rejecting all metaphysics, exercised a profound influence over subsequent English thinkers like John Stuart Mill and Herbert Spencer. And in the sphere of political reform another Polytechnician, Victor Considérant (1808–1893), made an impact as a protagonist of a kind of socialism whereby people could associate in groups for productive enterprise. This active, enterprising, questioning school in its heyday was the object of emulation in both Europe and America. In Vienna, Zurich and Russia, institutions were modelled upon it, whilst in America, West Point and the Rensellaer Polytechnic in New York were, by the admission of their principals, profoundly influenced by its curriculum.

In view of the longevity and success of the École Polytechnique, one wonders what would have been the fate of France had another of the technical institutions created in 1794 enjoyed a similar existence: the National Aerostatic School at Meudon. Created to train balloonists for the French Army, it had the privilege of witnessing two balloon companies operating on the Rhine and the Meuse. Napoleon seems to have had little use for them and allowed the school to languish. Had he employed them at Waterloo, they might have forewarned him of Blücher's advance and saved the day.

The legacy of the École Polytechnique to France was a cadre of systematically trained engineers with a matchless theoretical background. It was from this cadre that G. E. Haussman (1809–1891) obtained the technical assistance that enabled him completely to remodel Paris and create a model for subsequent town planners. Gustave Eiffel (1832–1923), another product of the École Polytechnique, built bridges in Europe, Africa and Indo-China, carefully experimented on aerodynamics and wind pressures, employed factory-made parts for his structures like the Garabit Viaduct (1880–4) and symbolized his engineering skill with a 1,000-foot tower of iron, erected for the Paris Exhibition just 100 years after the French Revolution had begun.

Another pioneer venture in Paris was the establishment of a combined museum of applied science, public technical library and teaching, testing and research station in the old Benedictine Priory of St. Martin des Champs in 1799. The basis of this remarkable institution, called the Conservatoire des Arts et Métiers, was the collection of automata made by Jacques de Vaucanson (1709–1782). In this collection there was an old mechanical silk loom which Vaucanson had made whilst Inspector of Silk Manufactures. After the Conservatoire had been established, J. M. Jacquard (1752–1834) rummaging in the collection, put together Vaucanson's old loom, improved it, and by 1804 had produced his own. This became the basic automatic weaver of the nineteenth century. It was an inspiration to Charles Babbage, the English mathematician, when he began to work on a mechanical computer a generation later.

THE PRODUCTION OF WAR MATERIAL IN BRITAIN

Soldiers, whether British, Prussian, Russian or Swedish, need uniform and blankets. All the countries were supplied with equal facilities by Benjamin Gott (1762–1840) of Leeds whose great factory at Bean Ing, established in 1792, was driven by a Boulton and Watt engine and was employing 1,000 people by 1800. His application of cotton factory techniques to the woollen industry started a boom in Leeds. 'Houses, nay, whole streets', wrote a visitor to that town in 1800, 'are building almost every year, and if we may hazard from external appearances the trade and manufactures of this town seem in their effect almost equally lucrative to a Peruvian mine.' The population of the town increased fivefold between 1771 and 1821.

An even greater Peruvian mine opened up in the iron foundries, busy providing chains, anchors, nails and guns as never before. This voracious hunger for ordnance supplies, sharpened by the increasing duties on bar iron, filled the order books of men like Richard Crawshay (1739–1810) of Cyfartha and enabled him to supplant Wilkinson as the largest ironmaster in the kingdom and probably in Europe.

Birmingham prospered and supplied two-thirds of the arms used by British troops in a war that lasted, with brief intermission, for a whole generation. Trade in guns had become so great that by 1813 a private proof-house was established in the town. And in addition to guns, chemical products, from sulphuric acid (used for brass founding, metal-pickling and refining, at this time) to the glass containers in which it was transported, flowed from factories like that of James Keir, F.R.S. (1735–1820). His alkali works offer a point of departure for those anxious to trace the connection between chemical engineering and society. Like Gott, Keir was a friend of Boulton and especially Priestley who, until 1794 when his laboratory was burnt by an angry mob, was a minister in the town.

Boulton was in his element. He was selling what all the manufacturers wanted—power—and even medals: 20 tons of them were sent from his factory to the French Revolutionary government before it undertook hostilities against Britain. And when Boulton and Watt's monopoly expired in 1800, there were a number of engineers who were straining at the leash to put their own power plants on the market.

THE DEVELOPMENT OF STEAM POWER IN BRITAIN

Both the supply and mobility of armies depended on horses. The shortage of horses and fodder in England accelerated the use of mechanical power in collieries, while the absorption of 140,000 men in the navy and 350,000 in the army (and the necessity of maintaining them) provided arguments for introducing steam power into factories. These factories grew from the basic technologies like textiles, to paper making and printing. Liberated from threats of litigation by the expiry of Watt's patent in 1800, Trevithick hastened to experiment on his own projects and on Christmas Eve, 1801, got his first steam car under way up a steep hill. Though its boiler could not keep up enough steam and though it was inadvertently destroyed through being left unattended with the fire undrawn, it made history. Steam propulsion had arrived. Trevithick now built a second steam car which he sent to London and another which he ran on rails in Wales. It cracked them—they were only flat cast-iron—so it was used to drive an ironworks hammer instead. In 1805 he sent two railway locomotives to Newcastle. In the following year he set three steam dredgers to work on the Thames.

London fascinated him. He conceived the idea of a tunnel under the Thames and actually constructed 1,000 feet of it before it was abandoned. He also rented a piece of land in Euston Square, fenced it, and charged spectators a shilling admittance to see and travel on a locomotive which he had brought up to the capital. He patented a floating dock in 1809. In 1810 he returned to Cornwall and set to work redesigning an engine with higher pressure than ever, that became world famous as the 'Cornish Engine'. He serviced it to such a peak of efficiency that it, in turn, consumed less coal than Boulton and Watt's engines. He applied it to agriculture for threshing, grinding, sawing and ploughing. He might have become the greatest figure in steam engineering had he not accepted an invitation to go to Peru and make his fortune draining its silver mines.

THE STEAM ENGINE ON WHEELS

As Trevithick was busy with the first models of his high pressure Cornish engines a young plugman at Dewley Colliery near Newcastle was learning, at the age of 18, to read and write at a night school. By 1802, this industrious northern youth became an engine man at Willington Ballast Hill, where he took up the repair of clocks and

A Chappe optical semaphore of 1792.

Henry Maudslay (1771–1831), whose works at Westminster Bridge Road was a nursery of mechanical engineers.

watches. His name was George Stephenson (1781–1848) and one of his friends at this time was a nearby engineer's apprentice called William Fairbairn (1789–1874).

Stephenson's life was at first singularly unfortunate. His wife died when his only son was not yet 3 years old. His father became an invalid. He himself was 'drawn' for the militia and had to find money for a substitute. At one time he seriously considered emigrating. The anodyne for his personal distresses was work and when in 1808 he obtained with two other men a contract to work the engines of the Killingworth Pit, he would spend his weekends dismantling them in order to understand the principles on which they worked. In four years' time he found himself the engine-wright of the colliery.

Stephenson's talents extended. He designed a safety lamp to enable colliers to work in gas-saturated areas but found that Sir Humphry Davy had been working successfully on the same problem. He also concentrated on the improvement of steam locomotion.

At nearby Wylam Colliery the proprietor had been experimenting with a locomotive that ran on the cog and rack principle—an idea first put into practice by John Blenkinsop of Leeds in 1811—but after two attempts consulted his own 'viewer', William Hedley. Hedley, two years older than Stephenson, hit on the right idea: steam engines did not need rack rails and toothed wheels but could obtain sufficient friction by a smooth rail and wheel rim. Hedley's patent of 13 March 1813 led to the laying down of a smooth rail track at Wylam in 1814, where Hedley's engines were also making use of his second discovery: the steam blast, whereby the exhaust steam was returned to the chimney to produce a greater blast.

Hedley's experiments intrigued Stephenson who, in 1813, began building engines which drew a load of 30 tons at 4 miles an hour up an incline 1 in 450. Subsequent experiments led him to advocate the elimination (as far as possible) of gradients. In 1821 when he was appointed engineer for the projected Stockton and Darlington Railway, he further suggested that the rails should be of wrought rather than cast iron. As soon as work began on the railway, Stephenson opened his own engine works at Newcastle, where was made the first engine that passed over the lines. It can be seen today on the platform at Darlington Station.

THE STEAM ENGINE AFLOAT

In 1802 William Symington, son of a Scottish mining engineer, built himself an engine and then a boat called the *Charlotte Dundas*, which in 6 hours tugged 2 seventy-ton barges for 20 miles along the Forth and Clyde Canal. Robert Burns, the poet, had seen his early experiments; Robert Fulton and Henry Bell saw this one. Fulton returned to America where, in 1807, he opened the first regular passenger steamship service between New York and Albany with the *Clermont*, engined and boilered by Boulton and Watt. Henry Bell built the first passenger-steamer, the *Comet*, which carried passengers between Glasgow and Helensburgh on the Clyde, making its first trip in August 1812. Its 4-horse-power engine was housed amidships in a brick structure, because the vessel itself was of wood. If Fulton had successfully inaugurated steam navigation in the New World, Bell had similarly inaugurated it in the Old.

The navy was heavily committed to the sailing ship throughout the war and repulsed Henry Bell in 1800, 1802 and 1813, causing him to approach other countries. Samuel Whitbread, a keen promoter of new causes, asked in the House of Commons in 1815 why the new steam boats were ignored by the Admiralty and confessed himself astonished that 'the improvements which to a wonderful extent had been made in all the private concerns of the country, were so slow in finding their way into public establishments, and especially the dockyards'. In the following year the First Lord, in a parliamentary reply, gave it as his opinion that steam power was only fit to tow men-of-war out of harbour 'when they would be prevented from sailing by contrary winds'. Naval engineering was not as uniformly unprogressive. Colonel Mark Beaufoy had established a Society for the Improvement of Naval Architecture in 1791, which collapsed through lack of funds. The British Navy, hitherto content with copying the French, was also stimulated by the efforts of Gabriel Snodgrass, the surveyor to the East India Company. The Barham Report of the year following Trafalgar saw the beginning of the professional education of shipwrights. A school was opened in 1811 at Portsmouth under Dr. Inman, a Cambridge mathematician, who trained some forty-one students before it was closed down in 1832. Some of its students were later to win distinction: Isaac Watts, the chief constructor, Lloyd, the engineer-in-chief, and Creuze, the chief surveyor of Lloyds.

THE BRITISH ARMY

The British Army were a little more progressive and on 25 April 1787 the engineers were designated as a Royal Corps with the motto: 'Ubique'. In 1791 the Ordnance Survey was established by the Duke of Richmond, then Master-General of the Ordnance. It still produces the most artistic and informative maps in the world. When Britain was committed to the long land campaign in the Spanish Peninsula against Napoleon, John Fox Burgoyne fortified the lines of Torres Vedras, the 'Maginot Line' behind which Wellington withdrew for strategic reasons. William Reid bridged the rivers in Wellington's advance. Because of Wellington's reports on the defective military engineering in the field, a special School of Military Engineering was established at Chatham with Charles Pashley (1780–1861) as its first director, a versatile, energetic man who contests with Aspdin the honour of inventing Portland Cement. This school was to have a hand in the military development of ballooning, mechanical traction, telegraphs and camouflage.

Wellington's extensive campaign in the Spanish Peninsula elicited not only the skill of John Fox Burgoyne (1782–1871) but that of William Reid (1791–1858) who was to build the bridges of boats that enabled the British army to cross the Spanish rivers. Both these military engineers, subsequently knighted, were to play their part in engineering enterprises for the next half century: Sir John Burgoyne as Inspector General of Fortifications and Sir William Reid as Chairman of the Executive Committee of the Great Exhibition of 1851. Reid's enthusiasm for technical training led him to suggest to John Weale (1791–1862) the publication of a series of rudimentary technical treatises.

New weapons were forged in this war, notably a service rocket, associated with the name of William Congreve, who joined the Woolwich laboratory (founded in 1695), having been a pupil of the academy (founded in 1741). In 1791 Congreve succeeded, after sixteen years of experiment, in perfecting a rocket which was known by his name. It was tried both on land and sea. The British Army used rockets at the Battle of Bladensburg, to open the road to Washington, and thereby added the words 'the rockets' red glare' to the national anthem of the United States. A rocket company served at the battle of Leipzig in 1813, where its noise and glare threw the French into great confusion. Knighted by the Czar, Congreve was appointed to succeed his father as Comptroller of the Royal Laboratory and devoted the years after the war in trying to improve his rockets, as well as the steam engine.

AMERICA AND THE RISE OF PRODUCTION ENGINEERING

Fear, first of the British in 1794, then of the French in 1798, was the bellows that kept a military fire burning in the United States of America. To provide the necessary weapons Eli Whitney (1765–1825) a former cotton ginner, contracted to produce 10,000 muskets in 2 years by using jigs. This technique would enable unskilled workers to make interchangeable parts in great numbers. It saved time, money and skill. The Government were impressed, even though he took 8 years to fulfil his contract instead of 3, and in 1812 they gave him another order for 15,000 muskets. His technique of mass-manufacture was followed by Simeon North (1765–1852) a former scythe maker, who provided pistols for the War and Navy departments.

Whitney's mechanical skill had already initiated a second revolution in America. His cotton gin, enabling cotton to be cleaned by machine instead of by hand, was widely pirated. Cotton growing boomed in the Gulf States. From 8,000,000 pounds produced in 1794, 70,000,000 were being produced in 1805. And, thanks to the furtive emigration of one of Richard Arkwright's apprentices from England in 1789, water-powered spinning mills were available to cope with it. This Englishman from Belper, Samuel Slater, together with Sylvanus Brown, his mechanic, virtually initiated the United States machine tool industry.

When the British did actually declare war on America in 1812 the military academy at West Point, established in 1794, existed on paper only. Though intended from 1802 to be a school of engineering, the first in the country's history, it left the students to shift for themselves, whilst the superintendent was absent for most of the time, building harbour defences. Indeed, by the time war broke out the officers and cadets had been dispersed to various services in the army. Re-opened in 1813 under Sylvanus Thayer as superintendent and remodelled on the lines of the French École Polytechnique, it began to discharge its rôle with great efficiency: a rôle which was in turn to be altered by the Mexican War and the Civil War.

The war also put in a sound position the gunpowder works of Eleuthère Irénée Du Pont (1771–1834) near Wilmington, Delaware. This works had been established ten years earlier on the model of the Royal French Mill at Essonne, where Du Pont had been a pupil of Lavoisier before emigrating to America

The first mill in the world to turn raw cotton into cloth was built during the 1812 war by Paul Moody and F. C. Lowell (1775–1817).

Lowell's son, John (1799–1836), used his father's wealth to found the Lowell Institute where it was hoped to provide instruction enabling the sterile and unproductive soil of New England to yield crops.

The American War of 1812 not only witnessed the use of rockets by the British but also provided further striking examples of the progress of engineering. For H.M.S. *Shannon*, under Captain Broke, and equipped with gunsights, range quadrants and azimuth scales, captured the U.S.S. *Chesapeake*. On the American side, the first steam-driven warship, with a catamaran-like double hull and a 16-foot paddle wheel between each part was designed by Fulton for use against the British.

BRITISH MACHINE-TOOLS AND MASS MANUFACTURE

A former chief engineer of New York, designer of a gun foundry and planner of the defences of Long and Staten Islands came to Britain in 1797. His name, Marc Isambard Brunel (1769–1849), was to become even better known here than in France (the country of his birth) or America (the country of his first adoption) because of the machinery he devised for the mass manufacture of ships' blocks. These block machines were made for him by Henry Maudslay (1771–1831) for installation at Southampton. They enabled 6 men to do what 60 men had done before, effected a saving of £17,000 in a year and played no inconspicuous part in enabling Britain to keep that vital command of the sea in the war against Napoleon. Since a seventy-four-gun ship of the line required some 1,400 blocks, the importance of Brunel's work needs no emphasis. It was important also for Maudslay, who in 1810 was able to set up in partnership with Joseph Field an engineering works of his own which was to be the 'school' of most of the eminent mechanical engineers of the high Victorian age.

The man who organized Maudslay's and Brunel's work in this field was Sir Samuel Bentham (1757–1831), the Inspector-General of Navy Works from 1795 to 1807. Younger brother of the perhaps more famous Jeremy Bentham the political philosopher (with whom he co-operated in fitting up the Panopticon or model school and prison), Samuel Bentham had spent many years in Russia, where he not only managed a shipyard at Kritchev but commanded a flotilla. As a specialist in organized production techniques he was also responsible for the production of ships' biscuits at the Victualling Office in Deptford which, according to the *Book of Trades, or Library of the Useful Arts* (1804), was 'curious and interesting'. The dough was worked by a machine and distributed on an assembly line to a moulder, a marker, a splitter, a chucker and a depositor who worked 'with the regularity of a clock'. It was an assembly line that was to excite attention from Peter Barlow thirty-two years later when he described *Manufactures and Machinery in Britain* (1836).

Brunel's associate in this block-making machinery, Henry Maudslay, had entered the service of Joseph Bramah (1748–1814) in 1789 and worked with him for nine years. During this time Bramah built with a hydraulic press—a device for converting a number of comparatively small impulses into a steady continuous pressure. Bramah's minor mechanical innovations were a water closet, a burglar-proof lock, the ever-sharp pencil, a machine for numbering banknotes and a planing machine. He was also anxious to use the screw as a propulsive force for vessels.

Maudslay left Bramah and in 1798 set up on his own, first in Marylebone, then in Westminster Bridge Road, making machinery for the elder Brunel to manufacture ship blocks. Patents for calico printing, small steam engines and a differential, enabling lathes to be power driven, followed. In partnership with Joshua Field after 1810 he improved marine engineering, made great improvements in the lathe and devised a measuring machine which would register ten thousandths of an inch.

Robert Owen (1771–1858), amongst his many contributions to social thought, stressed the importance of human relations in industry in 1813:

'If then', he wrote, 'the care as to the state of your inanimate machines can produce such beneficial results, what may not be expected if you devote equal attention to your vital machines, which are far more wonderfully constructed?'

Owen's insights were to enable him to become a guiding star in the firmament of the workers now gathering in the industrial towns.

THE DEVELOPMENT OF CIVIL ENGINEERING IN BRITAIN

What Maudslay was to do for mechanical engineering, Thomas Telford (1757–1834) was doing for civil engineering. He was appointed in 1793 to connect the Mersey, the Dee and the Severn by a canal. This, the largest yet undertaken, was to be followed by others even larger. The 'hard' way of the eighteenth-century engineer enabled him to turn his hands to roads, canals or bridges. Born in

Scotland, he lost his father early and at the age of 15 was apprenticed to a mason of the Duke of Buccleugh at Langholm, where his cheerful industry attracted the attention of a local inhabitant who made him free of her library. He wrote poetry for the *Edinburgh Magazine*. At 23, Telford removed there, where much rebuilding was going on.

Later Telford moved to London where he was employed as a stone-hewer on Somerset house. His first contract was for a house for the commissioner of Portsmouth Dockyard. One of his Scots friends invited him to recondition Shrewsbury castle, which he did so well that he was made surveyor of the public works for Shropshire. Brindley's aqueducts were liable to fracture and collapse by frost, because the lining consisted of moist earth. Telford hit upon the idea of a cast iron trough-liner, which not only made it possible to use less masonry but strengthened the structure. His aqueduct, crossing the Dee at Pont-Cysylltau, was a large undertaking in the Vale of Llan-gollen: 121 feet above the river, it carried a waterway 11 feet 10 inches wide and stretched for over 338 yards on nineteen arches. Sir Walter Scott told Southey that 'it was the most impressive work of art he had ever seen'. Four miles away another Telford aqueduct at Chirk, spanned the Ceiriog. This, only two-thirds of the length and little over half as high, was nevertheless just as artistic and useful. The first took 10, the second, 5 years to build. Before they were completed, Telford's imaginative use of iron in bridge building had been exploited over the Severn at Buildwas and he was asked to build another in London.

His heart lay, however, in the improvement of Scottish roads and harbours. The Government was anxious to make a waterway through the great glen of Scotland partly to stop emigration, partly for strategic reasons, and commissioned Telford to make a survey. As a result the Caledonian Canal was undertaken and ultimately finished in 1822 with Telford as chief engineer. He also made 920 miles of new road, realigned 28 miles of military road and built over a hundred bridges which, in eighteen years, transformed the highlands.

The social and economic results of Telford's work in Scotland were considerable. In these eighteen years, 3,200 men had been taught the use of tools and, as he said, 'These undertakings may be regarded in the light of a working academy from which eight hundred men have annually gone forth improved workmen.' In his opinion, the country had been advanced 'at least a century'. Furthermore, the harbour works like those at Wick, Aberdeen, Peterhead, Banff, Leith, Edinburgh and Dundee advanced the commerce and fisheries of Scotland.

Yet the really major work of this period, the Caledonian Canal with its eight locks ('Neptune's staircase' as Telford called it) which took twenty-one years to build, proved a virtual white elephant.

In the midst of such major national surgery, Telford found time to build canals between the Baltic and the North Sea (the Gotha canal), to rebuild the roads between Carlisle and Glasgow, Shrewsbury and Holyhead and (as a natural consequence) a bridge over the Menai Straits. This was on the suspension principle. Begun in 1819 with a grant from Parliament, it rested on sixteen wrought iron chairs each weighing 23½ tons and its success was such that he was later asked to bridge the estuary of the Conway and the Clyde at Glasgow.

Telford was asked to design the Liverpool to Manchester Railway but refused, preferring to concentrate on bridging and docks. Among his achievements were bridges at Tewkesbury, Gloucester, Edinburgh; Fen drainage (the North Level); and St. Catherine's Docks in London.

Napoleon's lightning strikes were possible because P. M. J. Tresaguet (1716–96) built the best roads in the world at that time. His astounding intelligence system was created by Claude Chappe (1763–1805) and his brother Ignace (1760–1828) who laced the French Empire with levered semaphores, situated on towers. Awarded the title of *l'Ingénieur Télégraphe*, Chappe, with his brother, connected Paris to Lille (1794), to Strasbourg (1797), Brest (1798), Lyon (1799) and enabled Napoleon to sustain his conquest of Italy by linking Lyon to Turin, Milan (1805) and Venice (1809). The *Histoire de la Télégraphie* which Ignace published in 1824 is a thrilling account of how the vast French imperial conquests were linked by the communicating towers (spread ten or so miles apart) of the Chappe system. Not the least of the tributes it received was that from British engineers, who linked London to the Channel Ports by the same means.

What a contrast English roads presented! 'It is a matter to be wondered at' reported the third of a series of committees of the House of Commons on the roads 'that so great a source of national concern has been neglected'. The crowds of men and horses moving from place to place, the ever increasing tonnage of carts bearing iron goods of every kind, the mounting number of stage coaches, the expansion of the postal service; these, amplified by the complaints of the Irish M.P.s who had to travel to London, made the roads a national concern. But national action was powerless against the Turnpike Trusts (which maintained their own engineers). So the

Chairman of the Board of Agriculture, Sir John Sinclair (1754–1835) was fortunate in obtaining a memorandum on roads from John Loudon McAdam.

John Loudon McAdam (1756–1836) was constructing model roads whilst at school at Maybole and, at the age of 14 when his father died, was sent to America in the care of an uncle. There he amassed a fortune as an agent for the sale of prizes. Surviving the peace with enough to retire he purchased an estate in Ayrshire that lay on an old high road, where he spent thirteen years as a magistrate and road trustee and experimented in coal for manufactures. In 1808, at 42, he was appointed agent for revictualling the navy in western ports and migrated to Falmouth. In the next sixteen years he travelled over thirty thousand miles to pursue his investigations and spent £5,000 in experiments on roads. He found that they should be made of broken stone, cambered and well flattened by the passage of traffic. His top dressing consisted of small cubical fragments, none weighing more than 6 ounces. In 1815 McAdam became surveyor general of the Bristol Turnpike Trust roads where he put his theories into practice. *A Practical Essay on the Scientific Repair and Preservation of Roads* (1819) and his *Present State of Roadmaking* (1820) drew such attention to his work that three years later a Parliamentary Committee was set up to examine his technique. They were completely converted. His most capable son, James McAdam, was appointed general surveyor of roads in 1827. McAdam himself was voted some £10,000 in view of his personal expenditure in the past. Until his death nine years later he lived in Hoddesdon in Hertfordshire, watching his process adopted by most of the world. Southey, loyal to the memory of his friend Telford, thought that 'Macadamising the streets of London is likely to prove quackadamising', but Jeremy Bentham thought well of him and commended the perpetuation of his name in popular speech.

In 1815 the Bristol Turnpike Trust, of which McAdam had been a leading member for some years, consolidated 149 miles of road under his jurisdiction and offered him the surveyorship. His 'system'—based on accommodating the road to the traffic, standardized road engineering procedures, a 10-inch-thick road on rural subsoil with a surface of stones mixed with sand or earth, honest surveyors and clear instructions—spread throughout the country. In little over two years his system was adopted by eleven other trusts embracing 700 miles of turnpike roads in fifteen counties. By 1823 his three sons (William, James and John Loudon Junior) were responsible for eighty-five trusts comprising 2,000 miles of turnpike road and another thirty-eight were administered in accordance with his system.

Such extensive surgery, though it scarred the face of Britain, yielded up a rich store of information that was eagerly collected and synthesized by William Smith. As a resident engineer of the Somerset Coal Canal Company he set up in practice in Bath in 1798 as a 'Civil Engineer and Military Surveyor and Land Drainer' obtaining from his work such insight into stratigraphy and geology that he compiled a great map of the English strata. His geology researches were forwarded by Richard Crawshay, the iron king of South Wales. The Duke of Bedford wished him to train his stewards. Smith was forced by financial troubles to sell his collections to the British Museum. He showed that coal could be won beneath the magnesian limestone.

THE GENERALIZING OF APPLIED SCIENCE

Telford, Stephenson, the Earl of Dundonald and many like them, took little thought of generalizing their activities. It was left to an American citizen of English descent, Benjamin Thompson, Count Rumford, to launch the Royal Institution in 1799 to diffuse the knowledge and facilitate 'the general introduction of useful mechanical inventions and improvements'. This diffusion and facilitation was to be achieved by 'courses of philosophic lectures and experiments' which would teach 'the application of science to the common purposes of life'. The first teacher was Sir Humphry Davy (1778–1829) who invented the arc light, the first electric furnace and shares with Stephenson the honour of inventing the miner's safety lamp. This last, together with concomitant needs and improvisations, helped the rise of coal production from 10,000,000 tons in 1810 to 16,000,000 tons in 1829.

Equally significant was the formation, in 1805, of the Society of Millwrights at the Museum Tavern in Bloomsbury. Rennie and Congreve were among the members and Bryan Donkin was the treasurer.

But it was H. R. Palmer, in the year 1818, who convened the small group, consisting of William Maudslay, Joshua Field and three others at the King's Tavern in Cheapside, as a kind of revolt against the Society of Civil Engineers that was run by the Rennies. With William Maudslay in the chair, this meeting gave birth to the Institution of Civil Engineers. Palmer said:

'An engineer is a mediator between the philosopher and the working mechanic, and, like an interpreter between two foreigners, must understand the language of both. . . . Hence the absolute necessity of his possessing both practical and theoretical knowledge.'

Such 'theoretical knowledge' had been in the mind of William Nicholson (1753–1815) who had established the first English periodical devoted to general science that was independent of the academies. Before establishing it in 1797, he had been commercial agent for Wedgwood, and whilst editing it, he ran a school in Soho, acted as a patent agent, and designed waterworks for, amongst others, West Middlesex, Portsmouth, Gosport and Portsea Island. *Nicholson's Journal* was such a success that a rival journal was started a year later in 1798—the *Philosophical Magazine*. This, in 1813, absorbed *Nicholson's Journal*. Many a notable discovery was first given to the world in its columns. By 1809 Nicholson began to devote himself to the *British Cyclopaedia, or Dictionary of Arts and Science* for he had done notable service in this field too by publishing dictionaries of practical and theoretical chemistry. His successor in this latter field, Andrew Ure (1778–1857), a professor at the first technical college to be founded in the United Kingdom (Anderson's University at Glasgow, founded in 1799), gave lectures to working men which, when the war was over, inspired Charles Dupin, the French engineer, to suggest that similar courses should be given at the École des Arts et Métiers in Paris.

By 1820, after the little group convened by Palmer had been meeting for two years, one of the recruits, William Provis, proposed that Telford be invited to become president of their Institution of Civil Engineers. Telford took two months to make up his mind, but when he accepted, he threw his weight behind the little group with the result that, unlike all the previous groups we have met, it acquired a corporate existence, with premises of its own in Buckingham Street in the Strand and later at Cannon Row, Westminster.

Telford stressed the importance of the institutions being kept alive by the voluntary efforts of its members. 'In other countries', he said, alluding to France and Germany, 'similar establishments are instituted by government, and their members and proceedings are under their control.' He gave it his magnificent library, with its collection of French engineering works. He instituted, through another of his pupils, Joseph Mitchell, the habit of recording the substance of papers, and encouraged it to meet weekly.

Nine years after its first meeting the Institution asked Thomas Tredgold to define the term 'civil engineer', 'in order that this description and these objects may be embodied in a Petition to the Attorney-General, in an application for a Charter'. His reply dated 4 January 1828 is comprehensive and eloquent:

'Civil Engineering', he wrote, 'is the art of directing the great sources of power in Nature for the use and convenience of man; being that practical application of the most important principles of natural philosophy which has, in a considerable degree, realized the anticipations of Bacon, and changed the aspect and state of affairs in the whole world. The most important object of Civil Engineering is to improve the means of production and of traffic in states, both for external and internal trade.' He concluded, 'we are every day witnessing new applications, as well as the extension of the older ones to every part of the globe'.

Those 'great sources of power in Nature' were to be intensively exploited in the generation to come.

CHAPTER 12

The Age of the Mechanical Engineer, 1815-1857

LIVERPOOL: FUNNEL OF TRADE

The Napoleonic wars diverted British export trade. By 1830, one sixth of it was going to America, and from America over three-quarters of all the raw cotton consumed in the United Kingdom was imported. Liverpool, as the funnel of this trade, had other pressing transport problems too: the ebb of emigrants to the American frontier and the flow of immigrants from Ireland to Britain to work as 'navigators' digging the canals. All sorts of interests, like coal owners and iron founders, wanted access to the Mersey.

Canals were too inadequate and too slow for such traffic. The success of Stephenson's steam railway from Stockton to Derby in 1825 encouraged several promoters to approach him with the idea of a railway linking Manchester and Liverpool. Telford himself preferred steam carriages which were free to go where a good road lay, but unfortunately the roads were just not capable of sustaining them. The canal interests were up in arms against the proposed railway. Even the engineers were at one anothers' throats about it. The fiercest opposition came from farmers and landowners, whose threats made surveys difficult, if not impossible to make. Opposition was sustained in Parliament, and when the Bill for its construction was introduced, Stephenson's methods were subjected to a cross-examination, designed to expose their impracticability. For a time the opposition was successful, but after a fresh survey by the Rennies, work began.

The promoters themselves were by no means convinced that locomotives were the answer to the problem of traction. Most of them favoured fixed stationary engines, spaced along the line, to haul the trucks. To settle the question a competition was held at Rainhill on 1 October 1829. The conditions stipulated that the locomotives should sustain an average speed of 10 miles an hour with a steam pressure not exceeding 50 pounds per square inch. They had to run backwards and forwards along the line at an average speed of 10 miles an hour, till at least 70 miles had been covered.

On 6 October 1829 four locomotives forgathered: the *Novelty*, built by John Braithwaite, the *Sanspareil* by Hackworth, the *Perseverance* by Burstall and the *Rocket* built by Stephenson. The *Rocket's* subsequent victory was greeted with rhapsodies. Even the *Scotsman* wrote that it 'had established principles which will give a greater impulse to civilization than it has ever received from any single cause since the press first opened the gates of knowledge to the human species at large'. By one of the ironies of fate, the politician who had been most active in promoting the parliamentary bill for the railway —William Huskisson—was killed on the day when the line was officially opened on 15 September 1830.

THE MACHINE TOOLS INDUSTRY

Five years after the Liverpool–Manchester Railway opened, James Nasmyth (1808–90) settled in Manchester. Railways and the textile industry had created a great need for slide lathes, drilling, boring, planing and slotting machines, and Nasmyth soon had to build a bigger works (finished in 1837), known as the Bridgewater Foundry, at Patricroft, near the junction of the Liverpool–Manchester Railway and the Bridgewater Canal.

'These are indeed glorious times for the Engineers', he wrote as his foundry was being built, 'I never was in such a state of bustle in my life, such quantity of people come knocking at my little door from morning till night. . . . The demand for work is really quite wonderful and I will do all in my power to bring about what I am certain is the true view of the Business viz. to have such as planing machines and lathes etc., etc., all ready to supply the parties who come every day asking for them.'

He was 'quite up in the clouds about the prospect it opens'. The standardized machine tools he was able to produce, outlined in a catalogue, did not require a skilled workman but a 'well-selected labourer'. As A. E. Musson has pointed out, these automatic machine-tools heralded the modern machine age, and his factory was a pioneer of 'assembly lines'. For Nasmyth, whilst the buildings were under discussion, proposed that they should be *all in a line*. 'In this way', he wrote, 'we will be able to keep in good order.'

Nasmyth also invented, in 1839 (patenting it in 1842), an even larger machine tool: the steam hammer. This made possible the quick, cheap forging of larger units and made it possible for Nasmyth's firm to begin manufacturing railway locomotives—ultimately the chief product of the firm which succeeded his. In fourteen years he had built 109, in addition to some small high-pressure engines. One of these small engines was even bought by Boulton, Watt and Company. Hydraulic presses and donkey pumps grew naturally from these powerful little prime movers.

THE MAUDSLAY NURSERY

Nasmyth was but one of the pupils of Maudslay Sons and Field. a firm virtually launched during the Napoleonic War (in 1810 at Lambeth) when they made machinery for Brunel's blocks and for Congreve's rockets. Now they turned to make table engines for the Prussian Technological Institute, Berlin, drainage and pumping engines, engines for rope haulage, air pumps and engines for the Constantinople mint, time balls for various towns, tank vessels and, towards the end of the century, 'Great Wheels' for Earls Court, Blackpool, Vienna and Paris, and high pressure engines for the Upper Goodavery Navigation Works in India.

Like Telford, Maudslay also trained men and apprentices who rode the waves of industrial expansion that surged through Britain after the Napoleonic Wars.

One of them, Richard Roberts (1789–1864), who had set up his own works at Manchester in 1816, invented a screw-cutting lathe, a gas meter, a planing machine for metal and a self-acting mule. He was to pioneer a Jacquard machine for punching holes in iron, evolved to overcome a strike on the Menai tubular bridge built by Robert Stephenson (1803–1859).

Another pupil, Samuel Seaward (1800–42), established a Canal Ironworks at Millwall and Poplar in 1824, turned to marine engineering and in 1829 was running the Diamond Steam Packet Company. His swing bridges, dredges, cranes, mills, tubular boilers, telescopic funnels and propeller couplings were famous.

Yet another pupil was David Napier (1790–1869) a cousin of the Clyde shipbuilder. Napier built printing machines for, amongst others, T. C. Hansard, the parliamentary printer. He went on to make arsenal equipment and bullet-making machinery for foreign governments.

Often these pupils joined up with each other. Thus William Muir (1806–1888), a second cousin of William Murdock, became Maudslay's draughtsman and in 1840 joined another Maudslay pupil, Joseph Whitworth (1803–87), who had set up at Manchester in 1833 and constructed many machine tools including a 'jim crow' or self-acting planing machine which cut at both strokes. Whitworth standardized the thread of screws, as well as perfecting the method of producing a true plane surface.

The last Maudslay pupil who must be mentioned is John Clement, who was the engineer to Charles Babbage (1792–1871) the great pioneer of mechanical computation.

BABBAGE AND BODMER

Babbage had obtained a government grant for making a calculating machine based on 'the method of differences' in 1823, but disagreed with the engineer. He devoted thirty-seven years of his life and most of his fortune to this machine. Babbage was interested in production engineering and in 1832 published his *Economy of Manufactures*, after a study of foreign factories. Two years later he founded the Statistical Society.

Babbage's experiments on his famous computer, though apparently unsuccessful, were said by Lord Rosse to have more than repaid, in improvements to machinery and tools, the money spent on it. The leading engineers of the day, Donkin, Rennie, Brunel and Maudslay, served on the committees of the Royal Society which examined the bills of the mechanical engineer who was trying to make it for him.

Another machine-tool maker, J. G. Bodmer (1786–1864), worked on water wheels, steam power and textile machines with equal zest. He laid out a factory, Chorlton Mills, Manchester, in which processing and handling were integrated. J. W. Roe sees in him the pioneer of the moving crane, the endless belt and the travelling grate (a quasi-automatic stoker): all put into operation in Lancashire where he flourished as a consultant. Bodmer was, in a very real sense, a pioneer of the assembly line.

THE RAILWAY BOOMS

The first of the railway booms which called forth the multiple skills of the mechanical engineers began in 1835. In the subsequent two years 88 companies were floated, capitalized at approximately £70,000,000. Their shares displaced canal securities as standard investments. At the same time, 71 mining, 20 banking and 11 insur-

The spans of the Britannia tubular bridge over the Menai Straits being manoeuvred into position. A lithograph by G. Hawkins.

A portion of Babbage's difference engine, showing the fine quality of the machining.

(By courtesy of the Director, Science Museum, South Kensington)

Photograph and signature of Charles Babbage (1792–1871)

ance companies were floated, out of a total number of 300 enterprises being promoted at that time. Most of the 300, in the depression that followed, sank without trace, except the railways. The investment mania kindled by this resulted in the emergence of the type of promoter personified by Dickens in *Martin Chuzzlewit* (1843): Mr. Tigg Montague of the Anglo-Bengalee Disinterested Loan and Life Assurance Company. As a result there was passed the Joint Stock Companies Registration and Regulation Act of 1844, which insisted on sound regulations for the promotion of companies and on publicity through the whole course of their proceedings.

The second railway boom broke the depression that followed 1837. On 1 January 1843 there were less than 2,000 miles of railway open to traffic. Six years later there were 5,000 miles and another 7,000 had been sanctioned. The total capital involved rose from 65 to over 200 million pounds.

Economically, this had three results. The first was a democratization of the money market; the second, its expansion geographically; the third an acceleration of the growth of negotiable investments. Everyone was anxious to buy railway shares and *The Times* remarked that 'the clergy were almost wholly forsaking scripture for script'. To supply the demand, provincial share markets grew in Leeds, Wakefield, Bradford, Halifax, Huddersfield, Leicester, Birmingham and many other inland towns.

With the share markets grew the contractors, Thomas Brassey (1805–70) and Morton Peto (1809–89), especially. Brassey, originally a land surveyor and quarry manager, was persuaded to undertake some small construction contracts by George Stephenson, and in the subsequent thirty-five years of his business life he successfully undertook 170 different contracts, involving 8,000 miles of railway in Canada, the Crimea, Australia, the Argentine and India. His counterpart, Morton Peto, built railways both in England and abroad, as well as Nelson's Column. He took his men around in societies and groups, providing them with books, teachers, savings books and benefit societies. Both contractors kept large staffs of scientific and commercial men to whom they could refer difficult problems, thus carrying forward another development of the great canal constructors, like Telford.

PROBLEMS OF THE PERMANENT WAY

From Telford, too, the railway engineers inherited techniques and improved on their inheritance. One of Telford's assistants, J. B.

MacNeill (1793–1880), set up in 1834 as a consulting engineer in Glasgow and London. As a great admirer of Navier, the French engineer, he published two years later, a translation of *Means of Comparing the Advantages of Different Lines of Railway*. MacNeil applied the skill of a civil engineer to problems of the permanent way. His system of sectio-planography, or of displaying embankments cuttings and gradients in one view, was adopted by the House of Commons as standard form for all railway plans presented to them for sanctioning. Such a name did he win that when Sir Robert Kane (1809–90), an early Irish chemical engineer, suggested that a big technical school should be founded in Dublin, the council of Trinity College hurriedly established a chair of civil engineering in 1842 and offered it to MacNeill. He held it for ten years.

Railways posed new problems of bridge building and here Robert Stephenson (1803–59) was to the forefront. He built the High Level Bridge over the Tyne at Newcastle in 1849 and the Britannia Tubular Bridge which crossed the Menai Straits on the Irish route. The latter, near Telford's bridge built a quarter of a century earlier, Stephenson built with the help of William Fairbairn and Professor Eaton Hodgkinson who spent months in calculating the strengths of the various types of metal tubes, as the Admiralty would not allow arches. What emerged from their experiments were three central towers supporting four giant tubes each 472 feet long and 1,400 tons weight. The resident engineer, Edwin Clark, has left us an account of the dramatic raising of these four tubes more than 100 feet above the waterway. Two of Bramah's jacks were used, operated by steam pumps of 40 horsepower each. The Britannia Bridge also utilized Fairbairn's steam rivetter.

THE INSTITUTION OF MECHANICAL ENGINEERS

A more personal problem to railway engineers was their status *vis a vis* the old canal engineers, who formed the crusty core of the Institution of Civil Engineers. Near the Lickey incline at Blackwell on the Bristol and Birmingham railway in 1846, some railway engineers were watching some locomotive trials. The story goes that it began to rain and they took shelter in a platelayer's hut where they discussed the refusal of the Institution of Civil Engineers to admit to membership the doyen of their profession—George Stephenson—unless he submitted evidence of his capacity as an engineer in the form of an essay. They—and Stephenson—felt this was a professional as well as a personal affront, and so they resolved to establish an institution of their own.

The engineer of the Bristol and Birmingham Railway, J. E. McConnell, convened a meeting at his house at the foot of the Lickey incline. To it came, amongst others, C. F. Beyer. A circular was issued inviting mechanics and engineers 'engaged in the different Manufactories, Railways and other Establishments in the Kingdom' to meet on 7 October 1846 to form an Institution of Mechanical Engineers, through which 'by a mutual exchange of ideas respecting improvements in various branches of Mechanical Science' they might 'increase their knowledge and give an impulse to Inventions likely to be useful to the World.' The meeting was well attended, the idea of the Institution was agreed upon, and it was formally founded on 27 January 1847. Needless to say, George Stephenson was elected president, and to the original group of promoters was added the weight of Joseph Miller, who joined Beyer and McConnell as a Vice-President.

The Railway and the City

Another set of problems was posed by railways in towns. That between Camden Town and Euston Square was solved by Charles Fox (1810–74) who built it within a covered way and retaining walls. In such a fashion was the metropolitan railway born. Fox's firm was the first to carry out the systematic manufacture of railway plant and stock. Though he undertook works in all parts of the world—the first narrow gauge line in India, the Berlin water works, and railways in France, Canada, Australia, South Africa—his best known and most characteristic undertaking was the Crystal Palace. In this he was associated with two other railway engineers, Sir William Cubitt (1785–1861), who built the Croydon atmospheric railway, and Joseph Paxton (1801–65).

As an essay in prefabrication alone the Crystal Palace was unprecedented. In 17 weeks, 18 acres of Hyde Park had been covered by 900,000 feet of glass fastened to a graceful web-like structure of 3,300 columns and 2,300 girders. It was, as *Punch* exclaimed in wonder, a Crystal Palace. Bradbury, the proprietor of *Punch*, was an intimate friend of Paxton. So was Charles Dickens. Dickens, in fact, thought so much of him that he dashed up to Derbyshire in October 1845 to discuss with Paxton the possibility of founding the *Daily News*. From being just the head gardener at Chatsworth, Paxton became a real power in the land. His duke treated him as an intimate friend rather than as a servant, took him abroad on tours, and came to rely more and more on his advice. For Paxton was shrewd as well as imaginative. He saw the infinite possibilities of railways, and plunged what money he earned into the steel networks that spread so rapidly over the face of England in the roaring forties. In fact Dickens once wrote, 'Paxton has command of every railway interest in England and abroad except the Great Western.'

These railway interests led Paxton to London on 11 June 1850 to see the chairman of the Midland Railway. The chairman, who was also an M.P., showed Paxton round the House of Commons. In doing so, he was so struck with Paxton's remarks about it and the proposed Exhibition building, that he took Paxton along to see one of the Exhibition commissioners. After a general conversation, Paxton offered to submit detailed plans of what he considered would make a good building.

He had to be quick about it, for the scheduled date of opening was 1 May the following year. But whatever might grow under Paxton's green and ubiquitous fingers, grass never had time to grow under his feet. Within ten days he was back with all the details of what was to become the Crystal Palace. And what was all the more remarkable about his achievement was the almost casual way in which he spent those ten days. He kept an engagement to watch a tube of the Menai Bridge being floated into position and attended a meeting of a railway court at Derby. It was while he was actually listening to the story of an absent-minded signalman at Derby that he produced the doodle on a piece of blotting paper which was his design in embryo. It embodied all the improvements he had made whilst serving as the Duke of Devonshire's head gardener at Chatsworth—large glass panes, mass-manufactured sash bars to hold them in position, iron columns which both drained and supported the structure, the curvilinear roof—and many more. And they were all linked by a master idea borrowed directly from the structure of the leaf of the Victoria Regia Water Lily. By the beginning of August the contractors were at work. So many were the distinguished people who tried to see what was afoot behind the hoardings (which the contractors set up before beginning their work) that when the Duke of Devonshire asked for special permission to see behind them, Lord Granville dryly remarked, 'If any exception is to be made for him, it can only be as Paxton and Company.'

When the Great Exhibition was opened by the Queen on 1 May

1851, it was Paxton who walked, close on the four Pursuivants of Arms, at the head of the procession. Five months later, after a phenomenally successful show, the three of them (Fox, Cubitt and Paxton) were knighted by the Queen.

Fox moved the Palace to Sydenham, where it became the true conservatory of the Victorian period. It survived damage by wind (in 1861), by fire (in 1866) and by water (in 1880). It even survived Paxton, who died in 1865. It was finally destroyed by fire in 1936. Within its walls were held the first automobile show and the first meeting of the Aero Club. In nursing it, Paxton was led to advocate an even more ambitious project—nothing less than a 12-mile arcade to relieve the congestion of London's traffic. Like the Palace, it was to be built of iron and glass. It was to house shops and houses, a roadway and eight lines of railway. But the committee to whom he submitted the plan were not convinced that its cost—some £34,000,000 —could be met out of the rentals on the shops it would contain, and the scheme was shelved.

Its structural basis can still be seen in Victorian railway stations. It foreshadows the ferro-vitreous structures that are today linked with the names of Gropius, Le Corbusier and Frank Lloyd Wright.

STEAM ON THE LAND

The steam engine also saved the fens where the drained lands had sunk beyond the capacity of windmills to evacuate the water. Arthur Young, the eagle-eyed reporter of the agricultural scene at the close of the eighteenth century, feared that it only needed a couple of floods to reinstate frogs, coots and wild duck on 300,000 acres of the richest land in England.

Joseph Glynn led the way by draining Deeping Fen and Littleport Fen by means of steam pumps. He was followed by William Wells (1818–1889) who cleared Whittlesea Mere so rapidly that a fenman was left stranded in his boat and fish were left gasping in the ooze.

Steam provided stronger power for mole drains, for threshing, and for haulage. Indeed, the development of the road traction engine was vastly facilitated by the demands of the more progressive farmers. Perhaps the man who most notably endowed the English farmer with mechanical power was John Fowler (1826–64) whose steam ploughs and cultivators became deservedly famous.

STEAM ON THE SEAWAYS

The great reciprocating complex of industrialization sustained and accelerated by the railway was still further extended by the steamboat. In this, several railway constructors had a hand, notably I. K. Brunel, son of Marc Isambard Brunel, who had built the Great Western Railway on the broad gauge of 7 feet. Brunel, with the same dogged insistence that had led him to discard Stephenson's gauge of 4 feet 8½ inches for his railways, insisted that a steamship could sail to New York without refuelling. He was proved right by the *Great Western*, a ship of great power and longitudinal strength, built of oak. He suggested it in October 1835 at a meeting of the Directors of the Great Western Railway at Radley's Hotel in Bridge Street, Blackfriars. In a dramatic race across the Atlantic with the *Sirius* in April 1838, the *Great Western*, powered by engines built by Maudslay Sons and Field, arrived with a quarter of her fuel left, whilst the *Sirius* had burnt everything combustible on board. The *Great Western* crossed the Atlantic sixty-seven times in eight years. It was followed by the *Great Britain*, built of iron. This vessel needed an intermediate paddle shaft of such dimensions that it threatened to 'gag' any existing tilt hammer available. So the engineer, Francis Humphreys, communicated with James Nasmyth who, in half an hour, designed his famous steam hammer to cope with the forging. But the great intermediate paddle shaft was destined never to be used, for Brunel decided to convert the *Great Britain* to screw propulsion. As such, it was the first iron ship and the first screw ship to cross any ocean.

THE RISE OF MARINE ENGINEERING

While Brunel was the brilliant impressario, others were at work in the wings attracting little attention. Among them were the cousins Napier, David (1790–1869) and Robert (1796–1876), who really deserve the title of founding fathers of marine engineering. David supplied the engines of the first Glasgow–Belfast ferry. Robert established in 1823 a small works at Camlachie in the east end of Glasgow on the Clyde to make engines. Robert's first engine powered the 54-ton cargo boat *The Leven*. Further orders compelled him to extend his works to Washington Street (1828) and Lancefield (1835). From building engines he began to build boats. The *Berenice*, built in 1836 for the East India Company, attracted the attention of Samuel Cunard, a merchant at Halifax, Nova Scotia, who was then founding the British and North American Royal Mail Steam Packet Company. The Cunard Company was a generous customer and fifteen of Robert Napier's iron vessels, built for them alone, established the tradition of the 'Atlantic Greyhounds'. French companies, anxious to obtain speedy, first-class ships, came to him too, and his *Pereire* and *Ville de Paris* both made new records for the Atlantic crossing.

THE GREAT EASTERN

Disgusted by the railway mania, Brunel turned, in 1852, to the building of a steamship for the East Indies and Australia which should be six times as large as any ship afloat, some 18,915 tons. Begun on 25 February 1854, this great iron ship was double-hulled, and propelled by sails, paddles and screws. Designed to carry 4,000 people, its construction and launching posed so many problems that their solution carried British engineering across another threshold. To begin with, the engines designed to turn the 58-feet-diameter paddle-wheels and 24-feet screw were powerful enough to drive all the cotton mills in Manchester. Its bunker space was such that it was to be capable of steaming nearly round the world without re-fuelling. Its designer, John Scott Russell, was chosen by Brunel for his work in connection with the organization of the Crystal Palace Exhibition, but the relationships between them were such that the problems of its construction multiplied.

Of the constructional techniques and engineering novelties embodied in this great ship, called, from its intended employment, the *Great Eastern*, three call for special mention. The first was the use of an experimental tank by William Froude (1810–79) in which problems of stability and resistance could be studied. Telford, as we have seen, used one at the Adelaide Gallery, but Froude's was the real precursor of that at the National Physical Laboratory which was built in his name. The second was the steam stearing gear designed by J. M. Gray (1832–1909): probably the first successful application of the hydraulic-servo or follow-up mechanism now so popular. The third was the major problem of actually launching this giant sideways from Scott Russell's yard at Millwall into the Thames. The hydraulic presses employed burst their casings under the strain, and to build stronger ones, Brunel called in the Tangye Brothers. So well did they do the job that their newly established firm in Birmingham never lacked orders. As Richard Tangye acknowledged: 'We launched the *Great Eastern* and she launched us.'

The *Great Eastern* launched Tangyes but it killed Brunel. Overworked, disappointed and heavily involved in the losses incurred in its building, he died on 15 September 1859, as it was on its trials in the English Channel. It was the most gigantic engineering experiment of the age. Though its original function was never discharged, it was chosen, for its steadiness and manœuvrability (it could turn in its own length) to lay the Atlantic Cable in 1865.

THE TELEGRAPH AND CABLE

The telegraph and cable was born of the marriage between the laboratory science of Faraday and Sturgeon and the practical techniques and necessities of the steam age. The semaphore used in the Napoleonic war was satisfactory enough for the Government to discourage Francis Ronalds, who in 1816 had devised an electric telegraph that worked over 8 miles through an insulated wire embedded in a garden in Hammersmith. The railways were also satisfied with the semaphore (they use it to this day), and discouraged W. C. Cooke, who offered them an electric telegraph in 1837. When, however, Professor Charles Wheatstone joined forces with Cooke, in the same year, a line was laid between Euston and Camden Town. The practical utility of another line laid between Paddington and Slough was given national prominence in 1843 when it helped to capture a murderer. In 1846 the Electric Telegraph Company began to operate the railway telegraphs and several companies were formed to operate other services for the public. On 13 November 1851 the submarine telegraph was successfully laid between Dover and Cap Gris Nez by the Brett brothers, and to mark its successful completion a salvo was fired simultaneously from the French and British coasts. The booming of those guns had scarcely died away before a bigger boom began in cable laying. Cable enterprises supplanted railways as popular investments. J. W. Brett was received by the Emperor of the French, who asked him to link Corsica and Algiers by telegraph, which he duly did. These single threads became a network, one of Brett's engineers being so enthusiastic that he spent his honeymoon laying a cable to Ireland.

In 1854, Frederick N. Gisborne, an English engineer, went to New York where he persuaded Cyrus Field to form a cable company. British capital followed, taking up 229 of the £1,000 shares, leaving only 88 for the entire United States, and all these were owned by Cyrus Field himself. Field's attempts to lay an Atlantic Cable began in 1857. He arranged for H.M.S. *Agamemnon* and the U.S. *Niagara* to meet in mid Atlantic and share out the cable, but the cable broke after 200 miles. In 1858 they set sail from England and America, this time successfully overcoming six breakages. The cable thus laid operated for three weeks before it broke again and another six years elapsed before he tried again.

In 1864, thanks to William Thomson's mirror galvanometer and the *Great Eastern*, an Atlantic cable was laid without a fault. The first had to be abandoned in mid Atlantic; the second, at a cost of $3,000,000, was successful. Cables gave the *Great Eastern* a second lease of life. It laid Reuter's Franco–American Cable—some 2,584 nautical miles and an 'all-red' line from Britain to India, where hitherto messages by land line took an average of eight days to transmit.

The steam engine and the steam ship thus incubated, amongst many other engineering devices and techniques, the first real practical application of electricity. In doing so, they encouraged firms and groups which were to do much to create the motor which was to ultimately compete with steam on the railways and supplant it in the workshops. Such a firm was that of the Siemens brothers, which grew with the cable boom. In 1874 they built a special cable ship in collaboration with William Froude called the *Faraday*. This laid the sixth Atlantic cable, which was the first to link Great Britain with the United States of America without going through Canada. Two years later, Sir Charles William Siemens could say with justification:

'Submarine telegraphs are specifically English enterprises. I might go further and say that every submarine cable which is now working is, almost without exception, the produce of this country and has been shipped from the Thames.'

But the force behind the Siemens firm was German technical training, which we shall explore later.

CHAPTER 13

New Horizons in Britain, 1815–1854

THE BENTHAMITES

Samuel Bentham's brother, Jeremy (1748–1832), conceived a formula for calculating with scientific accuracy the exact amount of pleasure that would be derived from any action. This formula or 'felicific calculus', 'the greatest happiness of the greatest number', was to be applied in social policy. From his brother Samuel, Jeremy also derived the idea of a 'panopticon' or combined prison or school, whereby the head could witness every action by every one in the institution. Jeremy Bentham's disciples had, to an even greater degree, his faith in the calculus of motives. His friend, James Mill, hoped that one day the workings of the human mind 'would be as plain as the road from Charing Cross to St. Pauls', whilst his secretary, Edwin Chadwick, was one of the first to set in motion the criticism of laws and methods of administration which were to sweep away obsolete laws. Chadwick, as much as any one man, cleaned up the industrial towns by revealing social conditions through the Benthamite technique of investigation by Royal Commission. Royal Commission reports were further exploited by two emigrés living in England at that time: Karl Marx (1818–83) and Friedrich Engels (1820–95). One lived in Soho, the other in Manchester. Both ransacked the blue books produced by a generation of Royal Commissioners to document, with formidable scholarship, their thesis of the class war and the ultimate triumph of the proletariat. Not a little of their energy was devoted to assimilating current science. Indeed, their claim was that they had made socialism scientific instead of Utopian. Yet it was Marx who paid the most generous tribute to the half century of industrialism that had taken place up to 1848, when he said it had drawn even the most barbarous nations into civilization and rescued 'a considerable part of the population from the idiocy of rural life'.

MUNICIPAL ENGINEERING

Early Victorian towns had to be seen and smelt to be appreciated. The oldest necessities for people living in cities are a constant supply of fresh water, facilities for disposal of excrement, and the dissipation of malodorous results of earning their living. All these problems generally needed a servile class until the nineteenth century, and in the hasty expansion of towns, had not been satisfactorily solved. The death rate, which had been falling steadily, suddenly began to rise in 1830. At Birmingham and Bristol it nearly doubled between 1831 and 1841 (14·6 to 17·2 per thousand, 16·9 to 31 per thousand, respectively), whilst in Leeds, Manchester and Liverpool the increase was almost as alarming (20·7 to 27·2, 30·2 to 33·8 and 21 to 34·8 per thousand, respectively). Today (1956) it is 11·7. The towns had, in other words, begun to fester, and the lives of those within them were to be graphically described by Engels in his *Condition of the Working Class in England* (1844): as the cumulative effect of early labour, long hours, bad sanitation and inadequate administration.

London was a stinking case-history. The circle surrounding Jeremy Bentham gave willing and encouraging help to an eager young doctor at Whitechapel, Southwood Smith, who exposed the 'resurrection men' then exhuming corpses for sale to medical students. Jeremy Bentham bequeathed his own body for Southwood Smith to dissect and for the rest of Southwood Smith's life the skeleton remained in his consulting room; now it rests in University College, London.

As a good Benthamite, Southwood Smith's sole occupation was social medicine. He would work from four or five o'clock in the morning till eight or nine at night. He became in 1832 a member of the famous Factory Commission, set up to inquire into the health of children in factories. Many drawings illustrated the official report, made largely at his instigation. Six years after this—in 1838—an outbreak of fever occurred in Whitechapel. Southwood Smith was appointed to conduct an inquiry. It was the chance of a lifetime. He extended the scope of the inquiry to cover twenty metropolitan boroughs and unions, and it took four years. His report on this question is historic. Over 10,000 copies were published. It was a classic example of the way in which scientific examination of the causes of ill-health became the foundations of legislation.

The lesson that epidemics were really products of bad sanitation was driven home. Mortality statistics reinforced the argument, for they showed that bad sanitation alone was responsible for over

139

40,000 deaths a year. Southwood Smith helped to form associations to improve the houses of the working classes, and one of them— 'The Health of Towns Association'—worked to secure Parliamentary action in the matter. A State Board of Health, with a staff of inspectors, was advocated, and for it Southwood Smith with Edwin Chadwick worked like beavers, throwing off tracts and addresses to working men, urging them to take the initiative to improve the condition of affairs.

Smith might have been a good medical officer, but compared to Chadwick he was a bumbling amateur in administration. Where Southwood Smith wrote tracts, Chadwick composed memoranda and drove Whig ministers, often against their will, to act. Reeking sewers and undrained towns were most offensive to one who had written the section on a minister of health in Bentham's *Constitutional Code*. A third report was needed and Chadwick himself on behalf of the Poor Law Board of which he was secretary, drafted a *Report on the Sanitary Condition of the Labouring Population of Great Britain* (1842). He saw that 'the chief remedies' consisted in:
'applications of the science of engineering, of which the medical men know nothing; and to gain power for their applications, and deal with local rights which stand in the way of practical improvements, some jurisprudence is necessary of which engineers know nothing. . . . Aid must be sought from the science of the *Civil Engineer*, not from the physician, who has done his work when he has pointed out the disease *that results from the neglect of proper administrative measures*'.

The administrative measures were Chadwick's, the engineering ideas he obtained from John Roe, engineer to the Holborn and Finsbury Commission of Sewers. Together they insisted that sanitation should be hydraulic, arterial and water-carried, which meant that water had to be laid on to every house to flush away the excrement, and to carry away sewage in pipe drains (instead of the brick arches then in use). Sewers continued Roe, should be egg shaped with a good fall to speed the velocity of the water and an efficient system of sewage disposal at the end of the process. Chadwick even read Liebig's *Chemistry of the Soil* to make his point that the excrement of one man yielded 16·41 lb. of nitrogen a year. Waste material in the sewage might well prove valuable, and present day utilization of even the methane gas at the Mogden Sewage Works in London shows how effectively he made his point.

Chadwick's great sanitary report, published in 1842, recommended that water companies, landlords and Commissioners of Sewers should be replaced by a Central Board. Parliament was moved to appoint a Royal Commission to examine the health of towns. This Commission's report was virtually an expansion of Chadwick's accompanied, as we have seen, by educational activity by a number of pressure groups in which Southwood Smith played a part. Chadwick himself was occupied with the practice of burial in towns, on which he published another revealing report in 1843.

So, after so much work the first State Board of Health was set up in this country on 5 August 1848 with Southwood Smith and Chadwick as its leading members. His reports continued to be published, some of them obtaining an international reputation. Persistently, patiently, he collected his facts, and marshalled them into the order that alone could make his doctrine orthodox. It was short lived, this Board of Health. Its activity was resented—the word 'snooping' was used of it and so, after six years' work, it was abolished, and *The Times* remarked that 'the English people would prefer to take the chance of cholera, rather than be bullied into health'.

Chadwick's plan involved a centralized administration, and in the campaign against it much play was made with some fractures of the earthenware pipes which were used in towns following his plan. Instead of arming him, civil engineers like Joseph Bazalgette (1819–91) argued with him. Bazalgette became, in 1855, the chief engineer to the Metropolitan Board of Works, and in the course of the next twenty years carried out the construction of the 83 miles of sewers draining the 100 square miles of the largest city in the world. This was the first work of real magnitude on which Portland cement was used. He and other engineers like Captain Douglas Galton and Thomas Hawksley were helped by the invention of Dr. Angus Smith (1817–84), a former assistant of Lyon Playfair, who patented a technique of coating heated metal pipes with coal oil to resist corrosion. Angus Smith became a chief inspector under the Alkali Act (to which we will later refer). He was one of the first of many 'pure' scientists whose excursions into the field of sanitary engineering were to have good effects. The contribution made by Thomas Graham (1805–69) to the colloids of sewage, the use of creosote from the emergent chemical industry, and the later contribution of Koch and Pasteur, culminated in the chlorination of drinking water at Maidstone in 1897. All these developments in turn made their demands on the chemical industry.

In spite of the raging quarrels between administrator and engineer, progress was made. In 1852 the compulsory filtration of all river water used for domestic purposes was secured—thus making general a wholesome practice instituted by James Simpson at Chelsea in 1829. British water supply became so good that twenty years later J. P. Kirkwood (1807–77) was sent from America to investigate it, with a view to purifying American water, notably in Poughkeepsie. By then further improvement had been effected. And Chadwick, from the sidelines, pushed his case. By 1866 local authorities were compelled by Act of Parliament to appoint sanitary inspectors whose responsibility was for water and sewerage. Three years later, in 1869, a Royal Commission looked into the whole matter of public health: water, drainage, smoke, food and burial grounds. This led to the establishment of the Local Government Board. The codification of all these Acts took place in an Act of 1875: the Magna Carta of public health which prevented the rivers from becoming open sewers by compelling local authorities to set up sewage works. As a result, smallpox, typhoid, typhus and cholera virtually disappeared.

THE STIGMATA OF INDUSTRIALIZATION

The herding of the working populations in towns, badly planned and hastily built, evoked a number of protests. The first of these was Charles Hall's *Effects of Civilisation* (1805) which posed the sharp paradox of early capitalist society, whereby eight-tenths of the population received one-eighth of the produced wealth while two-tenths who produced nothing at all received seven-eighths of the produced wealth. Hall's protest was widely read. One of his friends and helpers, John Minter Morgan, was anxious to see the rise of better planned towns. Morgan christianized the stern environmentalist gospel of Robert Owen that man could only be improved by improving his material surroundings. A friend of Morgan's, James Silk Buckingham, who had served as a member of parliament for one of the new industrial towns—Sheffield—came out in 1848 with the first complete and concrete scheme for a garden city in his book *National Evils and Practical Remedies*. Buckingham's work was to inspire, in the next generation, Ebenezer Howard to initiate the first garden city in England at Letchworth.

It was the employees of the engineering industries who gave fresh vitality to the Trades Union movement. The millwrights at the close of the eighteenth century had been particularly resistant to change and this had led their employers to initiate, through the House of Commons, the legislation of 1799 and 1800 which made combinations amongst workmen illegal. In spite of the relaxation of these Acts in 1825, the Trade Union movement was poor and unorganized. But with the advent of the Steam Engine Makers in 1824, the Journeymen Steam Engine and Machine Makers and Millwrights in 1826, the Associated Fraternity of Iron Forgers in 1830 and the Boilermakers in 1837, engineering workers began to nucleate. Led by William Newton (1822–76) and William Allen, his assistant, these friendly unions of mechanics built up a sound financial, administrative machine that by 1851 became the Amalgamated Society of Engineers. As the Webbs have pointed out, it 'served as a model for all new national trade societies, whilst old organizations found themselves gradually incorporating its leading features'. By 1919 the A.S.E. had become the A.E.U.—the Amalgamated Engineering Union.

STEAM INTELLECT SOCIETIES

Charles Dupin the French mathematician and naval engineer, visiting Britain in 1816, was much struck by two brothers who worked in a bakehouse which they had equipped with portable gaslights. In the intervals of baking bread, these two made a steam engine which they used in turn to make 'des machines et des instruments de physique'. Their uncle (who apparently owned the bakehouse) was neither amused nor interested; according to Dupin 'il préfère de beaucoup la boulangerie et la pâtisserie à la gazométrie et l'astronomie'.

These baking brothers, who symbolized for Dupin the incubation of technological zeal in Britain, were neither an isolated nor a new phenomenon. As early as 1717 the weavers and artisans of Spitalfields were working out mathematical problems on their slates. In 1789 a group of young men in Birmingham began to earn for themselves the title of 'The Cast Iron Philosophers' from their habits of discussing mechanics and natural philosophy with foundry workers and others. In 1800 the geologist, Thomas Webster, was discussing, at the newly founded Royal Institution, 'the application of science to the common purposes of life'. In the autumn of the same year George Birkbeck, a 24-year-old medical graduate of Edinburgh, enrolled seventy-five workmen for a course of lectures 'abounding in experiments, and conducted with the greatest simplicity of expression and familiarity of illustration, solely for persons engaged in the practical exercise of the mechanic arts'.

143

The steam-intellect societies had begun. Technology-hungry artisans were, as Dupin rightly said, a portent. One of them, Timothy Claxton, was determined to form a 'Society of Ingenious Working Mechanics . . . to study and improve the Arts of the Kingdom'. It began in 1817 and lasted till 1820 when Claxton went to Russia. Four years later he appeared in Boston with an air pump and some apparatus.

The anastomozing threads of all this activity came together in 1823 when a mechanics' institution was founded in London. The literal founder was a Fleet Street journalist and patent agent called J. C. Robertson, editor of the *Mechanics' Magazine*. Yet the London Mechanics' Institute was based on George Birkbeck's earlier work in Glasgow, the effects of which M. Dupin so much admired. And the London Mechanics' Institute was the real starting point of the mechanics' institute movement in Britain. Nine years after Birkbeck's death in 1841 there were 700 mechanics' institutes in Britain alone, catering for a membership of 120,000. Over 500 of these were in England. Over a quarter of these were in Yorkshire and another ninth in Lancashire. The outstanding institution in Yorkshire was at Huddersfield; it is now a technical college.

Some of those who lectured in mechanics' institutes were really trying to put the industrial revolution in perspective, to provide some antidote to the division of labour which factory work entailed. Birkbeck himself translated Dupin's book on mathematics applied to the fine arts. Henry Dircks of Liverpool and C. F. Partington laboured in their spare time on seventeenth-century technological pioneers like Hartlib and the Marquess of Worcester. Others, like James Hole, the secretary of the Yorkshire Union, looked out from existing institutions to an 'Industrial University': 'Inefficient as they may be as a means of industrial instruction', wrote Hole, 'they afford the best instruments for introducing an improved system.' That 'improved system' was to occupy the efforts of scientists and statesmen for the rest of the century, and at each stage apathy, indifference, hostility and misunderstanding dogged their efforts.

Dupin's prediction to the mechanics of Paris that in the next twenty years 'the conquests of science and industry in Great Britain' would 'surpass those of all the generations which had given so much prosperity to her people' was, as we know, fulfilled. The Yorkshire Union was virtually a Royal Institution for the working classes, as virile as those for whom it catered. These 'steam-intellect societies' (as they were called by the novelist Thomas Love Peacock) were more

The Adelaide Galleries, *c*. 1830, showing the tank where Telford conducted his experiments. A lighthouse and steamship can be seen, together with a bridge.

The Polytechnic, Regent Street, as it originally stood. (From E. M. Wood, *The Polytechnic and Quintin Hogg*)

James Nasmyth (1808–1890), one of Maudslay's most distinguished apprentices, who applied steam power to the hammer.

Matthew Boulton (1728–1809), whose capital and commercial acumen did so much to make the steam engine a success.

an expression of working class interest in science, rational knowledge and the elevating effects of education to offset the degrading division of labour, than places where any adequate training could be obtained. It is notable that a popular writer and speaker, who caught up and articulated the spirit behind them, was Samuel Smiles, secretary of the Leeds and Thirsk Railway, whose *Self-Help* (1859) was a virtual gospel for the West Riding where he first worked. And Smiles' reputation was also made by his extremely readable accounts of the men who had fabricated the Victorian world in his *Lives of the Engineers*.

THE GROWTH OF SCIENTIFIC GROUPS

Charles Babbage, the designer of the 'difference' engine, had attended at Berlin in 1828, a meeting of the Deutscher Naturforscher Versammlung, a body which had been gathering annually in Germany to discuss scientific matters since 1822. His admiration for it was shared by Sir David Brewster, who took steps to establish a British counterpart by associating himself with John Phillips, the Secretary of the Yorkshire Philosophical Society (which itself only dated from 1821) to form the British Association for the Advancement of Science.

From its first annual meeting at York, in 1831, the British Association became 'an impulsive, directive force' (to use the words of the first secretary). In a stirring address, he continued, 'What Bacon foresaw in distant perspective, it has been reserved to our day to realise.' 'Mechanical Arts' was one of the six sub-committees formed in 1831. By 1836 it had become 'Mechanical Science' under the distinguishing letter G, denoting one of the seven sections into which these sub-committees had grown. By the twentieth century Section G was renamed 'Engineering'.

Yet the British Association took little interest in the possibilities of flight. As late as 1894, in a discussion on Maxim's flying machine, Kelvin described it as 'a kind of child's perambulator with a sunshade magnified eight times'.

SIR GEORGE CAYLEY

One of the leading promoters of the British Association in 1831 was Sir George Cayley (1773–1857), he had also helped to found both the Yorkshire and the Scarborough Philosophical Societies. He also tried, but unsuccessfully, to found an aeronautical society, first in 1816, then in 1837, and again in 1840. After 1840 the Duke of Argyle took up the task and ultimately succeeded in founding the Aeronautical Society of Great Britain.

Cayley was in every modern sense the inventor of the aeroplane. He defined the basic aerodynamic forces operating on a wing as early as 1799, and the little aeroplane he scratched on the back of the disc which is now in the Science Museum, is the modern aeroplane. He made two full-size gliders. As Orville Wright observed in 1912 'he knew more of the principles of aeronautics than any of his predecessors, and as much as any that followed him up to the close of the nineteenth century'.

Cayley was anxious to popularize and encourage a popular interest in science and engineering. He helped in the establishment of the Adelaide Gallery, near the Lowther Arcade, where the National Gallery now stands. It took its name from the consort of King William IV, and was opened in 1830. Here in 1830 Telford worked on tractive forces with a model canal 70 feet long and 4 feet wide and model boats some of them driven by clockwork. Here in 1834 Charles Wheatstone determined the velocity of electricity with an electric mirror and Faraday, four years later, determined the character of the shock of the gymnotus. As a visitor remarked:

'Clever Professors were there, teaching elaborate science in lectures of twenty minutes each. Fearful engines revolved and hissed and quivered. Mice led gasping sub-aqueous lives in diving-bells. Clockwork steamers ticked round and round a basin perpetually to prove the efficacy of invisible paddle wheels. There were artful snares laid for giving galvanic shocks to the unwary.'

On 16 May 1837 William Sturgeon presided here over the inaugural meeting of the Electrical Society of London; and with Gassiot drew up a set of Rules, establishing weekly meetings at the Adelaide Gallery. The Electrical Society issued a journal which, after six years' existence, came to an untimely end. Sturgeon's vision is seen in his observation that:

'As a universal agent in nature and space electricity takes the first rank in the temple of knowledge; but while many other branches of science have possessed the powerful and well directed aid of associated men and funds, that of electricity has been left to the fitful energies of individuals.'

'As a result', he concluded, 'thousands of important experiments as well as much acute reasoning has been lost.' After the Society lapsed, Sturgeon went to Manchester, where he superintended a similar gallery and became a friend of James Prescott Joule. It was at the

Adelaide Gallery, in 1843 (which became the Royal Gallery of Popular Science when Queen Victoria ascended the throne), that W. S. Henson made experiments on model power-driven aeroplanes, the field which Cayley himself so much adorned.

A second institution promoted by Cayley was the Polytechnic Institution. Though granted a Royal Charter in 1839 it suffered from a lack of money. Subscribing industrialists wishing to control it were resisted by Cayley who appointed 'a good scientific board confined by no aristocracy or orthodox men who sit like an incubus on all rising talent *that is not of their own shop*'.

THE DYING AMATEURS OF SCIENCE

The term 'scientist' now emerges, first used by William Whewell in his *Philosophy of the Inductive Sciences* in 1840 (vol. i, p. 113). Whewell, the professor of moral philosophy at Cambridge, was a great friend of Michael Faraday. Faraday's own electrical discoveries at the Royal Institution, which produced the first dynamo, were in a sense an offspring of his original interest in electrolysis. Faraday gave Christmas lectures at the Royal Institution, and Whewell was a Cambridge don. Electricity was to owe much in its practical application to a Salford brewer, James Prescott Joule.

Slightly deformed, with symptoms of spinal trouble, Joule started in his father's office at the age of 15 where he acquired an interest in science (like Priestley who first studied carbon dioxide in a brewery) and came to appreciate the relationships between the temperatures and pressures of gases, between pumping and heating, simply by watching the brewery at work. Twice a week he studied under John Dalton, president of the Manchester Philosophical Society from 1817 to 1844. Dalton, who discovered the law of chemical combinations and tabulated the atomic weight of various elements, imbued Joule with his own love of experiment, so that, when only 19 years old, Joule had turned one of his father's rooms into a laboratory, and published his first paper. When on holiday near Lake Windermere he once blew off his own eyebrows with a pistol while trying to measure echo speeds. He also experimented on a lame horse with electricity.

It was to Sturgeon's *Annals of Electricity* that he sent the results of his early experiments. In the fourth of these he gave us the guiding motive of his research:

'I can hardly doubt', he wrote, 'that electro-magnetism will ulitmately be substituted for steam to propel machinery. If the power of the engine is in proportion to the attractive force of its magnets, and if this attraction is as the square of the electric force, the economy will be in the direct ratio of the quantity of electricity, and the cost of the engine may be reduced *ad infinitum.*'

He was considerably encouraged when William Sturgeon became superintendent of the Victoria Gallery of Practical Science at Manchester, a superintendence which soon failed owing to lack of money.

Joule first announced his discovery of the first law of thermodynamics in 1843. This was that the quantity of heat capable of increasing the temperature of a pound of water by one degree Fahrenheit is equal to a mechanical force capable of raising 838 pounds to a height of one foot.

Joule did not need the prestige of a university chair to elevate his name in the world of science. In 1847, at 28, he gave a lecture in the reading room of St. Ann's Church, Manchester, expounding his discovery of the law of the conservation of energy. This lecture was reprinted in the *Manchester Guardian* for 5 and 12 May in that year. He also read papers to the British Association at Cork in 1843, at Cambridge in 1845 and at Oxford in 1847, and it was at the last named of these meetings that he met a young man—just 22 years old —William Thompson (Lord Kelvin as he later became), with whom he was to exchange many ideas. Before he was 30 he was elected to honorary membership of the Royal Academy of Science at Turin, and two years later to the Royal Society.

By 1849 Joule had established the indestructibility of energy, the mechanical equivalent of heat and the existence of an absolute zero of temperature. He provided the material for the mathematicians to develop the theory of heat, and by showing that the efficiency of heat engines depended on the ranges of their working temperatures, he was in fact paving the way for the development of hot air engines.

His friend, young William Thompson (later Lord Kelvin), suggested that apparatus should be constructed to measure the temperature of gases before and after compression through a small hole. The Royal Society provided some of the money, and Joule set up in his brewery, where he conducted experiments which led ultimately to low-temperature refrigeration.

In 1854 Joule sold the brewery and moved his apparatus to his house in Oak Fields, Whalley Range. Joule was so immersed in experiments that he would not take his meals properly—he would iust run in and out of the house gobbling them up. In 1861 he moved his apparatus again to Thorncliffe, Old Trafford, but one of his neighbours complained of the noise made by Joule's engine. Despite such discouragements, Joule continued to work and his efforts are recognized in the commemoration of his name as a physical unit of power.

THE BEGINNINGS OF PROFESSIONAL TRAINING FOR ENGINEERS

But amateurism was not enough. In 1832, a year after his appointment as Professor of Physics at the University of London, William Ritchie (1790–1837) stated to the Senate that he was prepared, without interfering with his courses in natural philosophy, to give courses in civil engineering. In this he was supported by Thomas Coates, the Clerk to the Council, who remarked three years later:

'At present Civil Engineering can only be learned at enormous expense in the office of a professional man, who requires a fee varying from £1,000 to £500 for five years' pupilage. Nevertheless it is a science which can especially well be brought into a classroom. . . . The University would be a fitter place for teaching Civil Engineering than a separate institution.'

King's College set up the first department, University College, the first chair. At King's the prospectus of a course in civil and mechanical engineering was issued in 1838, sponsored by J. F. Daniel (1790–1845) the Professor of Chemistry and a Copley Medallist and Charles Wheatstone (1802–75) the Professor of Experimental Physics and pioneer of spectrum analysis and telegraphy. In the following year a lecturer in manufacturing art and machinery was appointed and in 1844 the department was renamed the Department of General Instruction in the Applied Sciences. The number of students rose from 31 to 71 in 10 years and by 1868 the authorities told a Parliamentary Select Committee on Scientific Instructions, 'We find our young men greatly valued and sought after for engineering appointments.'

Meanwhile, University College founded three chairs in engineering. The first, in civil engineering, was established in 1841 and to it was appointed C. B. Vignoles (1793–1875), a leading civil engineer. Vignoles had experience in both America and Russia. His suspension bridge over the Dneiper at Kieff was the largest in the world at the time of its construction. In 1846, the year in which the Institution of Mechanical Engineers was founded, a chair of mechanical engineering was established at University College, to which Eaton Hodgkinson (1789–1861) was appointed. Hodgkinson was an authority on the strength of cast-iron beams and, with William Fairbairn, helped to test the rectangular tubes for Stephenson's bridge over the Menai Straits. To a third chair, that of machinery, established in the same year, Bennet Woodcroft (1803–79) was appointed. Woodcroft was the inventor of 'tappets' for looms and a consulting engineer in London before taking up his appointment. He had worked on patents and industrial history and his published work is still used. By 1859 the reputation of the chair of civil engineering was both recognized and enhanced by the appointment of William Pole (1814–1900), who had served as first Professor of Engineering at Elphinstone College, Bombay, from 1844 to 1847, and who had assisted such great engineers as J. M. Rendel and Sir John Fowler. Pole's versatility was legendary. Fellow of the Royal Society and a doctor of music, he wrote as easily and authoritatively on engineering, music and whist. In the intervals of acting as honorary secretary of the Institution of Civil Engineers, he also acted as examiner for musical degrees.

Private professional schools and professional training, however, was very much a private affair, *ad hoc*, and established for immediate purposes. Four institutions at this time illustrate this.

Putney House, converted into a College for Civil Engineers was opened at Putney House in 1834. The railway and gas engineer, Samuel Clegg (1814–56), taught here and published a treatise on coal gas before becoming, in 1849, a Professor of Engineering at Chatham. Another teacher, Edward Frankland (1825–99), went, in 1851, to Owens College, Manchester. The principal, Morgan Cowie, was well aware of the significance of the college and in *A Letter to His Grace the Duke of Buccleugh*, published in 1847, wrote:

'If our country is to maintain itself as the leader of civilisation and the first in the progress of nations . . . the rising generation must be prepared for a struggle which can only be successful if they have acquired familiarity with the principles of science and their application to the wants and necessities of man.'

The students of Putney College began, in 1855, to foregather, led by Henry Palfrey Stephenson (1826–90), a gas engineer, to read papers to one another. Stephenson gave the first—on 'The Plurality of Worlds'—in which he 'advocated the extreme probability of the heavenly bodies being inhabited, especially the planets, by rational people similar to the inhabitants of this World'. Another early member was John Aird (1833–1911), who built several gas and water works at home and abroad, and is chiefly known for the construction of the Assuan and Assiut dams on the Nile, for which he was made a baronet. In 1857 the Putney Club changed its name to The Society of Engineers. One of those who read a paper to the reconstructed society on 'Steam on Common Roads' was Alfred Yarrow (1842–1932). He had, in 1859, also helped to found the Civil and Mechanical Engineers' Society at Radcliffe in Lancashire. This amalgamated with the Young Engineers' Scientific Association in 1861. Though originally composed of marine engineers (Yarrow himself was later to achieve world distinction as a maker of light marine craft utilizing prefabricatory techniques) the society also embraced railway engineers, and their president during the years 1863–5 was W. H. Maw, editor of *Engineering*. In 1910 the Civil and Mechanical Engineering Society amalgamated with the Society of Engineers, retaining the latter name.

The urgent need for chemists was such that those who, like Dr. William Gregory and Dr. John Gardner, had had the good fortune to study in Germany under Liebig, seized the occasion of Liebig's visit to England to campaign for a German-type laboratory in England. Liebig was asked to nominate a director and, after some delay, A. W. Hofman (1818–92) was appointed in 1845. Hofman agreed to come for two years, but stayed for twenty, during which time he placed the teaching of chemistry on a new footing.

A similar college of chemistry was founded at Liverpool in 1848 by J. S. Muspratt (1821–71), son of the founder of the alkali industry in Lancashire.

The fourth foundation was begun in 1846 at Queenwood in Hampshire, a 'model community' founded six years before by Robert Owen and built by Hansom, the designer of the famous cab. The community experiment failed and George Edmondson took it over as a college. John Tyndall (1820–91) supervised the engineering department and taught mathematics there. His wife saw it as:

'an odd mixture of an ordinary junior boys' school and a technical college for older boys and young men—the whole being under one and the same roof'.

According to Edward Frankland, it was the first English school to introduce 'the practical or laboratory teaching of science'. Amongst their pupils who later won distinction were Henry Fawcett (1833–84) who became Postmaster-General in 1880 and James Mansergh (1834–1905) the civil engineer who was to win a name as a constructor of water and sewage works. The college lasted until Edmondson's death in 1863.

To these four private ventures should be added two others, those of John Owens at Manchester and Henry R. Hartley at Southampton.

Both became institutions where a meagre scientific training was to be had. Though Owens' College did not open its doors until 1851 and the Hartley Institution until 1862, the motives behind the legacies of both men stemmed from the eighteen-forties.

THEORY OF PRIME MOVERS

Four important books appeared at this time which synthesized progress to date. The first was by a pupil and subsequent partner of Rendel, Nathaniel Beardmore (1816–72): one of the first British hydraulic engineers. His *Manual of Hydrology*, first published in 1852, became the text-book of the profession. The second was Joseph Glynn's *Rudimentary Treatise on the Power of Water with an appendix on centrifugal and rotary pumps*, to which reference has already been made. It first appeared in 1853 and was widely read. The third was Rankine's *Steam Engines and Prime Movers* (1859). W. J. Macquorn Rankine (1820–72) was a pupil of Sir John McNeill and spent his time on water projects. He succeeded Lewis Gordon as professor of civil engineering and mechanics in 1855. The fourth was William Fairbairn's *Mills and Millwork* (1862). Of this perhaps it is best said that it should be read by any student of the history of engineering.

THE SCHOOL OF MINES

The deep pits of Northumberland and Durham were, by the 1840s, yielding over a quarter of British coal. By 1849 Monkwearmouth, the deepest, went down to 1,716 feet. Lancashire, too, was sinking deeper pits and the more powerful winding engines needed were fitted with wire ropes (patented in England in 1835) and two-decker metal cages. As the coal was more rapidly extracted the railways carried it away. Wigan was even exporting coal to America. Deep pits demanded ventilation and here, in 1849, the South Wales Coalfield showed the way by mechanical means, without recourse to ventilation by furnace. In 1850 a Mines Act required all mines to be inspected by qualified men. As a result, Henry de la Beche formed, out of the Geological Survey established in 1832, a School of Mines in Jermyn Street which, amongst its other teachers, included Lyon Playfair. This was the first step to be taken by the Government to ensure an adequate professional training for those engaged in responsible posts in an engineering industry. Events were to accelerate this trend as the century wore on.

CHAPTER 14

The Crimean War and After, 1854-1857

When the Turkish Fleet was attacked in Sinhope harbour by the Russians and practically blown out of the water on 27 November 1853 the French and English were deeply outraged. Four thousand Turks were killed under the guns of an Anglo-French Fleet that had sailed through the Dardanelles. When England and France went to war with Russia on 29 March 1854 the memory of this swift and tragic destruction was fresh in the naval constructors' minds, and for the first time they built vessels sheathed in armour plate. France originated the idea and on 30 September 1855, in the attack on Kinburn to menace the rear of the Russian army in the Crimea, three French armoured floating batteries were employed. These three French ships had their baptism of fire and were struck between sixty and seventy times without complete penetration, only dents of 1½ inches deep being made in the plates.

In 1855 Britain followed with the *Trusty*, *Thunder* and *Meteor*, all with wooden hulls and armoured bottoms, and by 1856 was building iron hulls too, though none of the British floating batteries was completed in time for the Crimean War. One of them, the *Terror*, was built by Palmers of Jarrow with plates rolled by Samuel Beale & Co. (now Parkgate Iron Works) some 4 inches thick. (About this time the *Gloire* was built for the French Navy.)

To sheath these vessels, heavy industries were called into play. Thomas Turton and Sons (under the capable guidance of F. T. Mappin) installed the Nasmyth Hammer in their Sheffield works on 26 March 1855. John Brown bought, on 2 October 1855, a bankrupt works in the same town and transferred his business (hitherto scattered in five enterprises round Sheffield) and renamed it the Atlas Steel and Spring Works. It began to make boilers, armour-plate, springs, forgings and special orders like the plates for Charing Cross Bridge. In 1860, Brown personally visited a French naval base and

153

came back with the idea of rolling instead of forging the armour-plate. Needless to say, before he died he was sheathing three-quarters of the British Navy. Brown was also the first to introduce the Bessemer process in Sheffield in 1859.

BESSEMER

When the Crimean War began, Henry Bessemer (1813–98) was, perfecting a patent for giving rotation to an elongated projectile, and used to experiment in his own garden at Highgate with a cast-iron gun of his own manufacture. Bessemer's invention excited the interest of Napoleon III, who gave him *carte blanche* to go to Vincennes and experiment. But Bessemer found it easier to experiment at home. So, with unlimited credit at Baring Brothers (given him by Napoleon III), he returned home to manufacture two shot—one a 24- and the other a 30-pounder—for the French 4·75 guns. These he took to Vincennes and successfully fired them in the presence of Captain Minié (of the famous rifle). Minié observed that it would not be safe to fire a 30-pound shot from a 12-pounder gun, so Bessemer returned home to England again, and on 10 January 1855, took out a patent for fusing steel in a bath of molten iron in a reverbertory furnace. Steel was then £60 a ton. Bessemer cast a small model gun out of his steel and submitted it to the Emperor Napoleon, who issued orders that a furnace should be erected at the cannon factory at Ruelle, near Angoulême. Bessemer again returned to England, in order to get firebricks for his foundry. It was while he was experimenting with his furnace in England that he noticed how a draught of air decarbonized two bars of pig iron lying on the rim of his furnace, and that led to the discovery embodied in a patent he took out on 17 October 1855, for forcing air by fan into a closed cupola and removing impurities from melted iron. In December he patented what was in essence the Bessemer converter. The Bessemer process, with all that it meant, was announced in Sheffield on 9 February 1856.

The British Ordnance authorities (now reformed as a result of the Crimean Commission) in the person of Colonel F. M. Eardley-Wilmot, helped him by allowing him to roll bars at Woolwich. In August 1856, Bessemer read a paper to the British Association at Cheltenham on his process, and found that the Sheffield ironmasters were suspicious of his plan. Indeed, Edward Smith read a paper to the Sheffield Literary and Philosophical Society on 3 October 1856, which pronounced unfavourably against it. So Bessemer, in partnership with the Galloway Brothers of Manchester, and others, set up a foundry in Sheffield. But, as he himself said:

'No Sheffield manufacturer would adopt the process. Each one required an absolute monopoly of my invention if he looked at it at all. So I set up a works in the midst of the great steel industry of Sheffield. My purpose was not to work my process as a monopoly, but simply to force the trade to adopt it by underselling them in their own market which the extremely low cost of production would enable me to do, while still retaining a very high rate of profit on all that was produced.'

He certainly did. At that time, Sheffield manufacturers were selling steel at £60 a ton whereas Bessemer could buy Swedish iron at £7 per ton, and by blowing it for a few minutes in the converter, could produce material of equal quality at much lower cost. After fourteen years, the works (which had been greatly increased out of revenue) was sold for twenty-four times the subscribed capital of the firm. Notwithstanding this, they had already divided in profits a sum equal to fifty-seven times the gross capital. From the mere commercial working of the process, apart from royalties derived from the sale of the patent, each partner made in these fourteen years, eighty-one times the amount of capital subscribed. In other words, they enjoyed a dividend of 100 per cent on their investment for every two months of these fourteen years.

The Bessemer system, introduced first by John Brown and his partner, J. D. Ellis, together with William Bragge in 1858, necessitated new forges, converters and rolling mills. This in turn demanded fresh capital and in 1864 they became a limited liability company, being followed by Cammell Brothers and other companies in the same year. Cammells were early in the plating of vessels: the first ships they plated were the *Lord Clyde* and the *Royal Alfred* in 1864.

THE COMPANIES' ACT OF 1862

The repercussions on industrial organization, especially in the shipbuilding industry were decisive. The age of metal and machinery inevitably ripened the growth of large-scale industrial units. Shareholders in the *Great Eastern* (which was only floated by the flotation of several companies) went through the kind of traumatic experience that their predecessors had suffered in the railway mania of a decade before. As a result the Companies' Act of 1862 (25 and 26 Vict. c. 89) empowered seven or more associates to constitute a company, provided they subscribed to a memorandum of association. This Act facilitated the creation of companies on the model of John Brown's and Cammel, like the Staveley Coal and Iron Company, the Fairbairn Engineering Company of Leeds and Palmer's Shipbuilding Company. Even the private businesses became limited liability companies: Joseph Whitworth's in 1874, the Cunard Line in 1878, Robert Stephenson in 1886 and Samuel Cunliffe Lister in 1889. Nettlefold's, Tangyes and others had meanwhile done so.

THE EFFECT ON RAILWAYS

When Henry Bessemer suggested steel rails to John Ramsbottom (1814–1897) the locomotive superintendent of the London and North Western Railway at Crewe, Ramsbottom looked at him 'with astonishment, and almost with anger', saying, 'Mr. Bessemer, do you wish to see me tried for manslaughter?' Yet Ramsbottom's willingness to experiment converted him and the first Bessemer steel rails in the world were laid down at Crewe Station on 9–10 November 1861. Three years later the London and North Western Railway Company opened its own Bessemer steelmaking plant. This, as much as anything, urbanized the west end of the township and brought the number of employees up to 3,000. Ramsbottom's appointment at Crewe in 1857 'marked', as its most recent historian has said, 'a period of reorganisation, concentration and bold innovation'. L.N.W.R. carriage making was concentrated at Wolverhampton, engine construction at Crewe. From Crewe emerged the engines that were to be standard types for a generation—like the six-coupled goods, four-coupled passenger, both types injector equipped and with standardized parts. One example of Ramsbottom's work that survives today is the track water trough from which locomotives pick up water whilst running. The piston and rings that were named after him, and the duplex safety valve, are monuments of his energy.

Ramsbottom's successor, F. W. Webb, as Chief Mechanical Engineer of the L.N.W.R. from 1871 to 1903 carried the work of concentration and development still further. In 1873 he began to manufacture signals, and went on to establish a leather, soap and brick works. Even footwarmers and artificial limbs were manufactured here. And so efficient had construction of locomotives become that private manufacturers obtained and served an injunction on the L.N.W.R. to restrain them from catering for demand outside the company. Two years before Webb retired the company built its

4,000th engine, and the record assembly of one was timed at 25½ hours.

THE RISE OF ARMSTRONGS

It was just after the outbreak of the Crimean War that William Armstrong, whom we have already met as a hydraulic engineer, was commissioned to design submarine mines for the purpose of blowing up Russian ships sunk in the harbour of Sevastopol. His mines proved very successful and converted him from a hydraulic engineer to an armaments manufacturer. It is said that an incident in the battle of Inkerman on 5 November 1854 led him to devote his energies to the improvement of ordnance, and in the following month he suggested to Sir J. Graham the expediency of enlarging the ordinary rifle to the standard of a fixed gun to fire elongated projectiles of lead. The Duke of Newcastle, Secretary of State for War, interviewed him and authorized him to make six guns according to his plan. And so it was that Armstrong produced the first wrought-iron breech-loading gun to replace the bronze or cast-iron muzzle loaders then in use. He adopted the principle of shrinking on a wrought-iron jacket to a steel inner tube, to give the guns more strength. These steel tubes were made for him by the Vickers Company of Sheffield, which was then quite new. There is an interesting story of how the Vickers Company acquired this singular skill. In 1854 they acquired the patent of E. Riesse (which up to this time had been used in Germany) for making cast-steel bells, and working under this patent they developed the art of steel casting to such a pitch (the pun is not intended) that, when they were not casting the steel jackets of guns, they were casting the biggest bells in England. Indeed, one bell that they cast for the Great Exhibition of 1862 required the contents of 176 crucibles, weighed 4½ tons and was the largest steel casting yet made in England.

A third discovery was made by William Siemens and his brother Frederick in 1856: the regenerative furnace in which lost heat was recouped. An experimental furnace was built in Birmingham and a year later was being used for melting steel. Within the next decade it was being used to produce steel, by melting iron ore and pig iron. This was improved upon by two other brothers, who used to melt steel scrap and pig iron.

The Bessemer process depended on phosphorous-free ore from Sweden and could not exploit many of the English deposits. This was later overcome by Gilchrist Thomas who lined the furnace with an alkaline material, which, together with a flux, absorbed the phosphates. This enabled such towns as Middlesborough to utilize, after 1879, their phosphoric ores.

THE RISE OF THE SYNTHETIC ORGANIC CHEMICALS INDUSTRY

One accidental discovery of the year 1856 was made by an 18-year-old student at the Royal College of Chemistry, William Perkin. Experimenting in a laboratory at home to make quinine ($C_{20}H_{24}N_2O_3$) from allyltoluidine ($C_{10}H_{13}N$) by adding oxygen, and not knowing much about molecular structure, he heated it with an oxidizing agent consisting of a mixture of sulphuric acid and potassium bichromate, and got a red powder. Repeating the experiment with aniline, he got a black substance which when purified, dried and treated with spirits of wine yielded a brilliant mauve solution which had singular dyeing properties. Pullars of Perth were impressed by a sample he sent them, so he took out a patent on 26 August 1856.

With his father (a builder) and brother, Perkin established a works for the large-scale manufacture of this dye at Greenford Green, near London. To feed his plant he required benzene and nitric acid. His enterprise and initiative inspired others to employ pupils or assistants of Hofmann to make dyes.

The fashionable court of Napoleon III, a centre of conspicuous consumption, eagerly responded to the exciting new colours. French dyehouses bought them and two of the new dyes were called magenta and solferino to commemorate the French victories over the Austrians in 1859.

A Swedish engineer at St. Petersburg, who was evidently aware of the explosive effect of nitrated glycerine, tried to employ it during the Crimean War but failed. His son, Alfred Nobel (1833–96) returned to Sweden in 1859 for further experiments and found that nitro-glycerine could be exploded by a detonator. He began to manufacture it in a factory near Stockholm in 1864, the year of his first patent. But it was unstable enough to cause a number of fatal accidents. In 1866 Noble discovered that it became stable if absorbed in Kieselguhr. This new explosive, called 'dynamite' was a godsend to mining and civil engineering. When Nobel went on to provide gelignite and gelatinous explosives which were capable of alteration for specified tasks his interests expanded enormously. In five years the production of nitro-glycerine increased from 11 to 1,350 tons and Nobel built some fifteen factories throughout Europe. These supplied the explosives for the railway builders who were blasting their way through the Alps and the Rocky Mountains.

An even more significant trail—to the petrochemical industry—was blazed by Alfred's two brothers, Robert and Ludwig Nobel, who were the first to extract benzole and naphthalene from Baku oil in about 1880.

THE WAR PANICS

Between the end of the Crimean war in 1856 and the outbreak of the American Civil War there were two naval panics in England: one in 1858 on the deficiency of screw-driven ships, the other in 1859–61 on the shortage of iron-armoured vessels. These revived the question of training naval architects.

It will be remembered that the School of Naval Architecture had been established at Portsmouth in 1811 under Dr. Inman. This was closed down in 1832 by Sir James Graham. Then in 1843 schools for dockyard apprentices had been opened at Portsmouth, Chatham and Pembroke and later in four other towns. In 1848 the best of these pupils were gathered for higher training in a 'School of Mathematics and Naval Construction' at Portsmouth, but this, too, was closed down, also by Sir James Graham, in 1853.

The naval panics reopened the question and in 1860 Admiralty officials and naval contractors joined together to found the Institution of Naval Architects. In the same year the Scottish Institution of Ship Builders and Engineers was also formed. Three years later the Institution of Naval Architects proposed the establishment of a new College of Naval Architecture. One of the moving spirits was John Scott Russell who had been associated with the *Great Eastern*. The proposal was successful and in 1864 a Royal School of Naval Architecture and Marine Engineering was established at South Kensington. This received Admiralty students and fee-paying private students. The Admiralty students consisted of eight naval architects and eight marine engineers chosen by competitive examination from the dockyards.

Among the first eight students was W. H. White, later to be designer of Armstrong's shipyard at Elswick, and H. E. Deadman, his assistant. Other students included Francis Elgar, later to be Professor of Naval Architecture at Glasgow, and J. J. Welch, first Professor of Naval Architecture at Armstrong College, Newcastle-upon-Tyne. In spite of the apprehension of the private shipbuilders, the Germans thought that this third School of Naval Architecture 'gained an outstanding reputation not only in Britain but also . . . abroad'. Between 1869 and 1872 one of the teachers was W. C. Unwin, who in 1856 had become Fairbairn's scientific assistant, and ten years later manager of the Engine Department of the Fairbairn Engineering Company. Unwin also lectured to the Royal Engineers at Chatham between 1868 and 1871.

THE EXPANSION OF SMALL ARMS PRODUCTION

In 1855 the government called in Joseph Whitworth (1803–87), the most distinguished toolmaker in the world, whose introduction of the uniform system of screw thread that bears his name had made possible an enormous acceleration of constructional and repair work, to advise them on the machine tools necessary to re-equip the Small Arms Factory at Enfield. Whitworth began an elaborate investigation of the design of rifle barrels and projectiles, with the result that he developed a rifle more accurate than any previously made. This led him on to manufacture heavy guns, submitting the molten metal to a hydraulic pressure of many tons before it cooled.

The revolution in small arms production initiated by Samuel Colt (1814–62) at Whitneyville and Hartford, Connecticut, was itself accelerated by the Mexican War into which the United States had entered in 1846. Colt established a factory for the production of small arms in England which excited John Anderson, the inspector of machines in the Ordnance Department. 'No one,' he told a Parliamentary Committee on Small Arms in 1854, 'could go through that works without coming out a better engineer.' That veteran of assembly-line production, James Nasmyth, thought the same; 'perfection and economy such as I have never seen before. . . . You do not depend on dexterity—all you need is intellect.'

Anderson was sent to the United States in 1854–5 with two artillery officers and was still more impressed by the Springfield Armoury. From the ideas and machines in operation there he returned to introduce the Blanchard copying lathe to Woolwich, enabling irregular goods to be produced to a pattern. Furthermore, the Birmingham Small Arms Factory (consisting of twenty firms which had been selected to supply arms for the Crimean War) was established with American machines at Small Heath. A third innovation was the establishment of a small arms factory at Enfield which, begun in

1854, was soon turning out 2,000 guns a week. The Lee-Enfield rifle was to have a decisive impact on British fortunes especially in India.

That impact was not long in coming. For in the very year the Crimean War ended, the East India Company decided to equip its army with these rifles. Cartridges for them had to be heavily greased and had to be bitten to open the end and release the powder. The grease came from pigs and cows: one animal a particular abomination to Hindus, the other of particular veneration to Muslims. The company soon realized its mistake and hastily changed the grease for beeswax and vegetable oil. But the damage had been done; the tale had spread and in 1857 the Indian Mutiny broke out.

CHAPTER 15

India: Challenge and Stimulant

ENGINEERING UP TO 1857

Glimpses of the complex reciprocal influence of India and Britain upon each other in their long association have already been obtained. We have seen how the East India Company stimulated the growth of shipbuilding in Britain. The wars too were not without effect: Congreve is said to have been influenced by Tipu Sahib, whose 5,000 rocket throwers harassed and embarrassed their more conventionally armed British opponents at Seringapatam in 1792 and again in 1799.

So great was the need for engineers that the East India Company opened an engineering college at Addiscombe, near Croydon, in 1809. Taking over the house of Charles Jenkinson, the first Earl of Liverpool, they secured the services of Colonel (later Major-General) William Midge (1762–1800), then director of the ordnance survey, as commandant.

The harnessing of Indian rivers was one of the great achievements of Addiscombe cadets. One of them left Addiscombe in 1819 for Madras, and for the next forty-three years his famous anicuts on the Coleroon (1835–6), the Goodavery (1847–52), and the Krishna (1861), preserved millions from death by famine. When, sixteen years after the third anicut was finished, a great famine hit India in which over four million people perished, there were no deaths in the districts he had irrigated. Indeed, these districts were able to produce surplus food that saved three million lives. Well might the Madras Government say in 1858 of this cadet, Arthur T. Cotton (1803–99): 'If we have done our duty and have founded a system which will be a source of strength and wealth and credit to us as a nation it is due to one master mind, which, with admirable industry and perseverance, in spite of every discouragement, worked out this great result.'

162

Cotton's subsequent knighthood and return to England led to further political pressure being exerted in favour of further works.

The work so magnificently begun, has continued up to our own times. By 1947 nearly 70 million acres of land in India were under irrigation—three times as much as in the U.S.A., and greater than the aggregate amount of any ten countries in the world.

J. V. Bateman-Champaign (1835–87), another Addiscombe student, constructed the electric telegraph to India through Russia, Turkey and Persia, whilst J. F. D. Donnelly (1834–1902) as the first 'Director of Science' in the British Science and Art Department was to become architect of the first technological examinations held in Britain.

Nor were the lecturers at Addiscombe behindhand in promoting engineering progress. We have already met (p. 146), William Sturgeon (1783–1850), who began life as a shoemaker at Woolwich and was by 1824, thanks to his industrious study of electricity, appointed to the staff at Addiscombe. Sturgeon discovered in 1823 the soft iron electro-magnet, parent of the dynamo, and presented his apparatus to the Royal Society of Arts. His enthusiasm led him to establish thirteen years later, the *Annals of Electricity*, the first electrical journal in England.

The first steam engine was brought to India in 1820 by missionaries at Serampore, and the first steamer voyage effected five years later in 113 days (of which 50 were under sail). Travellers preferred the route via Alexandria and Suez which by 1838 took only 35 days, five days of which were spent travelling over the isthmus of Suez.

English railway engineers were very active in India. Charles Fox (1810–74), the builder of the Crystal Palace of 1851, constructed the first narrow-gauge line in India. J. J. Berkley (1819–62) was recommended by Stephenson as the Chief Engineer of the Great Indian Peninsula Railway. Of Berkley it was said that he 'succeeded not only in engineering matter, but in the more difficult task of educating men'. This was a tribute, among other things, to his encouragement of the Bombay Mechanics' Institute.

Before the mutiny, engineering had made great strides. Five hundred and thirty miles of the Ganges Canal, the biggest work of engineering in the world at the time, had been constructed by Sir Proby Cautley (1802–71), with the encouragement of James Thomason (1804–53), the Lieutenant-Governor of the North-Western Provinces. Thomason's enthusiasm for training engineers was such that, to ensure a sufficient supply of Indians to continue works like his Grand Trunk Road, he founded, in 1847 an Indian engineering college at Roorkee which bore his name and became a university a century later. Other engineering schools were established in the three presidencies to enable Indians to cope with the work initiated by the Department of Public Works (set up in 1854).

The first professor of Engineering at the Elphinstone College, Bombay, was 30-year-old William Pole (1814–1900), who helped in the survey for the great Indian railway. On his return to England he was employed in railway engineering and was consulted by de Lesseps on the proposal to dig a canal through the isthmus of Suez. He went on to become the consulting railway engineer to the Japanese Government and to hold the chair of Civil Engineering at University College, London. Pole's standing in the profession was high and his advice was sought on numerous engineering matters of the day from municipal waterworks to foreign harbours.

THE CONSEQUENCES OF THE MUTINY

If engineering advances helped to cause the mutiny in 1857 they also facilitated its suppression. The telegraph and railway enabled troops to be moved with great ease. And railway engineers took part in withstanding sieges, notably in the defence of Arrah. After the capture of Gwalior, the Mutiny came to an end and the process of westernization accelerated. Some of the reasons for this are worth exploring.

The Crimean War cut Britain off from her supplies of Russian hemp and so accelerated the cultivation of jute in India. On the other hand the discovery of aniline dyes by Perkin led to a severe decline in Indian cultivation of indigo. The rising standards of living in Britain and the improved shipping led to an increased consumption of tea and coffee grown both in India and Ceylon. The Mutiny itself stimulated the laying of the Red Sea and India Cable, with an energy which can be only described as feverish. As if this were not enough, a world economic crisis in 1857 led some British investors to regard continental enterprises with a somewhat jaundiced eye, and turn to India as a more profitable field of enterprise. Especially was this so in the period of reconstruction when railways, harbour works, dams and irrigation schemes went ahead with vigour. By 1870 4,000 miles of railway were built, of which J. H. Clapham has observed that 'perhaps, in the history of the human spirit, these Indian lines may prove to have been of greater moment than all those laid in America.'

THE SUEZ CANAL AND ITS EFFECTS

But there was one enterprise designed to link India and England more closely which was not given the official support it deserved: a canal across the isthmus of Suez.

Its progenitor and promoter was the Frenchman Ferdinand de Lesseps. Though snubbed by the Prime Minister in January 1856, he toured the British Isles in March 1857 to recruit commercial support. His success led to a debate taking place in the House of Commons in July, when Robert Stephenson, now an M.P., spoke so forthrightly against de Lesseps' canal that de Lesseps challenged him to a duel. Stephenson apologized and the apology was confirmed by the Secretary of the Institution of Civil Engineers. With the support of Said, the Viceroy of Egypt, and French investors, work was begun on 25 April 1859, and the town of Port Said founded.

The engineering skill of Voisin Bey and Laroche, his assistant, coupled with the mechanical trough-dredgers of M. Lavalley, made it possible to become, in the course of the work, relatively independent of the forced labour of Egyptian fellahin. The canal was formally opened by de Lesseps' cousin, the Empress Eugénie, on 17 November 1869, who led a convoy of 51 ships through it. Verdi wrote the opera *Aida* for the occasion. De Lesseps was fêted in London and decorated by Queen Victoria with the G.C.S.I.: an apt recognition of the tactical importance of the canal in the days of sea power. Four years later, when Ismail, Said's successor (after whom Ismailia was named) went bankrupt, his shares were purchased by Disraeli for £4,000,000. As Disraeli gloated to Lady Blandford, 'We have had all the gamblers, capitalists, financiers of the world, organized and platooned in bands of plunder, arrayed against us, and secret emissaries in every corner, and have baffled them all, and have never been suspected.'

The labouring fellahin of Egypt and the dredgers of M. Lavalley dug the grave of sailing ships, for the Suez Canal enhanced the building of steamships. Great wealth had been locked up in sail. Even greater wealth was now applied to build ships that would save insurance and compete for freight; for the next ten years British shipbuilders were very busy. In 1874 the S.S. *Britannic* was launched, engined by Maudslays, and the *Propontis*, which had a triple expansion engine designed by A. C. Kirk (1830–92), a former Maudslay draughtsman, now the manager of John Elder and Company, who was mainly responsible for introducing triple expansion engines into the Royal Navy.

It led to Egypt becoming a British responsibility. As *The Times* wrote on 27 November 1875 when the news of the British purchase of the shares was announced: 'The nation awakes this morning to find that it has acquired a heavy stake in the security and well-being of another distant land, and that it will be held by all the world to have entered upon a new phase of Eastern policy.' This 'security and well-being' led to Sir Benjamin Baker (1840–1907) designing the dams at Assuan and Assiut in the tradition of Cotton on the Goodavery. It also led to the emergence of an engineer officer as the commander-in-chief of the British Army—Kitchener of Khartoum.

COOPER'S HILL, 1871–1906

Two years after the Canal was opened the British Government established the Royal Engineering College at Cooper's Hill, near Staines, to replace Addiscombe, which disappeared with the East India Company after the Mutiny. As Professor of Mechanical Engineering, W. C. Unwin (1838–1933), aged 34, was appointed. Formerly a personal assistant to Sir William Fairbairn, Unwin had taken part in the trials of the Foy and Newall mechanical railway brakes (which preceded the equipment of all passenger rolling stock with continuous brakes), and in experiments on the fatigue of wrought-iron girders, and had been manager of the Engine Department of the Fairbairn Engineering Company.

Fairbairn had recommended him as a lecturer to the Royal Engineers at Chatham, and Unwin had also lectured on marine engineering to the Royal School of Naval Architecture and Marine Engineering at South Kensington. Now, at Cooper's Hill, this interest had full play. His original work and text books became famous: *The Elements of Machine Design* (1877) and *Treatise on Hydraulics* (in the *Encyclopaedia Britannica*, 1881) especially. The latter became the authoritative basis for the development of Indian irrigation and river control works for more than a generation, whilst the former was almost universally used in British drawing offices for half a century. He was to be a leading consultant on hydro-electric development in Niagara and South India, and was to play no inconspicuous part in connection with the development of the petrol engine. Unwin left Cooper's Hill in 1885.

Cooper's Hill continued until 1906, when it was closed because of the growth of engineering instruction in South Kensington and the provinces, a growth in which Unwin played his part, since he became Professor of the Central Institution of City and Guilds College in 1885, and when it was incorporated into the University of London he became the first London University Professor of Engineering.

THE INDIAN LABORATORY

Service in the challenging environment of India was almost a traumatic experience for British officers of intelligence and ability. For some of them, it sharpened their appreciation of engineering techniques and undoubtedly influenced their attitude when they returned to Britain again. Two outstanding officers leap to mind.

The first officer was Colonel Alexander Strange (1818–76) who, on returning to England from the trigonometrical survey of India in 1861, became so convinced of the importance of standard instruments that he persuaded the Government to establish, and then organized himself, a department for the inspection of scientific instruments in India for which he designed and superintended the construction of a massive set of standard instruments. Elected F.R.S. in 1864, he was a judge at the Paris Exhibition of 1867, and his experiences there led him to urge the British Government to establish a permanent commission on scientific questions, or better still, to appoint a Minister of Science. The appointment was only made in 1959. The intervening ninety-two years have only strengthened his case. He maintained, in characteristic fashion, that neither the governing classes nor the masses were qualified to discuss the matter intelligently. As *Nature* remarked in an obituary notice published on 23 March 1876: 'To him belongs the whole credit of having initiated in 1868 the movement which resulted in the appointment by her Majesty of the "Royal Commission on Scientific Instruction and the Advancement of Science"'. To that commission we must soon turn.

The second was Colonel R. E. B. Crompton (1845–1940), who as an ensign in India from 1864–88 introduced steam road haulage for the bullock trains then employed by the army. He was seconded as 'Superintendent of the Government Steam Train' and took charge of experiments to improve road steam traction which, owing to the Red Flag Act in England (see Chapter 20), had been severely curtailed. He was a great friend of R. W. Thompson, of Edinburgh, who equipped his steam traction engines with rubber tyres. On returning to England in 1871 he was to make his name as an electrical engineer and later as a pioneer of good roads.

CHAPTER 16

The American Civil War and After

Of all the nineteenth century wars, that which began on 12 April 1861, when the artillery of the ten southern cotton-belt states fired on the Federal Fort Sumter in Charleston Harbour, seems to have exerted the most intensively ripening effect of technology and engineering up to this time. Four years later, when Lee presented his sword in surrender to Grant at Appomattox Courthouse on 9 April 1865, chains of circumstance and necessity had been set in motion, and others accelerated, which together were to alter the centre of gravity of human living.

It was the first war in which nearly two million men served on one side and nearly a million on the other. To feed, arm, clothe, pay and destroy forces on this scale called forth extensions of ingenuity (which is engineering on a slapdash basis) of hitherto unknown dimensions. The steam-powered factory, the railway, the warship, the rifle, explosives, the telegraph and the balloon were instruments by which this was accomplished, and the forging of these instruments involved further engineering skills and institutional changes.

Technologically speaking, the Civil War was a lateral extension of the earlier struggle in the Crimea. The first general of the Federal Army of the Potomac, G. B. McClellan, whom Lee regarded as the best commander who ever faced him, was an engineer who had studied European military systems, visited the Crimea and brought back a new kind of saddle with which he equipped his cavalry. On his return from the Crimea he had become the chief engineer of the Illinois Central Railroad. Indeed, one of the reasons why he was dismissed from the Army of the Potomac was that he was 'too efficient and too scientific' to be continued in command. The times then needed cruder, more forceful characters like Burnside and, later, Grant. One of McClellan's chief critics—who published *The Army of the Potomac and its Mismanagement* (1861) and *Military Incapacity*

168

and What it Costs the Country (1862)—was Charles Ellet (1810–62), another American observer of the Crimean front. Ellet was the Brunel of America: a capable engineer, canal and bridge builder, and he had advised the Russians to employ ramboats. His advice, though neglected by the Russians, was adopted by the Southern States, whose ram, the *Merrimac* (or *Virginia*) overturned the *Cumberland* and *Congress* in Hampton Roads on 9 March 1862. The northern secretary of the navy, Gideon Welles, also rejected his advice, so Ellet obtained permission from Edwin Stanton, the Northern Secretary of War, to convert nine river boats to rams. With these he captured Memphis from the South, but unfortunately died in the process.

THE NORTHERN ARMY

In Edwin Stanton, the North were fortunate to have one of the great innovators in military history. Symbolically, six years before he took office, Stanton played a major role in breaking the monopoly which Cyrus McCormick was claiming for the manufacture of his reaper by defending John Manny, whom McCormick was suing for patent infringement. In Cincinnatti, and later in the United States Supreme Court, Stanton broke the arguments of McCormick's lawyers. Having heard him in court, Abraham Lincoln appointed him Secretary of War in 1861. So, having in peacetime made available mechanical reapers at a reasonable price, in wartime Stanton organized victory for the Northern armies.

From his first day in office, Stanton insisted on American made arms. He tried out private inventions, ensured that the government manufactured them on a large scale, and was so intensely preoccupied with mobilizing and supplying large forces that he created two new departments: for telegraphs and railroads. The Telegraph Department he put under T. T. Eckert (1825–1910) who played as big a role in the administrative organization of telegraphs as Morse had played in their perfection. The Railroad Department was entrusted to Herman Haupt (1817–1915), whose *General Theory of Bridge Construction* (1851) was a classic. Haupt was at the time the Civil War broke out, building the Hoosac Tunnel through the Berkshire massif between Boston and Albany. He left it to recruit his own labour force, which worked wonders. Amazed at the speed—some twenty days—with which Haupt rebuilt the Fredericksburg railroad, Lincoln wrote: 'That man Haupt has built a bridge across Potomac Creek on which loaded trains are passing every hour, and upon my word, gentlemen, there is nothing in it but cornstalks and bean-poles.'

'Cornstalks and bean-poles' perhaps better described the courier service for military news which Haupt established by sending telegraphers equipped with pocket instruments to make observations at the fighting front. These observations were transmitted to Stanton's own office at any hour of the day or night. Yet even Haupt seemed old-fashioned to some of his contemporaries. Thus Jacob H. Linville (1825–1906), bridge engineer of Pennsylvania Railroad, wrote in 1863: 'I went to Altoona with a young wife, with everything new to me. The bridges on the line were nearly all a wreck; I knew nothing of bridges, and had at my disposal nothing but Haupt's old book, all wrong. Orders came in to build new iron bridges, and I had to hustle with many sleepless nights, and days spent over old patterns and new plans and calculations.'

Bridge building went on at such a pace that in 1865 the Keystone Bridge Company incorporated to specialize in replacing wooden structures by iron.

The demands of the Army and rapid expansion of industrial life brought a considerable building of munition centres which, in turn, drained the farms of labour. So in 1863 the McCormick brothers put on the market a self-raking reaper that performed the most strenuous tasks hitherto undertaken by manual labour. It sold for 190 as opposed to 125 dollars for the earlier model. In that same year the income of the McCormick brothers exceeded a million dollars a year. With Cyrus abroad, his brother William purchased one million dollars' worth of real estate in Chicago, becoming the largest landowner in this thriving town. And how it throve is perhaps best illustrated from an extract from the *Chicago Tribune* on 8 October 1863: 'On every street and avenue one sees new buildings going up, immense stone, brick, and iron business blocks, marble palaces and new residences, everywhere the grading of streets, the building of sewers, the laying of water and gas-pipes, are all in progress at the same time.'

THE 'CHICAGO SCHOOL'

The rapid building of Chicago led by 1871 to another disaster: th e greatest fire of the century. Two-thirds of old Chicago was built of wood: the greater part of the new Chicago was built of iron, brick and stone. William Le B. Jenney (1832–1907), a graduate of the École Polytechnique in Paris, seized the opportunity to construct the

Home Insurance Company of Chicago (1883–5) and created another characteristic American artifact: the skyscraper. So many engineers and architects found similar opportunities to express themselves that a 'Chicago School' emerged: H. H. Richardson (the grandson of the famous chemist, Joseph Priestley), who had returned from study in Paris in the year the Civil War ended; J. W. Root and D. H. Burnham (who joined together in partnership two years after the Chicago fire), the successful exploiters of the steel frame; and Louis Sullivan (another graduate of Paris), whose tall office buildings began to stud American cities. The quatercentenary exhibition held at Chicago in 1893 symbolized the enormous strides which this 'School' was making in utilizing new materials and techniques in large block buildings. Such a command over new materials was to be even more brilliantly exploited by a young engineering graduate of the University of Wisconsin, who worked with Sullivan from 1889 to 1895: Frank Lloyd Wright. Wright was to do for the ordinary American house what his Chicago predecessors had done for public buildings: bring them into the sun.

THE DEVELOPMENT OF SKILLED LABOUR IN THE NORTH

Nearly every new industry began in America before the Civil War was fertilized by British skills. During the Civil War both a public and a private agency were established to recruit skilled labour on a large scale. Both were abandoned by 1868, as the enormous expansion of the American economy absorbed immigrants at a phenomenal rate.

Expelled by depressions in Europe, these immigrants needed ship and railway transport to carry them and mass-produced homes in which to live. Following them flowed capital to finance ventures to provide these facilities. The Union Pacific Railway (begun in 1862), the Northern Pacific (begun in 1864), and the Atlantic and Pacific (begun in 1866) all called for immense engineering achievements like Eads' St. Louis Bridge. The boom from 1866 to 1873 not only doubled railway milage but initiated new services, new machinery, new forms of business organization, opened up new areas, and sucked in a vast amount of new capital into the American economy.

ENGINEERING IN THE SOUTH

Lack of good railway vertebrae as much as anything lost the war for the South. Beginning the war with a total milage of 8,783 in a

171

country-wide total of 31,168, Southern trackage was not spread evenly from the Potomac to the Rio Grande. With no real trunk lines (each line was virtually a feeder to some waterway or local route) the average Southern stem seldom exceeded 200 miles, and the largest line under the control of a single company was the Mobile and Ohio, which ran from Mobile to Columbus, Kentucky (some 469 miles) and this was not complete until war began. There were only two major railway routes, one complete—the other unfinished. Both ran to Richmond, one from Memphis, Mississippi, through Corinth (which was on the Mobile and Ohio); the other, unfinished, ran from Pollard via Atlanta and Wilmington. Only twice in their course were these two lines connected laterally. They were iron fingers stretching towards each other, not everywhere had they joined hands. In one small railroad—the Roanoke Valley—two segments were actually of different gauges. Notable was the complete lack of railroads to Texas.

The Southern railways transported the first soldiers of the confederacy to their first battles (the battle on Manassas Plains, 21 July 1861, was a Southern victory because of the Manassas Gap railroad which enabled reinforcements to be hauled up when needed). But soon, owing to the shortage of iron, their railways had to furnish material for the army. So cannibalization of the tracks had to be organized, and on 22 January 1863 a commission was appointed 'to examine and advise on what railroads of the Confederate States the iron on their tracks can best be dispensed with.'

After 1861 not a single new rail was produced in the confederacy. There was not only no manufacture of steel and locomotive tyres, but the navy snatched up the production of the iron mill at Atlanta. Armed raids had to be organized to obtain iron, and a bar of railroad iron grew so valuable that Mexican Gulf Railroad wanted to tear up half its line and advertise it for sale. This was prevented by the government, but in the course of the war the governor was forced to confiscate track (e.g. of the Florida railway) itself.

It was Isaac St. John (1827–80), who suggested the removal of rail from unimportant lines, so that key routes might be maintained. He was one of the few engineers of distinction in the service of the South, and also served as superintendent of the Nitre and Mining Bureau. A recently published diary of the Civil War pays him great tribute: 'Colonel St. John,' wrote Robert Garlick Hill Kean, 'has developed the production of nitre from almost nothing to nearly a full supply. But for the loss of territory where the richest nitrous earths are found, he would have been wholly independent of importation. The loss of Tennessee has caused him to develop his works further in the interior, and in a few months his beds laid down near all the interior cities will be ripe. To his great energy, talent for organisation, and skilful invention in supplementing defective resources, the country owes as much as to any man in the service, whatever his rank or fame. Others have made good use of what the country afforded in resources. He has *created* when resources there were none.'

St. John worked under G. W. Rains (1817–98), who produced the gunpowder and organized the wholesale collection of nitre from limestone caves. From his efforts grew the Nitre and Mining Bureau of the War Department. This produced 2¾ million pounds of gunpowder for the South. His brother, Gabriel James Rains (1803–81), developed land mines, booby-traps and torpedoes, and one of his experiments involved the blowing up of an ammunition warehouse that cost the North some four million dollars.

The ingenuity of the South produced a number of midget submarines called Davids—one of which was the first to sink an enemy ship in action, the Federal *Housatonic* in 1864. Unfortunately it went to the bottom too, killing all her crew—which she had nearly done on five previous occasions. The torpedo—as an explosive charge on a small boat—was used by Lieutenant Cushing of the Federal Navy, who sank the Confederates' *Albemarle*, an ironclad of the *Merrimac* type.

The South was also experimenting with rockets. Robert Garlick Hill Kean, Head of the Bureau of War, went on 27 October 1861: 'to see a new rocket stand for shooting iron rockets, which are on the rifle principle, dispensing with the stick. The rocket is sheet iron with a shell in front, egg shaped. Part of the gases are vented through a hole directly in the rear about one-fourth the area of the cross-section; the residue, through small holes drilled obliquely through the iron end.' But of all the inventions produced by the racked ingenuity of the South, torpedoes were perhaps the most significant. They made them on a grand scale. One had a ton of powder, another a hydrogen cap and a third was fired by electricity. These fearsome monsters sank twenty fighting vessels of the North and, as the English journal *Engineering* wrote in retrospect on 14 January 1870, that 'the record of their work is of the greatest importance to ourselves'.

173

THE SEVENFOLD REVOLUTION IN NAVAL ENGINEERING

The built-up rifled gun, slow-burning powder, the fused shell, armour plating, iron construction, screw propulsion and the beginnings of turbine motors were, it is true, under active consideration in Europe. But the Civil War helped to make standard these, and other changes, which were hitherto either speculative, or at best experimental.

As head to the newly organized Ordnance Department of the North was appointed J. A. Dahlgren (1809–70), who had made the first triangulation survey and measured the first base line in the U.S.A. Dahlgren had served as professor of gunnery at Annapolis, and established the ordnance workshop at Washington Yard. His bottle-shaped 11-inch smooth bores—thick at the trunnions and diminishing to the muzzles—were to play their part in the war. For Dahlgren was an experimentalist as befitted a member of the American Association for the Advancement of Science. In 1857, he built an experimental gunnery ship, and during the war developed a 15-inch gun of his own. He also secured supplies of potassium nitrate, hitherto obtained from British India.

Guns need ships and the Civil War ripened the talents of John Ericsson (1803–89), brother of Niels Ericsson, the creator of the Swedish railway system. Ericcson had come to America at the invitation of the energetic Captain Robert F. Stockton of the U.S. Navy in 1841, who was impressed by his experiments with the Archimedes screw. Ericcson had spent thirteen years in England, where he had built high-pressure boilers, experimented with compressed air transmission and designed a surface condenser for marine engines. His engine, *The Novelty*, had run against Stephenson's *Rocket* at the Rainhill trials, and his screw driven vessel had crossed the Atlantic. He enabled American ships to ply the great lakes driven by screws. When the Civil War broke out in 1861, the North had to build a navy. Industry was called to expand production for vessels and guns, and Ericcson put forward a design for a low-hulled, armed *monitor* with rotating guns of large calibre. He promised to build it in 100 days and did so. It also succeeded in sinking the Southern ram *Merrimac* on 9 March 1862 at Hampton Roads. This victory led not only to further orders from the Northern States but to developments in battleship building all over the world. The warship had now become an engineering problem.

Iron ships (and monitors especially) need powerful engines, so a new Bureau of Steam Engineering was organized by the North under B. F. Isherwood (1822–1915), whose experience began on railways (he began life on the Utica and Schenectady Railroad) and who had worked on the great Croton aqueduct that brought water to New York. In the war with Mexico, Isherwood had served on the U.S. *Princeton*, the first propeller-driven boat in the U.S. Navy, and he had designed the first self-feathering paddle wheels in the U.S. Navy in 1852. His book *Engineering Precedents* (1859) embodying the results of ten years' work showed that he was admirably suited to his appointment. In this office, which he held till 1870, he conducted researches which led directly to the perfection of the steam turbine, for he showed that the ideas of Watt and Gay-Lussac concerning the expansion of steam had to be revised in the light of the increased ratio of expansion and cylinder condensation in the vessels built during the war. Losses, as he showed in *Experimental Researches in Steam Engineering* (1863, 1865), became larger whilst the additional work gained from increased expansion became progressively smaller. Nearly 600 boats were built in the North during the war. Isherwood's own creations were the commerce destroyers of the Wampanoag class, which when built were the fastest vessels in the world, travelling at $17\frac{3}{4}$ knots.

Many private shipbuilders responded to the needs of the North. J. B. Eads (1820–87) calls for special mention because of the chain of circumstance that was to make him great. When the war began he had already made a fortune by salvaging steamboats on the Mississippi by means of a diving bell of his own invention. Asked by Lincoln for his advice on the best ways of utilizing the western rivers for attack and defence, Eads proposed a fleet of armour-plated, steam-propelled gunboats. He built seven 600-ton boats on these lines in sixty-five days and twenty-five more incorporating various new devices. Such was his skill in deploying the resources of the four thousand men who worked on these in various parts of the country that in 1867 he was later given the job of constructing a bridge across the Mississippi at St. Louis. Twenty-seven of the leading civil engineers of America said it was impossible to build a 500-foot centre span with a clearance of 50 feet. Eads accepted the challenge and finished it in 1874. Its construction involved sinking a pier 136 feet below high water and 90 feet of sand. The medical problems encountered in this feat themselves stimulated new techniques in underwater engineering.

THE GROWTH OF STANDARDIZATION

Eads' biographer and friend, William Sellers (1824–1905) read a paper in 1864 to the Franklin Institute on 'A System of Screw Thread and Nuts', proposing a system of screw threads to supply the need for a generally accepted standard. This standard was adopted by the United States Government in 1868 and by the Pennsylvania Railroad a year later. It is still known as the 'Sellers or United States Standard'. Sellers was the Whitworth of America. In 1873 he took over the William Butcher Steel Company, renaming it the Midvale Steel Company. His company furnished the ironwork for the Centennial Exhibition in Philadelphia in 1876, and all the structural ironwork (except the cables) for the Brooklyn Bridge. One of his employees was Henry R. Towne, who was the father of production engineering and whom we shall meet again on p. 278.

THE RISE OF THE AMERICAN STEEL INDUSTRY

Floating gun batteries attracted A. L. Holley (1832–82), a graduate of Brown University, who had worked in the office of Corliss and Nightingale. At the time of the Civil War Holley was the technical editor of the *American Railway Review*, on the staff of the *New York Times*, the author of *American and European Railway Practice* (1860), in which he had been helped by his friend Zerah Colburn (whom we shall meet later). Having designed a locomotive for the Camden and Amboy Railway, his powers of quick assimilation, good hearing and bounding imagination were utilized by E. A. Stevens, who sent him to Europe for information on ordnance and armour. In 1862 Holley investigated Bessemer's process for steel, was impressed, and quickly returned to the U.S.A. to secure the active interest of Corning, Winslow and Company. In the following year he bought the American rights of the Bessemer Process. From now on Holley's biography becomes the history of steel manufacturing. He was commissioned to design steel works at Troy, New York, in 1867, and went on to design others in Chicago, Pittsburg and St. Louis. He became the consulting engineer of Cambria Steel, Bethlehem Steel and the Scranton Steel Works. As the foremost steel plant engineer in the United States he urged steel manufacturers to employ chemical analysis as well as physical tests to their products.

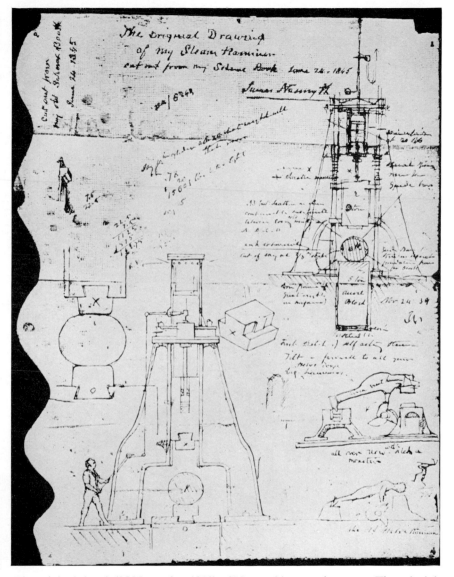

The original sketch (25 November 1839) of Nasmyth's steam hammer. The principle was actually adopted and manufactured by the French engineer Schneider at his Creusot works before Nasmyth could do it.

Michael Faraday (1791–1867), at the height of his powers in 1830. (From the portrait by Pickersgill)

SCIENCE AND THE GOVERNMENT

Before leaving the North there are three institutional innovations that call for special mention: the Navy Department's Permanent Commission, the National Academy of Sciences, and the land-grant colleges. The first was set up on 11 February 1862. To it all subjects of a scientific character could be referred for expert advice: projects like torpedoes and under-water guns needed such examination, and by 21 September 1865 no less than 257 reports had been made to the government. One of its leading members was A. D. Bache (1806–67), a great grandson of Benjamin Franklin, who at the age of 22 had become Professor of Natural Philosophy and Chemistry at the University of Pennsylvania, and later President of Girard College and Superintendent of the United States Coast Survey. His ally, C. H. Davis, was also one of the founders of the second body, the National Academy of Sciences in the following year.

The National Academy of Sciences was set up by Congress on 3 March 1863 without any discussion. Its committees dealt with such matters as the metric system (legalized in 1866), anti-rusting techniques for ships' bottoms, magnetic deviations in iron ships (it dealt with 27 vessels), the evaluation of hydrometers and work on charts. Some of its members were averse to meeting in wartime Washington, Benjamin Pierce, for one, complained that 'to meet in Washington is like a party of Poets meeting in a cotton machine, of angels in the palace of Beelzebub, or of imps in Abraham's bosom.'

The third institutional innovation—the land-grant colleges—needs setting in perspective. Up to 1850 almost the entire engineering cadre came from the West Point or the Rensselaer Polytechnic in New York (established in 1824). True, others were founded. Two railway magnates, Abbot Lawrence (1792–1855) and J. E. Sheffield (1793–1882), founded scientific schools at Yale and Harvard in 1847. Eliphalet Nott, at Union College, had also established engineering as a subject in the preceding year. The Brooklyn Polytechnic and the Cooper Union followed.

As the Civil War began, the Massachusetts Institute of Technology was taking shape. Its founder, W. B. Rogers (1804–82) was the son of a scientist and brother of three other scientists. Keenly interested in founding a polytechnic, he migrated from Virginia to Boston, and persuaded the legislature of Massachusetts to establish an Institute of Technology in 1861, becoming its first president. The term 'technology' was suggested by Dr. Jacob Bigelow (1786–1879), who had held the chair of application of science to the useful arts at Harvard, endowed by Count Rumford. Bigelow's Rumford Lectures, *The Elements of Technology* (1829), was one of the earliest books of its kind to be published in the U.S.A.

Combining as it did original research in applied science with the diffusion of knowledge, the Massachusetts Institute of Technology offered a model for the land-grant colleges that were established as a result of the Morrill Act of 1862. This measure, whereby the Federal Government granted land to the states to encourage the establishment of colleges of agriculture and the mechanic arts, had been blocked since 1856. It only passed because of the absence of the Southern delegation. The demand for engineers to build the great transcontinental railways and participate in the great expansion of the steel and oil industries after the war was so great that the engineering schools, especially of the land-grant colleges, multiplied. There were only 2 in 1840 but by 1870 there were 70.

Nor, in the evolution of American engineering, should the rôle of the Smithsonian Institution by forgotten. Founded by an Englishman, its first secretary, Joseph Henry (1797–1878), appointed in 1846, was a former tutor to the family of Stephen van Rensselaer and made his name as road surveyor in New York. He improved William Sturgeon's electro-magnet by using insulated wire wound in several layers round the iron core. Following the same lines as Faraday, he very possibly anticipated the discovery of induced current. Henry helped to organize the American Association for the Advancement of Science in 1847, and was president at its second meeting. As secretary of the Smithsonian he encouraged original research in the various fields of natural science and its publication in papers.

Henry made instruments for the balloonist, Thaddeus Lowe (1832–1913), who was later chief of the aeronautic section of the North, and was the first American to photograph armies from the air. The third secretary of the Smithsonian, Samuel P. Langley (1834–1906)—appointed in 1887, conducted an epoch-making series of experiments on the distribution of radiation in the solar spectrum by his bolometer which measured rises in temperature to a millionth of a degree. Langley carried the solar spectrum far beyond the then recognized limit in the dark regions beyond the red (i.e. the infra-red spectrum), and applied this to the study of the moon and heated bodies. Langley's other investigations centred round the possibilities of flight in heavier than air machines, and his laboratory investigations gave a great impetus to work in a field which had hitherto been regarded as fit only for eccentrics. On 6 May 1896 his fifth model, catapulted from a houseboat on the Potomac, flew for 3,000 feet without damage, and six months later his sixth flew even longer. These flights were the first sustained free flights of heavier than air machines. They are all the more notable in that the two European experimenters who were working on the problem at the same time, Ader in France and Maxim in England, gave up their experiments.

PATENTS

Some index of the acceleration of American engineering can be obtained from the fact that, from its establishment in 1790 to 1860, the U.S. Patent Office granted 36,000 patents, but from 1860 to 1890 it granted no less than 440,000: more than a twelvefold increase in less than half the time. Four hundred patents were taken out by one man alone: George Westinghouse (1846–1914), the son of a manufacturer of agricultural implements. His first, registered in 1865 when he was only nineteen, was for a rotary steam engine, the principle of which was applied to a water meter. At twenty-three, in 1869, he patented an air brake and formed the Westinghouse Air Brake Company. This made it possible for a driver to slow down and stop a train at will.

Increased speed made better signalling arrangements necessary and he both purchased and developed a number of patents which were exploited by the Union Switch and Signal Company, which he formed in 1882. In 1886 the Westinghouse Electric Company was founded. He built a model factory and model town for his employees. Perhaps he is best known for his introduction of alternating current for electric power transmission.

COMMUNICATION BREEDS NEW TECHNOLOGIES

Two big problems of the war were transport and communications, and so the pressure on expansion of railways and telegraphs during the Civil War was immense. These two fields produced three dominant figures of the American industrial boom that followed it.

At the outbreak of the Civil War, the driver of the first military train to enter the Federal capital of Washington was Andrew Carnegie (1835–1919), the private secretary of the Assistant Secretary of War in charge of transportation. Carnegie later superintended the evacuation of the defeated Federal Army after Bull Run, organized the telegraph department, and obtained such an insight into railroad needs that at the end of the Civil War he utilized his knowledge to become a salesman for the Keystone Bridge Company. It was in his capacity as a salesman that he became a friend of Sir Henry Bessemer and in 1873, at the age of 38, staked all his resources in concentrating on the manufacture of steel. His vast empire, rolled up with such organizational skill, was sold to the United States Steel Corporation in 1901 for 447 million dollars.

The second was Ezra Cornell (1807–74), who links the earlier phase of American engineering with the age of the telegraph. He had first settled at Ithaca as a millwright because of its proximity to the Erie Canal, building flour mills and tunnels for water power. He then built a pipe for Morse to lay a cable from Washington to Baltimore. Becoming interested in the magnetic telegraph, he helped to organize a company which ultimately became the Western Union Telegraph Company. Its lines, by 1860, reached from the Atlantic to the Mississippi River, and from the Great Lakes to Ohio. Cornell's name has been immortalized in the land-grant university with which it is still associated.

The hasty improvisations of the railway telegraph also stimulated Thomas Alva Edison (1847–1931). After the war ended he settled in New York and profited by the wild rush for news to patent a tape machine called a 'ticker'. Further improvements to the telegraph brought him enough money to start a laboratory in Newark on 1 October 1869. Two of his assistants, Siegmund Bergmann and Siegmund Schuckert were in this way fortunate to obtain the training that enabled them to return to Germany and found the famous works of these names. Edison went on from producing a printing and a quadruplex telegraph to improve the invention of the German, Philipp Reiss (made in 1861), and the Scotsman, Alexander Graham Bell (made in 1876), and produce a long-distance telephone by using a microphone of carbon granules. With the money for this invention (from Western Union) he bought a house and laboratory in Menlo Park. Having extended the human voice by the telegraph, he now managed to preserve it by the phonograph. This accomplished, Edison set himself to utilize electric current as an illuminant, and on 21 October 1879 he made the first carbon filament lamp. To provide the necessary power, he planned the Pearl Street Central Station which began operating on 4 September 1882. At the first International Electrical Exhibition in Paris in 1881 he had exhibited his incandescent lamp and his dynamo: the German rights were bought by Emil Rathenau and Oscar von Miller.

In 1886 the Edison Machine Works was established at Schenectady. This, three years later, became a unit of the Edison General Electric Company and in 1892 the largest plant of the General Electric Company. Schenectady became the headquarters of G.E.C. research. Here C. P. Steinmetz (1865–1923) worked as a consulting engineer from 1893 onwards and so did internationally famous chemists like W. R. Whitney and Irving Langmuir. This set a pattern for research laboratories in the new industries like electricity and chemical engineering that was to be copied all over the world.

THE RISE OF BUSINESS MACHINES

Perhaps the strangest link of events stems from the very heart of the war itself—the wounded soldiers. A young doctor (who only obtained his M.D. in 1860) called John Shaw Billings (1838–1913) was put in charge of a hospital in 1861 at Philadelphia, that was full of thousands of sick and wounded. There he made such a name for his unflurried efficiency that in April 1864 he was given responsibility for collecting statistics in the Army of the Potomac during the Wilderness Campaign. Invalided back to Washington, he began to build up the great Surgeon-General's Library and, with money turned in from hospitals after the war, vastly increased the number of volumes. He later virtually designed the Johns Hopkins Hospital, for which he selected William H. Welch and William Osler. His interest in statistics, stimulated by his part in the foundation of the American Public Health Association, led to his being put in charge of the vital statistics in the federal census of 1880. It was here that he met a young assistant of the School of Mines at Columbia, Herman Hollerith (1860–1929). Let Hollerith himself tell the story: 'One evening at Dr. B's tea-table he said to me, "There ought to be a machine for doing the purely mechanical work of tabulating population and similar statistics". Hollerith then went to the Massachusetts Institute of Technology in 1882 as an instructor in technology, and in 1884 went to the U.S. Patent Office in Washington, where he worked on the problem of perfecting mechanical aids in tabulating information. By 1890 he had invented machines working through punched hole non-conducting cards which would count items by means of an electric current passed through holes identically placed. It was first used in tabulating mortality statistics in Baltimore, New Jersey and in New York City. It was chosen for compiling the Census in 1890. It was similarly adopted for census returns in Canada, Norway and

181

Australia. Six years later he organized the Tabulating Machine Company to manufacture the machines and in 1911 that company was incorporated with others to become the Computing–Tabulator–Recording Company, later the International Business Machines Corporation, of which Hollerith was consulting engineer till 1921.

THE YANKEE STIMULUS TO BRITAIN

The impact of the Civil War on British engineering is no less noticeable. When the *Merrimac* fought the *Monitor*, *The Times* commented on the lesson it afforded to those who put their trust in wooden fleets—'Before this we had . . . 149 first-class warships, now we have two.'

Less than a month after the war began—i.e. by 9 May 1861—the first orders for arms reached Birmingham. Now Birmingham had only just organized its small arms industry as a result of the Crimean War. The Birmingham Small Arms Factory, consisting of twenty firms selected to supply arms for the Crimean War, had been equipped with American machines at Small Heath. American machines, 157 of them, including Blanchard's famous copying lathe, had been brought to England seven years previously by John Anderson, the inspector of machinery in the British Ordnance Department, as a result of a visit he had paid in 1854 to the factories of Samuel Colt at Whitneyville and Hartford, Connecticut (both enlarged as a result of the Mexican War of 1846). In the intervening seven years, these Yankee dragon's teeth had taken root in Birmingham. Small Heath was now to receive a second stimulus from the Civil War.

The Birmingham Small Arms Company, built out of the British engineering firms that received a boost from the Crimean War, were producing 142,819 rifles a year in 1861. This increased to 388,264 and 210,078 in the two following years. Altogether 1,078,205 rifles were produced in England for the combatants. Not without an excuse did William Siemens build an experimental plant at Birmingham in 1865 for making steel. This experimental plant made steel from iron ore and pig iron in a regenerative furnace, but the brothers Émile and Pierre Martin, at the same time, substituted scrap steel for the iron ore, and this, as the Siemens-Martin process, was destined to supplant the Bessemer process in America.

The civil war provided an indirect stimulus to civil engineering in England. The diminished exports of cotton to Lancashire led to the passage of a Public Works (Manufacturing Districts) Act in 1863—a piece of proto-Keynesian legislation to alleviate unemployment. This empowered distressed areas to raise loans of one year's rateable value of the area, repayable in thirty years at an interest of $3\frac{1}{2}$ per cent. As a result, between 1863 and 1865 ninety local authorities borrowed some £1,850,000 for public works. It was within the Act that no work was to be sanctioned that was not a work of permanent utility and sanitary improvement. So the constructional skill of Thomas Hawksley (1807–93) made the constant supply of water a standard feature of British towns and cities. Well might the Webbs claim that the Public Works Act was a measure of sanitary engineering rather than an emergency relief measure.

Whitworth's orders for Confederate guns led him to experiment with large sound castings of steel, using a hydraulic press instead of a steam hammer. This Whitworth steel, first produced in 1870, won such renown that in 1883 the Gun Foundry Board of the U.S.A. went to his works at Openshaw, near Manchester, and gave it as their opinion that the system there carried on surpassed all other methods of forging and said that their visit 'amounted to a revelation.'

Other case-histories could be cited. But from the engineering point of view we should be content with only one more: that of the docks at Cherbourg and Le Havre, built between 1865 and 1871 by the French to cater for the American trade. For this harbour no less than 20,000 tons of British Portland cement were used. This order proved such a stimulant to the Swanscombe Works of Messrs. White that they became the dominant firm in England. By 1900 they embraced the Associated Portland Cement Manufacturers Ltd., and in 1911 the British Portland Cement Manufacturers.

THE GROWTH OF TECHNICAL JOURNALISM IN BRITAIN

All these developments were copiously reported in English engineering circles, chiefly by an American Zerah Colburn (1832–70), who came to England when the Civil War broke out to edit a journal called *The Engineer*. By 1866 he had founded a second journal, *Engineering*, which he both owned and edited. Colburn's first visit to England was with A. L. Holley in 1857 when at the request of several leading American railroads, they had reported on English railway practice. Their report, *The Permanent Way and Coal-Burning Locomotive Boilers of European Railways with a Comparison of the Working Economy of European and American Lines and the Principles upon which Improvement must Proceed* (1858) became a standard work. Colburn had then stayed on in London to become editor of *The Engineer*, succeeding Edward C. Healey, a friend of Brunel, Stephenson and Fairbairn. After putting this on its feet he sailed on the *Great Eastern*'s first voyage to America, where he started a paper of the same title in Philadelphia. In 1861 he returned to England to take over the editorship of *The Engineer* for a second time. Both in *The Engineer* and in the Transactions of the Institute of Civil Engineers he wrote on current developments. His books on *Heat* (1863), *The Gas Works of London* (1864) and *Locomotive Engineers* (1864) were the results of a growing private practice. He stepped down from the editorial chair of *The Engineer* in 1864 but remained a private contributor. In 1866 he issued a second weekly journal in England—*Engineering*. This became an entrepôt of ideas. Till *The Times Engineering Supplement* was founded in 1905, it held with *The Engineer* the field for general and professional topics, although, of course, *The Electrician* fitfully illuminated its specialist field from 1861–4 and from 1878 onwards. Colburn's skill in exploiting the public demand for knowledge of current technological developments led Norman Lockyer, a former editor of a digest known as *The Reader*, to approach Macmillans with a proposal that they should issue *Nature*, which they have done without intermission since 4 November 1869.

One of Zerah Colburn's last editorials in *The Engineer* of 9 April 1869 stressed the significance of American engineering expansion and remarked: 'In the engineering and the engineering practice of the United States we can find a much closer parallel than is afforded in France or in Germany and if there exists there a method of professional education which has by long experience proved itself good it is thither that we may look for an example rather than in the Continental Schools.'

His words were to acquire added significance in the decades to come. They were reinforced from an unusual quarter by the French novelist Jules Verne, who sketched the exploits of an imaginary group of experimentally-minded artillery veterans of the American Civil War in *From the Earth to the Moon* (1865). The projectile (of aluminium) was, though fired from a 'gun', equipped with rockets. To its building and launching, Verne wrote, every country in the world contributed; except the British, who received the plan with 'contemptuous apathy'. That 'apathy' was soon to be disturbed.

CHAPTER 17

The German Exemplar

'T he nearest approach that has ever been made to realizing
Bacon's grand imagination of the New Atlantis—a university
for experimental knowledge of the Cosmos.' So *The Engineer*
on 25 March 1870 commented on the state laboratories of Berlin,
Leipzig and Bonn. The year was significant. Germany was to conquer
France and overthrow Napoleon III. As Disraeli remarked, 'this
represents the German revolution, a greater political event than the
French Revolution of the last century'. The German example was
given particular point as the Charlton factory of the Siemens
Brothers (transferred there from Millbank in 1866) expanded
rapidly.

The Siemens family offer a particularly good example of the pro-
jection of German engineering skill into England. The three brothers,
Werner (1816–92), William (1823–83) and Karl (1829–1906) constitute a formidable international trio influencing the engineering
development of Germany, England and even Russia.

Werner was a Prussian officer trained in the engineer and artillery
school at Berlin where he came under the influence of the great Peter
Beuth (1791–1863)—a technological evangelist. Beuth was a bachelor
wedded to his work. He paid two visits to England—in 1823 and
1826—to further his own knowledge, writing percipiently on what he
saw. He also travelled extensively on the continent. He sent members
of the Prussian Technical Institute on similar journeys. Three were
sent to the United States, another to England, and one to China.
Beuth organized the Government Technical Commission in 1819 and
as its director encouraged the preparation of technical books, the
delivery of lectures on technology and a liberal administration of the
patent law. In 1820 he established an Association for the Promotion
of Industrial Knowledge in Prussia which began with 367 members,
not one of whom described himself as a machine builder or engineer.

185

In 1821 he established an Industrial Institute in Berlin, mainly in civil engineering. Industrial exhibitions were held, an industrial library was organized, and a collection of foreign models built up. As Director of the Department of Manufactures and Commerce and of the Technical Commission, as Chairman of the Association for Promoting Industrial Knowledge and as Principal of the Industries Institute, Beuth acted as a foster-father to German engineering for a generation. The published transactions of the association reveal what a skilful intelligence service he established. By the time he had retired in 1845, industrial Germany was not only on its feet, but preparing to run. His industrial institute was the embryo of the famous Charlottenburg Technical University. The Institution of Civil Engineers in England mourned his passing and its secretary, Charles Manby, referred to his 'almost universal influence'.

It was Beuth's Technical Deputation in Berlin which granted Werner Siemens a patent for electro-gilding in 1842, which his brother William came to England to sell. Another brother, Frederick, was getting interested in the regenerative furnace, and yet another, Karl, was active in Russian engineering. At the Physical Society in Berlin (founded in 1845), Werner met a young mechanic called Halske, with whom, in 1849, after fourteen years' service in the army, he devoted himself to telegraph work.

William Siemens, a pupil of Weber, worked with Werner on a patent for electrolytic deposition of metals and marketed it in England in 1843, settling here a year later. He married Anne Gordon, sister of his friend Lewis Gordon, professor of engineering at Glasgow University. By 1856 with his other brother Frederick, he patented the regenerative furnace which was widely used in steel and glass works. Meanwhile, Werner in Germany had built up a great reputation as a telegraphic engineer. William was one of the first to apply electricity to locomotion and the Portrush Railway in 1883 was a pioneer venture. In all he took out one hundred and thirteen English patents.

The Siemens' great rival was Emil Rathenau (1838–1915) who returned from the U.S.A. with the German patent right for Edison's electric light and created the German Edison Company in 1883 which became the Allgemeine Elektrizitäts Gesellschaft or A.E.G. In self-defence in 1903 Siemens merged with the firm of Schukert. These two great rival cartels took the world lead in utilization of electricity and in 1913 produced 34 per cent of the world's electrical products.

THE GERMAN CHEMISTS

William Siemens was a pupil of Wöhler as well as Weber. Friedrich Wöhler (1800–82) first obtained metallic aluminium in 1827 and in the following year isolated beryllium. He is chiefly known for synthesizing urea from ammonia and cyanic acid. Since nearly half of urea is assimilable nitrogen it was to be extensively used as a fertilizer, a seasoner of wood, a base for soporifics and sedatives, a stabilizer for explosives, a nutrient for yeast, and a protein sparer for cattle. It was also to become a basis for the production of synthetic resins. Wöhler's illuminating contributions are perhaps best symbolized by his discovery of calcium carbide as a source of acetylene gas. Wöhler was associated after 1840 with an even greater German chemical teacher Justus von Liebig (1803–73) whose laboratory at Giessen became a nursery of accomplished scientists and the model upon which those scientists founded their own teaching laboratories. For twenty-eight years at Giessen and twenty-one at Munich, Liebig inspired investigations into the fundamental processes of animal and physical life, confirming amongst other things that plants obtain carbon and nitrogen from the carbon dioxide and nitrogen in the atmosphere. His discoveries, and those of his friends and assistants, were widely disseminated through the *Annalen der Pharmazie* which he founded in 1832. He was helped in this by Friedrich Wöhler, especially after 1840 when the journal became the *Annalen der Chemie und Pharmazie*. F. A. Kekulé (1829–96), one of Liebig's pupils at Giessen, was if not the originator, at least the formulator of the theory of the constitution of benzene, one of the most brilliant predictions in organic chemistry which led to the unravelling of the nature of the 'aromatic compounds'. Much of modern organic chemistry is a product of this theory.

A fourth German chemist, R. W. von Bunsen (1811–99), who lost the sight of one eye through an explosion and later nearly killed himself by arsenical poisoning in an experiment, studied the compositions of gases given off by blast furnaces. He found, with Lyon Playfair, that in England over 80 per cent of the heat of the fuel was dissipated with the waste gases. His *Gasometrische Methoden* (1857) indicate his famous methods of measuring such phenomena. He was the first to obtain metallic magnesium by electrolysis, to invent the carbon-zinc electric cell, the burner known by his name, and calorimeters for ice and vapour. His most distinguished contribution was the utilization of the spectrum for analytic purposes, through which he discovered the chlorides of caesium and rubidium. Bunsen started a practical course in chemistry at Marburg and influenced two great English chemists, Lyon Playfair and Henry Roscoe.

German chemical skill, exported to England, found excellent raw material on which to work. Liebig had worked on benzol (or bicarburet of hydrogen as Faraday had called it in 1825 on finding it in the decomposition products of whale oil) in 1835, and his pupil Hofmann had investigated other products of coal in 1843 and 1844. In the following year Hofmann came to England to take the Chair of Chemistry at the new Royal College of Chemistry. Hofmann's pupils and assistants literally created the synthetic dyestuffs industry in England.

The home of chemical engineering was Germany. After Perkin produced the first synthetic dye Hofmann returned to Germany with the invention. Then too in 1860 the rich potash beds in Saxony, Thuringia, Hanover and Alsace were being worked and Germany's production of sulphuric acid expanded enormously. Out of the many chemical companies of this time was to be formed the famous I.G. Farben. It was the state and industry which provided a laboratory for Paul Ehrlich who discovered salvarsan, a major chemotherapeutic drug and a cure for syphilis. Another major chemical discovery enabled Germany to wage the first world war: the Haber process for the fixation of nitrogen from the air. Armed with this, the explosives industry was independent of foreign supplies. And when war came the development of anaesthetics enabled German front-line soldiers to obtain quick and painless surgery.

EDUCATIONAL ENGINEERING

It was from professorial chairs rather than associations of professionals that German scientific and engineering advances were made. After the disappearance of a number of older inventors during the Napoleonic era, a new group of research oriented institutions was created. Berlin (1810), Breslau (1811), Bonn (1818) and Munich (1826) were supplemented by a number of technical high schools and military schools. Julius Weisbach (1806–71) was for thirty-five years, professor of mechanics and machine design at the School of Mines in Freiberg. His work was translated into English and used by Americans as early as 1848. His teaching, by means of an experimental laboratory, quickened interest in the design of machine parts that are subject to strain. F. Redtenbacher (1809–63) was even more influential. Holding a similar chair at the Karlsruhe Polytechnic, his work on springs, hooks and chains was widely used, and his influence on German engineering education was outstanding. His successor, F. Grashof (1826–93), had helped to launch the *Vereindeutscher Ingenieure* in 1856, and, as its editor, ensured that German engineers had a journal to carry the results of their work to a wide lay public. Sir Michael Sadler described the whole system as 'educational engineering' and remarked that it stood out in history as a classic example of the power of organized knowledge in furthering the material prosperity of the nation. It was from the Royal Polytechnic School in Berlin that John A. Roebling acquired the technical knowledge that was to make him a pioneer of suspension bridge construction in the United States of America.

Technical High Schools, based on trade schools, became true engineering colleges. Darmstadt (1822), Karlsruhe (1825), Munich (1827), Dresden (1828), Stuttgart (1829), Nuremberg (1929), Cassel (1830), Hanover (1831), Augsburg (1833), Brunswick (1835), and later foundations at Aachen, Danzig and Breslau were the forges where professional inquiries were tempered and tried. Some of these, such as Aachen, have been rebuilt and greatly developed since 1945.

OTTO AND DAIMLER

In Cologne in 1861 accounts of the gas engine reached a 29-year-old merchant's clerk, Nicolaus Otto. He was drawn to examine the idea: built himself one model, then another, reinforcing the experiments by further research into the literature. He realized that the mixture in his engine must be compressed, and that the explosion had to occur at top dead centre. He was not the first to consider the possibility of producing energy by putting the furnace into the cylinder. This stemmed from Huygens' and Papin's suggestions that the explosive power of gunpowder should be used to impel the piston. By 1794 R. Street was proposing to utilize a vaporized mixture of turpentine and air instead of gunpowder. W. Cecil described to the Cambridge Philosophical Society in 1820 a variant of this, whereby an explosive mixture of hydrogen and air was used to create a vacuum below a piston, allowing the atmospheric pressure to produce the working stroke, as in the earliest steam engines. In the same decade Samuel Brown was manufacturing gas engines for sale.

Whilst others were experimenting with the steam engine, improvements to the gas engine were going ahead. W. L. Wright in 1833 described an engine working with an inflammable mixture of gas and air. William Barnett, in 1838, stressed the importance of compressing the mixture before ignition. Jean Lenoir (1822–1900) began building engines in 1860 which worked with regularity and smoothness, and in 1862 Beau de Rochas laid down the principles of the four-stroke cycle: induction, compression, power and exhaust. It was Lenoir's engine that excited Otto.

Unfortunately Otto's power stroke was too forceful. So Otto tried to create a partial vacuum *under* the piston by the explosion of the gas mixture. After three years he joined with Eugen Langen, a scientifically trained industrial engineer, to manufacture gas engines. Langen was to supply capital, Otto the ideas. Langen, however, did more. Thanks to him the engineering problems connected with the power stroke were solved and when their engine was exhibited at the Paris Exhibition of 1887, it was awarded the gold medal. It did 80 to 90 strokes per minute. This success was widely reported, and a number of engines were ordered. To manufacture them a new company was formed at Deutz, near Cologne, with which Langen's name was associated. In three years it became a limited liability company, and as such needed engineers. At this stage they obtained the services of Gottlieb Daimler and William Maybach. Daimler stayed with them for ten years as technical director, developing techniques of mass production.

The more gas engines were produced, the more it was desired that they should have a fuel other than that supplied by the gas works. Langen had been told of the use of petroleum residues by Siegfried Marcus in Vienna. Marcus had been using these residues to drive a gas motor attached to a carriage but the Viennese police objected to the noise he made. Langen was interested in the application of the engine for propulsive purposes, indeed, he had tried to drive a tramcar in Liége by a gas engine. But disputes about patent rights distracted both Otto and Langen from going forward with liquid fuel. Instead they concentrated on perfecting their horizontal four-stroke engine of eight horse power, which was patented on 4 August 1877, and which ran at 150 to 180 strokes to the minute.

After ten years as the technical director of Langen and Otto's works at Deutz, Gottlieb Daimler decided to leave. So taking Maybach with him, Gottlieb Daimler set up a small workshop at Cannstatt to produce high-speed motors for propelling vehicles along roads. In November 1885 he was successful in getting one of his high-speed motors (900 revolutions per minute) to drive a wooden cycle. Motor transport had begun.

Before the first world war the firm of Adam Opel (1837–95) was fully converted from the manufacture of sewing machines and bicycles to that of motor-cars. Robert Bosch (1861–1942) returned from the U.S.A. in 1884 to become one of the largest producers of parts. In 1914 there were 64,000 motor cars.

THE COMPRESSION-IGNITION ENGINE

If German technical skill produced an internal combustion engine, German education in thermodynamics gave birth to the compression ignition engine. One German professor at the Technical High School at Munich pointed out to his class in 1878 that the steam engine only converted 6 per cent to 10 per cent of the heat of the fuel into effective work. Amongst his audience was 20-year-old Rudolph Diesel (1858–1913), who set himself from then on to realize the Carnot cycle. He worked for Linde in Paris and Berlin, experimenting in his spare time until, by the time he was thirty-five, he felt confident enough to publish *The Theory and Construction of a Rational Heat Engine to replace Steam Engines and Present Day Internal Combustion Engines*. The basis of this paper was that a gas, in this case the air, when subjected to rapid and powerful pressure becomes hot, sufficiently so as to cause the combustion of any suitable fuel which is injected into the cylinder at the critical moment. This blueprint excited his teacher and employer, who supported him (together with a machine manufacturing shop in Augsburg) in establishing a laboratory at Augsburg in 1893 where, in the course of the next four years he conducted six series of experiments which by 1897 produced a marketable engine. In this the air was compressed to 35 atmospheres by the piston and at the end of the compression stroke the fuel was injected in the form of a fine spray. (One of his prototype engines produced an explosion which not only wrecked itself but nearly killed him.)

The firms which backed him in Augsburg sent him abroad to sell it. He persuaded France to adopt it for submarines (his own country having refused) and witnessed its application in locomotives and lorries. He died in 1913 in mysterious circumstances, vanishing from the Harwich–Antwerp ferry.

In America, Adolphus Busch, the head of the famous brewing concern in St. Louis, promptly bought the American and Canadian rights of the Diesel engine for 1,000,000 gold marks and built a 60 h.p. unit at St. Louis in 1898, subsequently forming what he called the Diesel Motor Company of America. Since the term 'motor' was held to be a misnomer as applied to its product, the name was finally changed to the Diesel Engine Company of America. The heart of the Diesel engine is the fuel injection pump which injects fuel into the cylinder head against a pressure of approximately 500 lb. a square inch.

THE POLITICS OF POWER

One of Diesel's patrons was the Krupp firm. Bessemer steel was introduced into Germany by Alfred Krupp (1812–87) in 1862. With the great hammer which he had built to fabricate propeller shafts for ships, Krupp progressed from making large wheel tyres of cast steel to becoming the most notable manufacturer of cannon, subsisting first on foreign orders (especially from Russia) and then on orders from Prussia. Two years later when the Germans had won from France the rich potash deposits of Alsace and the equally rich iron resources of Lorraine, to work with the coal of the Ruhr, their engineering output accelerated rapidly.

In the four years after the 1870 war more iron and engineering works began in Prussia than in all the previous years of the century! In addition to Krupps other industrial giants were born. In 1871 August Thyssen (1842–1926) set up at Mülheim in the Ruhr, K. F. von Stumm-Halberg (1836–1901) in the Saar, and Prince G. H. von Donnersmark (1830–1926) in Silesia and the Rhineland. In 1871 Germany produced less than a quarter of the pig iron of the United Kingdom: in 1910 it was producing not only as much but nearly half as much again. This gigantic expansion had no parallel anywhere in the world, and it poured into railways, ships, and above all, armaments. Railway mileage nearly trebled in forty years. Shipping companies like North German Lloyd and the Hamburg–Amerika became the largest and most modern of the age. Armaments were sold all over the world to friend and enemy alike.

New inventions and the refinements of existing devices for speeding up production kindled an optimistic and almost naive belief in the capacity of technology to make gods of men. Kurd Lasswitz (1848–1910) expressed this view particularly well. In an essay in *Die Nation* (1898–9, p. 468) he wrote: 'What distinguishes the consciousness and indubitable cultural advance of modern humanity from the ancient and medieval is not so much the ethical and religious ideas; they were, in the main, already present in past ages. It is rather the conception of the power of man over nature, the conviction of

Chain-drums and checking gear used at the attempted launching of the *Leviathan*, or *Great Eastern*.

The iron-clad frigate, New Ironsides, and two Ericsson batteries going into action at Charleston.

the possibility of theoretically understanding and technically controlling nature. Modern man has attained maturity with the growth of the natural sciences.'

This control of nature intoxicated Lasswitz. He foresaw interplanetary travel, underground towns, pellet foods, and other startling developments. Like H. G. Wells he exploited the fantasies of science fiction to outline the goals, or Utopia, of his day. His *Auf Zwei Planeten* (1897) contains an imaginative account of the Martians' control of nature as a kind of foil for contemporary terrestrial failings. Amongst his imitators and successors was Martin Atlas, whose *Die Befreiung* (1910) centres round a new country, Penon, where energy is produced by a string of dynamos round the equator, and whose ruler adopts as his motto '*scientia redemptor mundi*'.

GERMAN INFLUENCE ON ENGLAND

Just how science was redeeming Germany was the subject of many pointed and piercing comments by the leaders of British science at this time. Lyon Playfair, a product of Liebig's Laboratory at Giessen, tried to rouse the British Association in 1885. A year later Werner von Siemens virtually created the Physikalische Technische Reichsanstalt by giving it land at Charlottenburg. This stirred Norman Lockyer who, as he witnessed its growth, commented freely on its significance in *Nature*. When the coping stone (if the image is not too repressive) was laid on the German system of technical training by an imperial decree establishing the cachet Dr. Ing. in 1899, Lockyer's desire to arouse his countrymen intensified. In 1903 he in turn delivered a homily to the British Association on 'The Influence of Brain Power on History', urging it to become a pressure group and persuade the government to establish British Higher Education on the German model. Getting no response, he formed the British Science Guild in 1905, which sustained a lively agitation up till the first world war.

That war, as we have seen, was fought by a Germany which was able, thanks to Haber, to make its own nitrates. It was also able, thanks to Paul Ehrlich (1854–1915) the discoverer of salvarsan, to employ drugs on a great scale. And in the wider spectrum of technological change, the influence of German brain power on history was nowhere more apparent than in the work of Heinrich Hertz (1821–94) on radio waves, W. K. Röntgen (1845–1923) on X-rays, and Max Planck (1858–1947) on Quantum theory. German science and technology was carried far and wide by scientific publishers like Springers of Berlin, just as their literature was made available by the famous Tauchnitz editions and the issues of Reclam Universal Bibliothek.

Bildung, the hallmark of the early twentieth century German, was a less impressive, though more important, reason for German national advance than the militaristic nonsense of the Junkers. This was clearly seen by many Englishmen. Sir Swire Smith, a colleague of Henry Roscoe on the famous Samuelson Commission during the 1880's, showed this when he published in a book with the revealing title, *The Real German Rivalry* (1916). Germany offers a classic example of the development of an economy through the complex interaction of innovators, entrepreneurs and bankers. Viewing the process from experience of life in an unindustrialized country (Egypt) and from the detached position as professor of economics at an Austrian university (Czernowitz), a young Austrian economist, Joseph H. Schumpeter (1883–1950), elaborated a theory of economic development which was to have a profound effect in the U.S.A., where he subsequently taught. Schumpeter, more than anyone else, stressed the dynamics of technology as the most important single cause of economic development in general and of the so-called capitalist system in particular. Indeed, he prophesied that from its very action some form of socialism must inevitably follow. The full truth of Schumpeter's view was not seen at the time, as his book, *The Theory of Economic Development* (1911), was not made available to English readers till two decades later.

After the First World War, a planned economy for Germany, based on some form of guild socialism, was envisaged by Walther Rathenau (1867–1922), son of the founder of the great German electrical combine, Allgemeine Elektrizitäts-Gesellschaft. As minister of reconstruction and later foreign minister, he put forward challenging ideas for remodelling the economy. Unfortunately, his career was tragically ended when he was murdered in June 1922. The Rathenau influence on the rebuilding of Germany died with him, but that of the Allgemeine Elektrizitäts Gesellschaft lived on. For from it sprang the ideas of Walther Gropius, who was profoundly to modify existing techniques of building all over the world.

Yet both German and American expansion in the electrical field were influenced by developments in the turbine. These we will now examine.

194

CHAPTER 18

The Generation of Electricity

THE DEVELOPMENT OF WATER POWER

One can best see the turbine in the light of the improvements to the waterwheel that had been steadily taking place whilst the reciprocating engine was rising in public favour.

'Few persons', remarked Joseph Glynn (1799–1863) 'have had occasion to use the steam engine more extensively. Yet', he continued, 'I have often felt that water power has been unduly superseded or neglected when it might have been usefully employed. Lately more enlarged views have been taken, and men who stand high in their profession have turned their attention to the subject and it has been more studied and better understood.' Writing in 1853 Glynn spoke with the authority of twenty-eight years of experience of using the steam engines to drain the fens, where 480 windmills had been replaced by 60 steam engines. Yet Glynn grasped the twofold role of the water wheel as a dual source of power and pumping: a breast wheel reversed was a draining wheel. As manager of the Butterley Iron Works in Derbyshire he built wheels which drained in Holland, Germany and British Guiana in the wet seasons, and irrigated in the dry.

THE FRENCH TURBINE

In France where critical work was taking place to secure greater efficiency from water power, J. V. Poncelet (1788–1867) had improved the undershot wheel by using curved floats and secured a maximum efficiency of 85 per cent. During his imprisonment in Russia during the Napoleonic campaign, he became interested in projective geometry, and as a military engineer at Metz and as professor of mechanics at the École d'Application he obtained the experience and opportunity to publish his *Memoire sur les roues hydrauliques à aubes courbes* (1826).

195

The name 'turbine' was coined by another French professor, Claude Burdin (1790–1873), who made waterwheels the subject of special instruction in the technical school at St. Etienne. Burdin had a rich stock of experiments on which to draw, for the study of the principles of hydraulics had developed on the continent in the latter part of the eighteenth century. Euler, Defarcieux, Papecino d'Antoni, Bossert and Nordwall were continental pioneers and there was of course the English Smeaton who, as we have seen, had worked on hydraulic power for the Royal Society.

One of Burdin's students, B. Fourneyron (1802–67), constructed a radial outward-flow water turbine. It was built at Pont sur l'Ognon, Sâone, France, in 1827, and in 1834 a second was constructed in Franche Comté to blow a furnace, working at times with a fall of nine inches. The idea spread, after Fourneyron turbines had been erected in the Black Forest. In 1841 N. J. Jonval introduced the parallel flow water turbine and in the following year a series of large waterwheels were installed at the Niagara Falls to supply power.

AMERICAN EXPERIMENTS

Meanwhile in America a reaction turbine had been built by Samuel B. Howd in 1838, but it was not until eleven years later that it was finally perfected by James B. Francis (1815–92), the son of an English railway engineer. Francis was employed at Lowell, Massachusetts, in dismembering, measuring and making detailed working drawings of a Stephenson locomotive for the Boston and Lowell railroad. He became chief engineer of the first locomotive builders in New England. In view of the large number of mills in Lowell, his employers asked him to measure the quantities of water drawn along the canal by the wheels. After working for four years on this problem, Francis reported in 1845. As a result his employers abandoned steam engine manufacture and devoted themselves to developing hydraulic machines, Francis being made chief engineer and general manager. In 1849 Francis designed a turbine, based on the Howd patent, with vanes redesigned to give a combination both of radial and axial flow: hence its name today, the mixed flow or Francis turbine. His book, *The Lowell Hydraulic Experiments* (1855) appeared two years after Glynn's persuasive little monograph, to reinforce his thesis.

THE CENTRIFUGAL PUMP

In France and America attention was concentrated on using the turbine as a source of power. In England attention was being concentrated on it as a centrifugal pump. Here J. C. Appold (1800–65) utilized the fortune he had made as a fur-skin dryer to experiment in mechanics, and devised a centrifugal pump to drain mines and fens. It was employed in Egypt and the West Indies where tracts of ground lay below the level of the natural outfall. In England it proved very versatile. One version drained Whittlesea mere and a great portion of the Bridgewater marshes; another version raised the thick viscous ink with which *The Times* was printed.

James Thomson (1822–92), the brother of the more famous William (later Lord Kelvin), a pupil of Lewis Gordon the first professor of civil engineering at Glasgow and also an employee of Fairbairn's, built a horizontal waterwheel called the *Danaide*, an improvement of the *Danaide* of Manouri Dectot. He patented an improved version in 1850, whereby it was mounted within a circular chamber and the water, injected under pressure, flowed tangentially into the circumference of the well and was led through the wheel to the centre by suitably formed radiating partitions. A number were designed by him for various factories. In 1851 he settled down in Belfast as a civil engineer to work on the properties of whirling fluids. This led to improvements in the action of blowing fans on one hand and to the improvement of centrifugal pumps on the other. These, and other improvements in the turbine, he described to the British Association when it visited Belfast in 1852. Jet pumps, too, were the subject of another paper of his. Thomson's centrifugal pumps were put to work on sugar plantations in Demarara. After a period of service as resident engineer to the Belfast Water Commissioners he became, in 1857, professor of civil engineering and surveying in the Queens University of Belfast and later succeeded Macquorn Rankine at Glasgow in 1872.

THE STEAM TURBINE

The reactive force of steam jets had been described by Hero of Alexandria in the first century A.D. The employment of a fixed jet of steam to drive a wheel by its action on blades projecting from the circumference was described by Giovanni Branca in 1629.

The invention of the dynamo revealed the limitations of the reciprocating engine. Baron Kempelen in 1784 and Richard Trevithick's 'whirling engine' of 1815 had, if anything, provided warnings of the dangers of departing from the piston principle. Yet the kinetic energy of the steam particles continued to attract Francis Bresson in

1852. Hall had proposed and Harthan patented a compound steam turbine in 1858.

It was an engineer at the Armstrong works at Elswick who was to save mankind wasting energy on the reciprocating engine. Charles Parsons (1854–1931) was the son of a President of the Royal Society and at the age of twelve had built a steam motor car. After studying at Trinity College, Dublin, and St. John's, Cambridge, he entered the Armstrong works at Elswick in 1877, to work on rocket torpedoes and rotary steam engines. In 1880 he tried to propel a vessel by a jet of air.

Armstrongs, his employers, were great hydraulic engineers. Since 1846 when W. G. Armstrong (1810–1905) had built up in the region of the three northern rivers of England, Tyne, Wear and Tees, a great industrial complex, they had been making cranes and hydraulic winding machinery for collieries, for lead mines and for printing. Where there was no water pressure, as in the construction of the Ferry Station at New Holland in Lincolnshire, he built accumulators to store water pressure by means of steam pumps. As a result of this invention hydraulic power became almost ubiquitous where cranes had to be worked, hoists and lifts operated, lock gates open and shut, and ships launched. The *Great Eastern* was perhaps the most notable example of this. Armstrong's success as a hydraulic engineer lay in his utilization of water as a means of distributing rather than obtaining energy.

THE PARSONS STEAM TURBINE

The water turbine served as a basis for the design of the steam turbine by Charles Parsons in 1884. Just as the water turbine was evolved from the waterwheel to use the kinetic energy of water instead of its pressure, so the steam turbine was evolved to utilize the kinetic energy of the steam. And just as water turbines developed along two lines—reaction and impulse—so did steam turbines. Parsons was impressed by the fact that the efficiency of water turbines was from 70 to 80 per cent. As he wrote: 'the safest course to follow was to adopt the water turbine as the basis of design of the steam turbine. In other words, it seemed to me reasonable to suppose that if the total drop in pressure in a steam turbine were to be divided up into a large number of small stages, and an elemental turbine like a water turbine were placed at each stage (which, as far as it was concerned, would be virtually working in an incompressible fluid) then each individual turbine of the series ought to give an efficiency similar to that of the water turbine, and that a high efficiency for the whole aggregate turbine would result; further, that only a moderate speed of revolution would be necessary to reach the maximum efficiency.'

His patent of 23 April 1884 (No. 6735) embodying this principle, entitled 'Improvements in rotary motors actuated by elastic fluid pressure and applicable also as pumps', ushered in a new age. It appeared, as it was intended to, at the very time when the newly introduced dynamo or electric generator needed driving, and Parsons, with this in mind, had taken out on the same day a patent for a generator.

The impact of the turbo-generator was immediate. Parsons left his partners in Clarke Chapman and Company (which he had entered in 1883) and formed his own company—C. A. Parsons and Co. Ltd.— whose Heaton Works at Newcastle-upon-Tyne soon became world famous. Unfortunately Clarke Chapman and Company retained his patent for an axial flow turbine and drum armature dynamo, and Parsons had to concentrate on radial inward and outward flow turbines. Newcastle was the first to install a Parsons set in a public power station.

THE TURBINE AND ELECTRIC POWER

The turbine principle vastly accelerated the production of electric power. Americans took a lead in establishing the new industry in England. C. F. Brush (1849–1929), a former telegraph engineer, established a branch of his company at Loughborough, and George Westinghouse established a works at King's Cross in London. The English rights to the Edison dynamo were obtained by Sir William Mather.

The British Westinghouse Electrical and Manufacturing Company, which succeeded Westinghouse's original subsidiary, began to manufacture Parsons Steam Turbines for power stations, and had as one of its directors the Hon. R. Clere Parsons, a brother of the inventor. At the close of the century they built a works at Trafford Park, which is still one of the largest engineering works in the world. The team of American builders imported to construct it excited comment from newspapers like *The Times* for the speed and efficiency with which they worked. The very bricklayers worked four times as fast as their English counterparts. Westinghouse enlisted Englishmen and sent them to Pittsburg for a course of technical training. These Pittsburg-trained Englishmen were known in their day as 'The Holy Forty', and included such distinguished engineers as

Miles Walker, who was later to become professor of electrical engineering at the University of Manchester, and A. P. M. Fleming, who was to become the head of the Westinghouse training department. In 1919 the British Westinghouse Company, together with the Metropolitan Carriage and Waggon Company, were taken over by Metropolitan Vickers.

DEVELOPMENTS IN LIGHTING

It was in the sphere of electric lighting that the real advances were made. A. N. Lodygin in Russia made lamps with carbon filaments in 1871. Not, however, until J. W. Swan carbonized a thread of dissolved cellulose in 1880 was electric lighting really practicable. Utilizing the work of Sprengel (who devised a pump in 1865 to evacuate a glass bulb) the glass bulb industry now developed rapidly.

Edison, who joined forces with Swan in 1883, provided the necessary current. Since the first generator of Faraday was discovered in 1831, a host of improvements had raised its efficiency by 40 per cent. Edison's bi-polar dynamo of 1878 was followed by his incandescent lamp a year later and his 'system' of central station power production in 1882. Edison designed the first central electric power station in England, installed at Holborn, London. Four years later Hopkinson's paper on dynamo electric machinery made it possible to calculate generating performance.

In 1882 the government endowed local authorities with the right to purchase, after 21 years, the private enterprise companies operating in their area. Though a subsequent act six years later doubled this period, there were considerable discouragements to private enterprise setting up lighting companies. So many of the companies were, in fact, civic affairs that in 1896 the Incorporated Municipal Electrical Association was formed.

Indigenous English firms were relatively few. One of the earliest was founded by a retired Indian army officer, Colonel R. E. B. Crompton (1845–1940), who had purchased for the Stanton Ironworks a Gramme dynamo and an arc lamp so that night shifts could be worked. His firm, Crompton and Company at Chelmsford, founded in 1878, introduced various improvements in the design of lighting equipment, which it installed in Windsor Castle, Buckingham Palace, the Law Courts and the Imperial Theatre at Vienna. Crompton designed and built a number of central electricity stations on the continent and in India and established one of the first public lighting undertakings—the Kensington Court Electric Company. As the engineer for nearly all of the low-tension systems of London, he was visited by Edison on 25 September 1889. Both Crompton and Edison were, however, to be outshone (if a metaphor may be excused) by a younger man—who disagreed with them both—Sebastian de Ferranti (1864–1930).

FERRANTI AND HIGH PRESSURE CURRENT

Ferranti was seventeen years old when he joined the Siemens works at Charlton in 1881, and began to assist Sir William Siemens in making steels. He supplemented his knowledge with evening classes at University College. A year later he produced the Ferranti alternator which, installed under the Cannon Street Railway Station, won him such recognition that he was able to start an independent business for the manufacture of electrical apparatus. His first contract was to supply the Grosvenor Gallery Electric Supply Corporation with a generator, to which he added transformers, meters, switches and other equipment. Finding that the noise and dirt of the Grosvenor Gallery was inconvenient, Ferranti now urged that the whole of London north of the Thames could be supplied from a power station at Deptford, which gave access to coal fuel and cooling water. This station he designed and built in two years (1888–90) as chief electrician to the London Electric Supply Corporation. Unfortunately, current opinion (if the pun may be permitted) ran against him: low power generator stations in the middle of small areas were thought more suitable. From 1892 onwards Ferranti returned to his inventions and his manufacturing business. A factory at Hollinwood, near Oldham, was built in 1896 which manufactured equipment for all types of electric lighting under Ferranti's own busy and inventive leadership (he took out over 176 patents in 45 years). The firm became one of the most progressive in Europe, and Ferranti's leadership was not confined to the firm. Ferranti constantly urged the importance of conserving coal and using electricity as a source of light, heat and power. The Electrical Age had really begun.

THE SOPHISTICATION OF THE STEAM TURBINE

Whilst Parsons was marrying thermal power to the waterwheel to produce electricity, Osborne Reynolds (1843–1912) had been working on a centrifugal multi-stage turbine pump in 1875, and developing a practical theory of lubrication. He patented this twelve years later. Reynolds, as the first occupant (appointed in 1868) of the chair of civil engineering at the Victoria University of Manchester (where the Whitworth Engineering Laboratory was established in 1885), built the first hydraulic model to work on the navigation of the Mersey. He showed how characteristics of the estuary, with respect to erosion and silting, could be reproduced in the model on a small scale. Similar hydraulic laboratories were soon constructed in England, France, Germany and Italy, to study hydraulic problems as well as the performance of particular structures under controlled conditions. Reynolds himself discovered his famous dimensionless quantity $VL\rho/\mu$, where V is the velocity of flow, L the length of the flow system, ρ the mass density of the liquid, and μ is the viscosity.

From 1889, when Parsons lost control of his axial-flow steam turbine patents, the work on turbines intensified. In Sweden C. G. P. Laval (1845–1913) patented an impulse turbine with a diverging nozzle in 1889, extending a principle first outlined by Branca two centuries previously. In France Auguste Rateau (1863–1930) designed in 1889 a multi-stage impulse (as opposed to a velocity compounded) turbine, and in the same year C. G. Curtis (1860–1953), an American, developed a velocity compounded turbine which proved capable of being employed with turbines of other types. B. Ljungstrom (1872–1948) in 1906 patented radial flow reaction turbines on a principle which Parsons had abandoned in 1894.

The Parsons turbo-electric generator lit up the Ormesby Iron Works in 1888. The Newcastle and District Lighting Co., in the following year, was the first public power station to purchase one of his sets and was followed by the Cambridge Electric Supply in 1891. In 1896 Pittsburgh installed the first Parsons turbine in America, and four years later Elberfeld installed the first Parsons 1000 k.w. turboalternator.

THE LIBERATOR OF NEW MATERIALS

By 1890 ambitious and imaginative extensions of hydraulic turbines were envisaged. It was even proposed to build a tidal power station on the Seine near Havre. Five years later, the Niagara Falls were tapped for generating purposes by using a tunnel 29 feet deep and 18 feet wide to convey water to ten 5000 h.p. double-outward-flow water turbines. In 1896 a hydro-electric station was built at Foyers in Scotland for the British Aluminium Company, and in the same year accumulators began to be used for taking excess peak loads. The Niagara Falls Power Company began to extend its operations and

deliver electricity to a wider area by means of high voltage transmission. Collieries, too, felt the benefit. In 1866 the first Guibal fan
for ventilation was installed at the Elswick Colliery, and within ten
or twelve years 200 were operating in various parts of Britain. The
advent of direct current, first used underground at Trafalgar Colliery
in 1882, led to new pumping machinery and safer illumination. Alternating current followed at Denaby eighteen years later, where it was
used for haulage, and Acton Hall used it for a coal-cutting machine.
By 1910 there was an all-electric colliery at Britannia, Monmouth.

The advent of electric power re-oriented the work of the chemical
engineers. The use of the electric arc to extract aluminium by Charles
M. Hall in Ohio in 1886 enabled engineers to utilize the third most
abundant metal on the earth's surface. It had been isolated some fifty
years before, but production had hitherto been far too costly. When
Hall's process was improved by Héroult in France it was cheapened.
Aluminium came at just the right time for the development of
aircraft and automobiles. Production in Britain began in 1896. In
1897 it was being used as a telephone line conductor in the Chicago
Stockyard. It was a godsend to the engineer who used it as a protective coating to steel structures, as a pigment in paint, as an alloy
with copper, magnesium, iron, nickel, zinc, silicon, manganese, and,
of course, in aeroplanes. Not only aluminium but magnesium, sodium, zinc and nickel were to be refined in this way. But perhaps the
most unforeseen result of cheap electric power was to enable the Germans to fix nitrogen from the air, thereby enabling them to produce
both the fertilizers and explosives needed to fight the first world war.

Electrothermal techniques enabled artificial graphite and silicon
carbide to be made in quantity. Here E. G. Acheson (1859–1939) made
striking advances in marketing carborundum. Calcium carbide could
also be made by heating lime and carbon in the electric furnaces of
Henry Moissan (1852–1907) and Thomas Willson (1860–1915).
And from calcium carbide came the acetylene which was not only to
be used in cutting and welding, but was to be important in the further
development of the organic chemicals industry. Hot calcium carbide
treated with nitrogen, provided cyanamide; an invaluable fertilizer,
and until the adoption of the Haber-Bosch process referred to in the
preceding paragraph, was a standard source of ammonia.

The electrical works themselves effected a revolution in factory
design as profound as that of the steam factories. In and around
Berlin the factories of Allgemeine Electrizitäts Gesellschaft were
designed by Peter Behrens (1868–1940) and those of Siemensstadt

by his pupil Walter Gropius (b. 1883). Both these men pioneered in the imaginative use of glass and concrete and utilized the concepts of Paxton to open up the twentieth-century town to light and air.

THE REVOLUTION IN TRANSPORT

Electricity also effected a revolution in transport. At an Electrical Exhibition staged at the Crystal Palace in 1882 F. J. Sprague (1857–1934), an American naval officer, served as secretary of a jury that considered gas engines and dynamos. Sprague was on unofficial leave, and only came because the exhibition afforded him a good opportunity to get abreast of current developments. He overstayed his leave by six months and only escaped court martial by preparing a voluminous report. On his return to America he left the navy to join Edison, but left him, too, on discovering that Edison was more interested in illumination than power. Sprague's development of a motor to operate on commercial circuits enabled him to contract for an electric railway at Richmond, Virginia, in 1887, the first of 110 such contracts he fulfilled before his company was merged with Edison's General Electric Company in 1890. From then on he turned to electric elevators, until that company, too, was absorbed by G.E.C. in 1902. He then turned to automatic train control. His genius lay in his capacity to bring together the best of a number of competing projects.

Nor should the work of George Gibbs (1861–1940), another product of the Edison 'stable' at Menlo Park, be forgotten. Gibbs was a pioneer of electrification in London, Liverpool, Paris and New York, and the first to use all-steel passenger cars. Many of the underground railways owe their original electrification to his stimulus.

The acceleration of transport was also visible on the sea through the installation of turbines in ships. The first was the launch *Turbinia*, which achieved a speed of $34\frac{1}{2}$ knots in 1897. By 1907 the Cunard Line had installed 70,000 h.p. turbine power units in the *Lusitania* and *Mauretania*. The British navy adopted them, too, as speedier, more economical and more manageable power units. The principle was to be extended and sophisticated, as we shall see, to aeroplanes.

THE SLOW ADOPTION OF ELECTRICAL POWER IN ENGLAND

The lack of large-scale industrial units in England as compared to Germany and America, the confident assurance of an industrial economy based on cotton, coal, iron, steel, and shipbuilding did not make for an easy adoption of the next great change in industry—the use of electrical power. Other inhibitory factors were the suspicions of state power on one side, and the restrictive legislation on the other. Labour was too cheap and available, and capital had too wide a field for investment outside Great Britain to tempt a too-rapid change-over to such a labour-saving power system.

Concentration on the telegraphic and illuminant phase of British electrical development had one indirect advantage: the work of J. J. Thomson and Rutherford on one hand and Sir Ambrose Fleming (who worked for a time in the London Edison Company) on the other. Fleming's two-electrode valve in 1904, followed by the American de Forest's grid in 1907, could be easily manufactured in lamp factories. From it stem, not only a new telegraphy, that of wireless, but radar, television and computor industries.

Thus the application of science to the common purposes of life has itself resulted in another step forward in modern physics, the full consequences of which are still to be seen.

In Britain working telegraph engineers contributed a great deal to both the research and the dissemination of electrical knowledge. Two examples spring to mind. One was an employee of the Great Northern Telegraph Company from 1870 to 1874: Oliver Heaviside (1850–1925), whose first article in *The English Mechanic* in 1872 was followed by others in the *Electrician* from 1885 to 1887. He retired to Camden Town to pursue experiments on his own, and his name is immortalized in the Heaviside Layer. The other was an earlier associate of the Society of Telegraph Engineers, H. R. Kempe (1852–1935), who spent the whole of his working life in the Post Office. In 1872 he founded the *Telegraphic Journal* which twenty years later became the *Electrical Review*. Four years later he published the *Handbook of Electrical Testing* which remained a standard text book until the twentieth century. Perhaps he is best known for his *Electrical Engineers' Pocket Book*, first issued in 1890, which five years later became an annual publication—covering all branches of engineering; fittingly still called by his name. As this chapter is being written the author has a copy of the 63rd edition on his table.

It was the telegraph that incubated the first professional association of electricians in 1871, when four British army officers, a naval officer and three telegraphic engineers formed a Society of Telegraph Engineers 'to advance Electrical and Telegraphic Science'. The three civilians were Wheatstone's son-in-law (R. R. Sabine), an associate of Siemens Brothers and Company (Louis Loeffler) and a veteran of the Atlantic Cable (Wildman Whitehouse). The first president was Charles William Siemens (1823–83). The new body was given every encouragement by the senior professional engineering body, the Institution of Civil Engineers, in whose buildings it met for many years. Beginning with 268 members, it survived the formation of the Physical Society (in 1873). In 1888 it changed its title to become the Institution of Electrical Engineers.

In 1889 the employees of the telegraph industry followed suit. Meeting at Manchester as the Amalgamated Society of Telegraph and Telephone Construction Men and the Union of Electrical Operatives, they decided to change their name to The Electrical Trades Union, obtaining the famous artist Walter Crane to design their banner. These two changes of name indicated that electricity had come to stay.

CHAPTER 19

The Processing of Molecules to 1914

Nature's pantry was not inexhaustible. As demands upon it grew, it became progressively less able to supply raw materials of the quantity and calibre needed to sustain life for an expanding population. Madder and indigo gave way to synthetic dyes, the silkworm to man-made fabrics, the pine trees to plastic, the hevea tree to artificial rubber, whale oil to petroleum. A myriad of natural products were either supplemented or supplanted by products of the laboratory.

To simplify the picture it would be perhaps best to begin with the cracking of nature's first line of reserves, that great chemical strongroom, coal, since many of modern chemical engineering techniques stemmed from this. Since chemists were, during the second half of the nineteenth century, interested in producing new compounds of carbon, they became interested in synthesis, and created a new field of study in organic chemistry. It looks a wilderness to us, and it was a wilderness to those who laboured in the field. As Friedrich Wöhler wrote to his great teacher Berzelius (1779–1848), who started the serious study of catalysis: 'Organic chemistry just now is enough to drive one mad. It gives me the impression of a primeval tropical forest, full of the most remarkable things, a monstrous and boundless thicket, with no way of escape, into which one may well dread to enter.'

In a relatively densely populated country like Germany the by-products of industry could not be projected into the atmosphere or run off into rivers. In finding uses for them a great aromatic chemical industry was built up.

COAL DISTILLATION

Subsequent improvements in the distillation of coal produced further products to improve the standard of living: naphtha (for lamps) and creosote (for sleepers and cross ties). The rising standard of public health needed tarred pipes for sewage. The London Gas and Coke Company prepared varnishes and pitch for the navy, roofing tar for the United States, and after 1833 they began to use their ammonia instead of decanting it into the Thames. Then solvent naphtha was also found valuable to use with rubber to make waterproofs. By 1856, as we saw in Chapter 14, the coal tar dye industry virtually began.

Attempts to professionalize this activity were made as early as 1849 as the *Illustrated London News* (of 26 May 1849) and the *Journal of Gas Lighting* (of 10 August 1849), indicate. By 1863 the British Association of Gas Managers had been formed under the famous sanitary engineer Thomas Hawksley (1807–93).

By the 1880's Carl Auer (later Baron Auer von Welsbach, 1858–1929) had perfected a gas mantle consisting of threads of cotton impregnated with solutions of the nitrates of thorium and cerium which on subsequent ignition formed an open cone of a kind of china made of the oxides of these two metals, which though very fragile emitted much luminous radiation when heated by a gas flame. This enabled gas manufacturers to concentrate on the heating value of the gas, since the mantles would give illumination. This harmonized with the increasing needs of industrial furnaces and domestic cooking (which needed gas to heat rather than illuminate) so gas producers raised the temperature of their carbonization plants. Unfortunately this led in turn to a decreased yield of tar at the very time when it was needed most.

Meanwhile the increased consumption of metals led to the construction of special ovens (as at Saarbrucken in 1854) to make metallurgical coke harder than that made by an ordinary gas company. The ammonia was recovered as sulphate of ammonia and the insoluble volatile products were used for heating. By 1861 the Siemens brothers enabled a firm to make gas for its own use by carbonizing coal in a fire brick chamber under which was a trough of water. Steam from the water penetrated the coal and was utilized. This process was consistently improved, notably by Dr. Ludwig Mond (and Thaddeus Lowe, an American) in England, who introduced more steam still, enabling more gas to be produced and the ammonia to be collected more economically. This ammonia, as will be seen, was needed for an important new purpose, the manufacture of alkali. Thus the coke ovens of the steel industry not only helped to resolve the shortage of industrial gas for the regenerative furnaces,

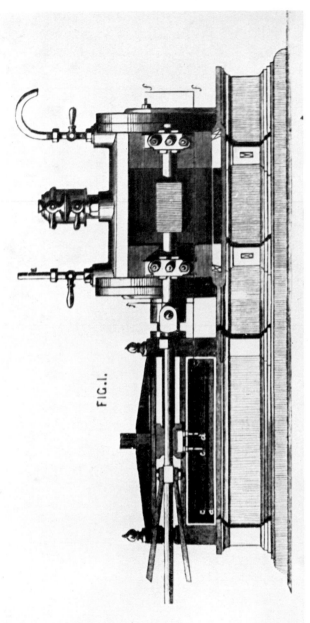

FIG. 1.

Lenoir's Gas Engine. The first gas engine was built by Lenoir in 1860. It developed $\frac{1}{2}$ h.p. at 110 r.p.m.

Sir Joseph Whitworth (1803–1887), pupil of Maudslay, who worked on Babbage's difference engine and was founder of the scholarships for training mechanical engineers. His uniform system of screw threads, first suggested in 1841, was in widespread use by 1860.

but eventually the increased production of ammonia made the Ammonia-soda process economically sound and so, very largely displaced the Leblanc process which hitherto had been the traditional basis for the heavy alkali industry of Great Britain.

THE RECOVERY OF WASTES: SULPHURIC ACID, CHEAP SODA AND SOAP

The smelly calcium sulphide residues of the Leblanc process had forced the government to take action. By the Alkali Acts of 1863 and 1874 the effective disposal of residues became compulsory, and inspectors (like Angus Smith and G. E. Davis) were appointed to see that it was done with the minimum offence to those living near an alkali works. One shrewd observer estimated that this disposal (in the sea or in old mine shafts) lost the country some 250,000 tons of sulphuric acid a year. This observer, Ludwig Mond (1839–1909), took out a patent to recover the sulphur from this waste. Later, he bought the English patent of a process for producing soda with ammonia and, with John Brunner, established a works at Withington in 1872 which revolutionized the alkali industry by producing a purer product at a lower cost than by the old Leblanc method. This ammonia-soda process was based on the principle of saturating brine with ammonium bicarbonate to produce ammonium chloride and sodium bicarbonate. Then, by heating the sodium bicarbonate, sodium carbonate was obtained, together with carbon dioxide and water. The appealing aspect of the process to our thrifty forebears was that both the carbon dioxide and the ammonia could be recovered and used again. It is rightfully associated with the name of Ernest Solvay (1838–1922), son of a Belgian salt-refiner who, though he was not the first to think of the idea, was the first to exploit it by means of breaking up the process into a number of unit operations. As Professor Newitt remarks: 'his process constitutes a landmark in the history of chemical technology in that it showed for the first time that continuous operation on a large scale was not only possible but also created conditions favourable to exact process control and high operating efficiency.'

So the old traditional firms who made soda by the Leblanc process, because of Mond's successful exploitation of the Solvay process, began to concentrate on exploiting their by-products, notably chlorine.

Mond went on to discover nickel carbonyl, a gaseous metal compound, which he exploited to manufacture nickel at Swansea from ore mined in Canada.

Ludwig Mond had developed physical chemistry in a works instead of a laboratory, giving England cheaper soda and cheap soap. The impact on domestic hygiene and the health of individuals was enormous. He made three-quarters of the nitrate of ammonia and all the nickel used by the British Army during the first world war, together with the first British gas and the filling for gas masks.

His son Alfred entered the House of Commons in 1906 and bought the *New English Review*. This took up John Masefield.

LABORATORY-MADE SUBSTITUTES

Explorations in the field of synthetic dyestuffs led to collateral discoveries utilizing artificial dyestuffs. Paul Ehrlich (1854–1915) argued that the right molecule should have a toxic effect on certain micro-organisms without harming the host. And so he developed arsphenamine as a cure for syphilis. Then again, the aromatic phenols, quinones and amines became progressively more important with the development of photography, which also demanded bromine (made from the salts at Stassfurt). It was also found that by acetylating salicylic acid its gastric effects were eliminated, and, in the form of aspirin, could be used as an analgesic and anti-pyretic drug. By 1879 two Americans, Ira Remsen (1846–1927) and Constantine Fahlberg had discovered saccharin. W. H. Perkin, the great English pioneer of coal-tar dyes, himself discovered how to synthesize coumarin in 1868, a basic ingredient of perfumes.

By this time the production of food itself was becoming a technology. Progressive urbanization had led to the application of chemistry to increase the productivity of the soil. Humphry Davy in 1813, by his *Elements of Agricultural Chemistry*, set the pace. John Bennet Lawes (1814–1900), from dressing bones with sulphuric acid, developed a system of artificial manure manufacturing mineral superphosphate. The fortune derived from this was devoted to an experimental station at Rothamsted whence with Sir Joseph H. Gilbert, a steady stream of information as to the growth of cattle issued. When the Thomas-Gilchrist process of making steel made its appearance in the seventies, the slag (containing the phosphorus which the process eliminated from the iron ore) was found to be another source of phosphate manures.

But the chemists reached out to simulate (even to produce) natural products. Hippolyte Mège-Mouriez, working on a false hypothesis, made 'butterfat' from beef suet which he called oleomargarine. His process patented in 1869 set off a rapid development of the industry. Gradually animal fats were replaced by vegetable fats and improvements in the hydrogenation processes. The rapid development of the industry was stimulated by the urgent needs of the cities, and it is symbolic that its discovery in France was a direct response to a social need. This need was at the same time met by the exploitation of refrigerants like ammonia to make ice. In 1874 frozen beef was being sold in Smithfield market and by 1880 all steamship lines to the U.S.A. were equipped with refrigerators.

OTHER SCARCITIES: PAPER, RUBBER, BONE AND EXPLOSIVES

Food and distraction for the mind was equally important to the urbanized populations of Britain and France. To provide it vastly increasing amounts of paper were needed by the steam presses of the Victorian age. Newer raw materials like esparto grass from Africa were tried, and in 1857 Thomas Routledge was using it to make paper at Eynsham, Oxfordshire, for the *Illustrated London News*. Within thirty years 200,000 tons were being imported for papermaking, and Routledge had begun to use bamboo. Bamboo, however, was impregnated with lignum and resin, which in turn had to be removed by sulphite of lime. Benjamin C. Tilghman, who was responsible for this discovery, began work on the problem in 1857. His experiments were carried forward by George Fry (1843–1934), the manager of a Swedish pulp mill, who came to England in 1883 as a consultant engineer designing chemical pulp mills: the first was built in Northfleet in Kent. This sulphite process, coupled with other acid and alkali processes, enormously accelerated the production of paper so that by 1886 it could be made from no less than 950 raw materials. Without this enormous supply of paper, the periodical and newspaper empires could not have even been contemplated. By the century's end the paper was being used for gas pipes, horseshoes, hansom cabs, and military hospitals. It had already been used to make coffins.

The Victorian steam engine needed harnessing as a prime mover. Belting, hoses, washers became so important that in 1857 Stephen Moulton introduced their manufacture from America. The connection between steam engines and rubber was as old as Walter Hancock (1799–1852), who with his elder brother had virtually founded the rubber trade in England. Thanks to the act of vulcanizing rubber (discovered by the American Charles Goodyear), belting, hoses and washers of vulcanized rubber were made available by Moulton's factory. In 1856 five American investors set up a factory for making vulcanized rubber in Scotland, and in a few months had incorporated themselves as the North British Rubber Company. So great was the demand that in less than ten years supplies of rubber from Brazil became exhausted, so the Marquis of Salisbury suggested that rubber trees should be planted in the British Colonies. H. A. Wickham was commissioned to collect a large number of seeds and take them to England. He collected 70,000, chartered a steamer in the name of the Indian Government, and sent them to Kew. In 1876 they were duly distributed to Ceylon, Singapore and Perak. Thus originated the plantations in Malaya and East Asia. But further demands on the hevea tree were made by the bicycle and later the motor car. This led to a search for synthetic rubber. In 1860 Greville Williams isolated isoprene, and in 1879 Bouchardat showed that a rubbery substance could be produced from isoprene by strong hydrochloric acid. In 1884 Sir William Tilden made isoprene from turpentine. Elastomeric engineering was on the way. The Germans were to concentrate on the preparation of methyl rubber substitute during the first world war, but its poor physical properties led to the production of the buna rubbers from coal and limestone and ultimately from petroleum. The second world war led to the bonding of rubber to metal—notably on tank wheels.

Another shortage of the 1860's was that of antlers, tusks, horns, and animal bone. Thanks to the Englishman Parkes, and the American John Wesley Hyatt, the British Xylonite Company and the Albany Dental Manufacturing Company were able to produce substitutes for ivory, tortoiseshell and mother of pearl. H. G. Wells, in company with many of the science students in the 1880's, wore celluloid collars. By 1890 A. Spitteler in Germany accidentally mixed sour milk and formaldehyde precipitating casein: the second plastic to be discovered. A third came from America where Dr. Baekeland in 1909 combined formaldehyde and phenol under controlled conditions of heat and pressure to provide phenolformaldehyde or bakelite.

A fourth shortage was of explosives for mining and engineering. From 1832, when Braconnot showed how an explosive substance could be created by the action of nitric acid on gun cotton, improvements were continuous. Schönbein's discovery of gun cotton in 1845 and Sobrero's of nitroglycerine two years later was improved by Alfred Nobel (1833–96), evaporating a solution of gun cotton and nitroglycerine in acetone. Nobel founded in 1873 a factory at Ardeer on the Ayrshire coast to make dynamite. This became a centre, not only for explosives but also for industrial nitrocellulose products, later to be important as a basis for paint, lacquers and leather cloth. Nobel's factory produced sulphuric acid for fertilisers and batteries and was to give its name to Ardil, a synthetic fibre from peanut protein. One of the early employees at this factory was Harry McGowan, later Lord McGowan and the Chairman of Imperial Chemical Industries.

The Nobel firms traded with the Birmingham ammunition trade, most notably with George Kynoch, who had begun making percussion caps in 1852 at Birmingham and in 1861 had moved to Witton. As a cartridge-maker Kynoch soon found he had to expand into the non-ferrous metals industry and laid down rolling mills in 1888. From cartridges were developed railway fog signals, humane killers and engine starting devices. It was from work on explosives that the plastics and man-made fibres industry was to receive a great stimulus.

THE RISE OF CHEMICAL ENGINEERING

Of all the shortages the greatest was that of trained men. Von Hofmann had, it is true, brought into England the chemical techniques of Liebig, and had, by training chemists at the Royal College of Chemistry, created a demand for them in industry.

Several German chemists like Kekulé, Caro, Martius, Griess and Witt came to work with Hofmann, but unless they could obtain work in breweries (as several of them did) they returned to Germany, where trained chemists were more appreciated than in England. When Hofmann himself returned to Germany in 1864 to take up a professorship, followed by Caro in 1867, Martius in 1870, and Witt in 1879, industrialists became alarmed. Dye skills went with the departing Germans, especially since Nicholson retired in 1865, Perkin in 1874, and Greville Williams in 1877.

Luckily in the year in which Hofmann returned to Germany and Perkin retired from business, Ivan Levinstein founded a dyestuffs works near Manchester at Blackley, to make magenta and later Blackley blue. He found that the manufacture of basic dyes involved the use of industrial alcohol, which was so expensive and difficult to obtain that he extended his field of operations. So interested was he in new research that he founded and edited the *Chemical Review*, the first chemical trade journal to be published in England. In it, he voiced an increasingly felt need for a native cadre of chemical engineers.

'We do not possess,' he wrote, 'what may be termed chemical engineers.' E. K. Muspratt, a former pupil of Liebig's, a member of the heavy chemical firm that bore his name, agreed that it was 'very difficult to find a manager who has a knowledge of engineering combined with a knowledge of chemistry. Such men must be educated, and it is only now, after the Germans and French have shown the way for forty years, that we are beginning to follow in the path.'

So with the active agreement of Ludwig Mond, a Society of Chemical Industry was formed at Owens College, Manchester, on 14 December 1880. Levinstein spoke for the dye-industry, Muspratt for the older heavy chemical industry, and Mond for the newer. A resolution was passed establishing a Society of Chemical Engineers, but the final argument was that 'Industry' should be substituted for 'Engineers'. Another promoter was one of Bunsen's pupils, H. E. Roscoe (1833–1915), the Professor of Chemistry at Owens College, who with Carl Schorlemmer (1834–92), had organized the first department of organic chemistry in England and provided a special course on chemical technology.

Launched in 1881 with 360 members, the Society of Chemical Industry soon absorbed bodies like the Newcastle Chemical Society (founded in 1865), the Tyne Chemical Society, and the Faraday Club of Widnes (founded in 1875). Its first secretary was the former alkali inspector G. E. Davis (1850–1907), who gave lectures on Chemical Engineering in the Manchester Technical School in 1887 (forerunner of the Manchester College of Science and Technology), which were later published in book form. As he said, 'to produce a competent chemical engineer, the knowledge of chemistry, engineering and physics must be co-equal. . . . Chemical Engineering must not be confounded with either Applied Chemistry, or with Chemical Technology, as the three studies are distinct. Chemical Engineering runs through the whole range of manufacturing, while Applied Chemistry simply touches the fringes of it and does not deal with engineering difficulties in the slightest degree; while Chemical Technology results from the fusion of the studies of Applied Chemistry and Chemical Engineering, and becomes specialized as the history and details of certain manufactured products.'

Another pioneer was Professor H. E. Armstrong, of the Central Technical College in London, who was also teaching chemical engineering in 1885.

The 'rationalization' of the chemical industry as a result of the creation of the United Alkali Company in 1890 resulted in Ferdinand Hurter, the chemist who improved the efficiency of the Leblanc process, being made the head of a central research and analytical laboratory: the first of its kind in Britain.

MAN-MADE FIBRES

Two years after the Society of Chemical Industry was founded, Sir Joseph Wilson Swan was looking for a satisfactory filament for his electric light bulb. Since Edison, his contemporary, was finding carbonized bamboo unsatisfactory, Swan, utilizing the existing knowledge of Hyatt and others, forced nitro-cellulose, dissolved in acetic acid, through a small hole in a coagulating bath of alcohol. He and his assistants, C. H. Stearn and C. F. Topham, then treated the resultant fibres with ammonium sulphide to neutralize the inflammable elements. With that imaginative insight which characterizes the successful applied scientist, he saw that what he had worked on as a potential filament for a light bulb could be a man-made fibre, and at the Inventors' Exhibition of 1885 table mats, crocheted from artificial silk, were exhibited by his wife. These mats afforded evidence that man-made fibres were now possible.

Two chemists trained at Owens College, Manchester, under Roscoe and Schorlemmer. now came to Swan's assistance. They were C. F. Cross and E. J. Bevan, who with Charles H. Stearn and C. F. Topham, began to work on the filaments which Swan so much desired. Cross and Bevan were consultant chemists in London and Cross was interested in textile research, especially in the lustrous product obtained when cotton was mercerized with caustic soda. Applying caustic soda to wood pulp, they discovered that it swelled when it was further treated with carbon bisulphide, and found it was converted into a new compound which dissolved in water or dilute caustic soda solution to a viscous yellow solution which they called viscose. Patented in 1892, viscose was used by Stearn in 1893 as a filament for lamps. Topham tried to spin it (as he had spun nitro-cellulose for earlier filaments) but failed. The failure was due, as Topham later discovered, to the need for the viscose to 'age'.

At this stage the large firm of Courtaulds came to the assistance of Cross and Bevan and employed them to investigate a modification of the nitrocellulose process that was in vogue in Switzerland. In 1904, Courtaulds purchased the patents of Cross, Bevan, Stearn and

Topham together with their Viscose Spinning Syndicate's workshop at Kew, and began to manufacture viscose yarn in quantity at Coventry. By 1913 they owned the only firm of its kind in America. At that time Britain was providing 27 per cent of the world output of rayon—some 6,000,000 lb. The other British rayon firm, Glanzstoff of Flint, began by using the cuprammonium process but changed to viscose by arrangement with Courtaulds. During the war Courtaulds acquired its plant and patents and by 1927 were producing 80 per cent of the total British output and 70 per cent of the U.S. output: one-half of the total world production.

Contemporaneously a Frenchman, the Comte de Chardonnet, had begun in 1878 to see if he could simulate the activity of silkworms. In 1884 he made his first fibre by dissolving natural cellulose and extruding it through small holes so that it solidified into a thread. Today the main sources of cellulose are wood pulp, cotton linters, or straw, but Chardonnet began with the mulberry-leaves—the silkworm's chief food. He dissolved them into cellulose pulp, then nitro cellulose, which he dissolved in alcohol and ether. Extruding the solution through a fine hole he obtained a nitrocellulose yarn which was treated to remove the nitrate radical. By 1891 he was operating an artificial silk factory at Besançon in eastern France and soon others were set up in Germany, Hungary and Belgium. In England the process had a brief life—lasting from 1898–1900. Though dangerous and expensive it lingered abroad and the last factory working was burnt down in Brazil in 1949.

A second false start in artificial fibres was the cuprammonium patent of Despeissis, which utilized cotton linters dissolved in copper sulphate and ammonia (hence the name) before extrusion. After extrusion the fibre was treated with sulphuric acid or caustic soda to remove the copper and ammonia. This too was expensive and had only a temporary popularity when used for stockings or ties. It never exceeded more than four or five per cent of total rayon output.

A fourth type of rayon was acetate rayon. This too was patented by Cross and Bevan and exploited later by a partnership which included William H. Walker, the professor of chemical engineering at the Massachusetts Institute of Technology. This partnership took out in 1902 the first patent in the United States for the production of a man-made textile yarn. It was also used by the Dreyfus brothers in France for the manufacture of film and other articles. In the first world war it was used most noticeably for aircraft 'dope'—cellulose acetate dissolved in acetone. Their factory at Spondon in England manufactured 10 tons a day. In 1918 the Dreyfus firm became the British Cellulose and Chemical Manufacturing Company.

THE RESEARCH TEAM

In 1889 as Cross and Bevan were bearing witness to the efficacy of the chemical training they had received at Manchester under Roscoe and Schorlemmer, Dr. William M. Burton was hired by the Standard Oil Company of America, to make marketable the products of the 'sour' oil from Lime, Ohio. Burton was also a graduate of a new type of institution, the Johns Hopkins University at Baltimore. In 1909 he initiated pioneer work on 'cracking'. In 4 years the first stills were operating at the Whiting Refinery and in 7 years the company was operating hundreds of them and had also licensed other companies to use them as well. American industry was to use these research teams as incubators of new ideas. As C. F. Kettering, the director of General Motors research laboratory remarked, 'who locks his laboratory door locks out more than he locks in'. Kettering's remark was to be later underlined when a younger co-worker of C. F. Cross, working in the laboratory of the Calico Printers Association at Accrington, discovered in 1940 yet another new fibre from ethylene glycol (which motorists know as anti-freeze) and terephthalic acid: terylene. His name was G. J. Whinfield.

Research teams both in the chemical and electrical industries were to be paced and stretched by the ever-more insistent demands of the internal combustion engine, which itself became a virtual laboratory on wheels. How this happened we shall now discover.

The rise of expenditure on research is dramatically illustrated by figures from the U.S.A. There the percentage of the gross national product devoted to it has risen from less than one quarter of one per cent in 1934, to over 1·5 per cent by 1954. Expressed in terms of sheer money this represents an increase from 200 million dollars (in 1934) to 5,400 million (in 1954). By 1958 it was 7,300 million and by 1959 10,000 million. Nor is the total likely to end there. The Task Force on the Impact of Science and Technology set up by the Republican party recommended that by 1976 it should be 36,000 million dollars, as part of a programme 'consciously and purposely to expand the scientific revolution'. This was the first time that an American political party had conducted such a study and issued such a report. Applied Science is now high politics, and as such, should be the concern of every educated man or woman.

217

CHAPTER 20

The Internal Combustion Revolution to 1914

C. F. Kettering represented a new phenomenon: the automobile engineer. Bernard Shaw universalized the type in 'Enry Straker, 'motor engineer and New Man' as he called him in *Man and Superman* (1905). When the leisured, gentlemanly Octavius sincerely declares his belief in 'the dignity of labour', Straker replies 'My business is to do away with labour. You'll get more out of me and a machine than you will out of twenty labourers, and not so much to drink either.' This was an understatement: today in America all the new prime movers provide the average American citizen with the equivalent of 100 energy slaves.

In name the automobile was French as were its collaterals, garage and chauffeur. In conception and creation, however, it owed much to Germans, British and Americans, as we shall see.

THE DECLINE OF THE STEAM CARRIAGE

Walter Hancock, whom we already encountered in the previous chapter as an elastomeric engineer, invented a steam engine in which the ordinary cylinder and piston were replaced by two flexible bags made of rubberized canvas, which were alternately filled with steam. It was as if the cylinder itself collapsed instead of the piston working up and down in it. Since it was so light, he tried it on the roads. Other more orthodox Hancock steam carriages plied for him from London to Stratford, carrying 12,760 passengers in 20 weeks. There were numerous other experimenters, for Julius Griffith of Brompton made a two-cylinder engine with a boiler of horizontal tubes extended across the firebox. Both engine and boiler were cushioned by spring suspension. Unfortunately the water could not be kept in the boiler tubes. Another road coach engineer was Colonel Maceroni (1788–1846), formerly A.D.C. to Murat, who had experimented with paddle

218

wheels, rockets, armoured ships and asphalt paving. His steam coach elicited favourable comment from *The Times* on 10 October 1834, but it landed him in debt and ultimately in gaol. Of the many such vehicles being built at this time, few of them seemed as suitable as that of Goldsworthy Gurney (1793–1875) to withstand prolonged tests. The War Office allowed him to take two army officers to Bath and back, which he did, much to the delight of the Duke of Wellington.

Gurney's steam carriage weighed no more than a 15-cwt. truck does today, and it embodied some novel features. One was the braking system—a couple of fearsome iron shoes which were lowered to the ground and skidded the vehicle to a standstill. The other was that it had no boiler, but instead a number of tubes in which the steam was heated. In case any of them burst under pressure, a large hammer was carried. A few taps with this and all was mended again. It was steered by its long rod (rather like the handle of a bath chair), the passengers sat in the middle, and the engineer (or stoker-waterman) in the rear. On 28 July 1829 it ran 15 miles in 65 minutes before the occupants were stoned by ostlers, horse dealers, drivers and the like. The stoker was injured and Gurney had to dismount with his passengers and literally fight to get the coach pushed to safety in a yard. Mercifully the magistrates soon arrived, and posted some constables around the yard, or else the mob would have surely broken it to bits. Warned by this, Gurney continued his journey the following day, thinking it advisable to draw the carriage the remainder of the way by horses, and did not get steam up again till he came to Bath.

The numerous pioneers of road steam carriages were working under heavy handicaps. With no mild steel, no artificial abrasives, no standardized screws, no cutting tools, no mineral oil (for lubrication or fuel) and no gas (for power), they were forced to rely on welded wrought iron, copper, and cast iron. Yet their efforts were remarkable and a Committee of the House of Commons described Gurney's carriage which made the first long journey of a self-propelled vehicle from London to Bath on 28–29 July 1829, as 'one of the most important improvements in the means of communication ever introduced'. The material deterrents to its improvements were reinforced by even more effective social barriers; for the jealous opposition of the horse and the railway interests secured the passing of strict legislation against the use of steam vehicles on the highway. By an Act of Parliament of 1865 all road vehicles were obliged to carry a crew of two, to run at not more than four miles an hour; and to be preceded by a man with a red flag. One operator, F. Hodges, was summoned so many times that he turned his carriage into a fire engine.

Progress in road locomotion was therefore diverted to manufacturing carriages for export and the ingenuity of engineers in Britain was channelled into perfection of the tricycle.

Work on the Combustion Engine

Though many British proposals for utilizing volatile fuels had been made before 1836 (Stuart in 1794, Cecil in 1820, Brown in 1826), after 1836 more pioneer work was done in France, Germany and America. In France Beau de Rochas laid down in 1862 the fundamental principles of the efficient workings of an engine using hydrocarbons built by his fellow countryman Lenoir in 1860, when he outlined the four-stroke cycle, and stated that the engine needed maximum rapid expansion with maximum pressure at the beginning of the power stroke. These principles, subsequently and independently applied in Germany by Otto and Langen in 1866, fired a number of experiments. In America George Brayton of Boston patented in 1872 the first engine to use petroleum as a fuel and Otto followed with his silent engine in 1876.

In England Dugald Clerk began to work on a two-stroke engine which he patented in 1881. With ships and stationary power plants in mind, others, like Priestman (1888), worked on oil engines, whilst the first patents for a compression-ignition engine were made by Akroyd-Stuart in 1890, three years before Diesel's. It is not amongst these that we must look for the progenitors of the automobile but amongst the British cycle engineers where real advances were being made.

British Cycle Engineers

The Michaux family of Paris, whose pedal velocipedes with wire spokes, tubular frames, mudguards and gears were winning races in Paris, excited Englishmen. One of them, Rowley B. Turner (1840–1917) brought a French machine back to England and rode it into the works of Coventry Sewing Machine Company in November 1868, where it caused a great stir. The Coventry Sewing Machine Company, taking advantage of the slack trade in their usual product, began to manufacture velocipedes for export to France. The Franco-Prussian War of 1870 however prevented this, and so the velocipedes were launched on the home market, where they sold so well that the company changed its name to the Coventry Machinists Company and began to manufacture them in earnest. A new industry was born, and the journal *The English Mechanic*, by featuring the new machines, played a very real part in its foundation.

This is not to say that bicycles were unknown in England before 1868, for in 1818 a 'hobby horse' was being manufactured by Dennis Johnson, a coachmaker, of Long Acre, for the regency bucks—hence its name, the 'dandy horse'. Twenty years later, cranks, pedals, seats and driving rods were added by Kirkpatrick Macmillan, a blacksmith of Dumfries (1810–78), who was prosecuted and fined for 'furious driving on the roads'. In 1846 Gavin Dalziel, a cooper of Lanark still further improved the velocipede, as it was now called. The French Pierre Lallemant had added rotary cranks and sold the patent to his employer, M. Michaux. It was at this stage that the Coventry Machinists Company took over the manufacture of velocipedes on the Michaux model.

Encouraged by eminent medical specialists like B. W. Richardson, bicycle riding became a pleasurable propaedeutic for townsmen. Manufacturers like Dan Rudge of Wolverhampton and Thomas Humber of Nottingham were quick to profit by the demand which followed. In 1874 there were twenty manufacturers, and by 1879 sixty. The most successful was James Starley (1801–81), of the Coventry Machinists Company, who having devoted eleven years to making sewing machines, turned to improving the manufacture of cycles.

Starley's improvements were decisive. The step behind (for mounting), the small hind wheel, and pivot centre steering, made his *Ariel* bicycle so popular that he left the Coventry Machinists Company and began to manufacture them for himself with W. Hillman. By 1876 he had patented the *Coventry* tricycle, with a double-throw crank, chain and chain wheels. This was followed by the *Salvo* quadricycles in which a bevel gear differential unit was incorporated in the axle to prevent one rider's superior strength pulling the machine to one side. This differential gear became standard for all tricycles or quadricycles.

The techniques of cycle manufacture were important for the evolution of the motor car. Wire-spoked wheels, chain drives, ball bearings and the pneumatic tyre (first fitted to a tricycle by J. B. Dunlop in 1887) were four of the many inventions that were adapted by the early manufacturers of motor vehicles. And it was for English

221

tricycles that continental motor inventors adapted their early oil motors. In November 1885 Daimler used one of his high-speed engines (900 revolutions per minute) to drive a wooden cycle.

THE LIFTING OF RESTRICTIONS IN ENGLAND, 1896

The invention of the pneumatic rubber tyre by John Boyd Dunlop in 1888 enabled the motor car to become a feasible proposition. In 1895 Evelyn Ellis imported a 4 h.p. Panhard from France and in the following year the speed and crew restrictions were lifted by Parliament. It raised the speed limit to 14 miles per hour, subject to the authority of the Local Government Board, which promptly reduced it to 12 miles per hour. To celebrate emancipation the motor car users organized a 'run' from London to Brighton, which is still annually celebrated.

Coventry, seat of the cycle industry, became the seat of the car industry when the manufacturer of Daimler cars began there in 1896. Cycle manufacturers like Humber and Rover turned over to car manufacture. One of those largely responsible for the lifting of restrictions on road transport was Sir David Solomons. As early as 1874 he had built an electrically propelled tricycle. As one of the promoters of the Self-Propelled Traffic Association in 1895, he organized the first motor-car exhibition at Tunbridge Wells in October of the same year. Having tasted blood, the Self-Propelled Traffic Association conducted further campaigns and secured the passage of the Motor Car Act of 1903. This recognized the 'motor car' as something quite different from a light locomotive, allowed it to travel (subject to the restrictive power of the local government board) at speeds up to 20 miles per hour, and laid down licensing regulations. The maximum weight of a motor car could not be increased from 3 to 5 tons.

The foundation of the English Daimler Company at Coventry in 1896 followed the first lifting of restrictions and was accompanied by the registration of Joseph Lucas as a public Company (it had been founded 21 years before). The second, or Motor Car Act of 1903, encouraged the foundation of many other firms. To satisfy the demands of bus companies like Thomas Tilling and Co., which began to operate in 1903, firms like Leylands, Crossleys, Maudslay and Morcoms soon rose. Wolseley's moved from making sheep-shearing machines to cars. Rolls-Royce was launched as a public company in 1906. In 1911 the Ford Motor Company began to produce the model T at Old Trafford. Singer, Rover and Morris began to emerge as noted manufacturers. By 1914, 193 different makes of car had been introduced. Some of them commemorate the early pioneers: the Earl of Shrewsbury and Talbot, Sir F. D. Dixon-Hartland of the Rover Company, the Hon. C. S. Rolls, H. T. Lock-King (who laid out Brooklands in 1907), Col. A. F. Mulliner (who built the first motor-car bodies in England), J. I. Thorneycroft (of the famous marine engineering works), and Montague S. Napier (who built the first car of repute in England).

Napier's offers a good example of the way in which a firm which since 1805 had been making cranes, bullet-making, coin weighing and hydraulic machinery, printing presses, peat compressors and telescopes, could be adapted to motor engineering. Montague Stanley Napier, a racing cyclist, was persuaded by another cyclist, S. F. Edge, to transform the tiller-steered Panhard into a competition winning motor car. Edge's mechanic, C. S. Rolls, entered the Paris–Toulouse race in 1900, and Edge himself won the Paris–Innsbruck race in 1902 in a 40 h.p. Napier to put British cars on the map. Napier's were pioneer users of large aluminium castings, wire wheels and six-cylinder engines. In 1902 they built the first goods lorry—a 3-tonner. Edge was bought out in 1912, and when the war came Napier's began to manufacture aeroplane engines in which after 1924 they decided to specialise. In 1931 there was a scheme to re-enter the field with W. O. Bentley. Abortive as this proved, nevertheless racing motorists like John Cobb adapted the Napier Aero engine for speed records.

THE MOTOR ENGINEER

The cyclists also gave birth to the first professional association of engineers. In January 1898, 29 engineers engaged in cycle production, at a meeting at Birmingham decided to form an association known as the Cycle Engineers Institute. The first paper read to it was a comparison of chain and toothed wheel driving gears. Several of the members were already members of the Institute of Mechanical Engineers. To raise the status of their fellow engineers, they based the rules of the new on those of the older institute and published *Proceedings* from 1899–1906. By 1904 there were over 17,000 motor vehicles on the road, 8,465 of which were private cars. So, in the same year the title was changed to the Automobile and Cycle Engineers Institute.

The first president under the new title was Herbert Austin of the Wolseley Company. Soon after it ceased to be a provincial association, when a London section was formed under Lieut.-Colonel Crompton in 1906 with headquarters at the Motor Union at Caxton House, Westminster, where Rees Jeffreys was installed as secretary.

This London section developed into the Institution of Automobile Engineers, which in 1910 obtained independent offices at 13 Queen Anne's Gate. Familiar figures appeared on the first council: Herbert Austin, Dr. Hele-Shaw, F. W. Lanchester and Victor Riley. Their discussions, recorded in their *Proceedings* from 1906 onwards, exhibit a lively preoccupation with the basic problems of automobile technology and even with works organization. The assembly line was discussed by Percy Martin's paper, delivered in 1907 on the 'Organization of a Motor Works'. This, as we shall see in Chapter 23, was to be revolutionized by Henry Ford.

The first to provide cars for the masses and not the leisured classes was Ransom E. Olds at Detroit in 1899. When the Oldsmobile factory in Detroit, the first especially designed for the manufacture of automobiles, was burnt down, R. E. Olds approached a cycle-part manufacturer, H. M. Leland, for engines. Leland, a product of the Federal Arsenal at Springfield and the Colt arms factories in New England, and a great precision engineer, became so interested in cars that in 1904 he became General Manager of the Cadillac Motor Company. Four years later a Cadillac won the Dewar Trophy of the Royal Automobile Club of Great Britain, setting new standards of engineering for the new American motor trade, especially after C. F. Kettering took over responsibility for their research work, perfecting a self-starter in 1911 that enabled women to drive cars with ease.

C. F. Kettering in America and F. W. Lanchester can be called the first of the mechanical engineers of the new century. Both conducted work that included all aspects of the application of internal combustion engines to roads and air. To manufacture Kettering's self-starter a company known as the Dayton Engineering Laboratories Company (DELCO) was established which during the war did valuable metallurgical research for the Allies.

The outstanding British motor engineer was F. W. Lanchester, whose development of the epicyclic change speed gear, pre-selection, wire wheels, electric ignition, direct driven top gear and worm transmission to the rear axle was decisive. Lanchester even designed machines to cut the worm gears. Going further, he laid down the

Otto and Langen's atmospheric gas engine 1866.

The 500 h.p. experimental gas turbine of the Hon. C. A. Parsons (1854–1931). This principle was to revolutionise the propulsion of ships and the generation of electricity.

true principles upon which the flight of an aeroplane depends. From his first paper in 1894 (which was never published) he sustained the vortex theory of sustentation, made explicit in his *Aerial Flight* (1907), and adumbrated by further original work, ideas fundamental to the later development of the science of flight and stability.

THE IMPROVEMENT OF ROADS

Cycles and cars need roads and both groups campaigned to make travelling easier and less uncomfortable. The Cyclists' Touring Club (founded in 1878) provided funds to form the Roads Improvement Association in 1886 which, with numerous pamphlets pressed improved techniques on the County Councils set up in 1888. But Parliament, which had been content to allow the Turnpike Trusts to become extinct, had an aversion to meddling any further with the road system. Singularly enough, almost exactly the same action was being taken by cycle manufacturers in the United States of America, where Colonel Albert Pope, of the Hartford and Columbia bicycle firm, organized a petition for better roads. Indeed Pope was so absorbed in road engineering that he secured the introduction of lectures on the subject at the Massachusetts Institute of Technology.

The Road Improvement Association was strengthened by the adhesion of the Automobile Association (formed in 1897), the Motor Union (formed in 1903) and the Motor Van and Wagon Users Association (formed in 1905) to form a powerful pressure group in the House of Commons to call attention to grievances. These were literally centred round dust raised by motorists who showered roadside market gardens and orchards till *The Lancet* in July 1907 referred to it as 'this great modern plague'. There were few roads like the 100-mile stretch from London to Bath, which were watered every day from end to end with pumps set up every two miles. Dust-laying competitions were organized by the Road Improvement Association —one in Berkshire in May 1907 was widely attended. It was even suggested that a pipe should be laid from Brighton to London to moisten the streets with sea water. The French Government, true to the Tresaguet tradition, took the lead in organizing the first International Road Congress to discuss the adaptation of roads to modern conditions. This, held in Paris from 11–18 October 1908, was the first of eight that were held at four-yearly intervals.

As a result of the agitation the Development and Road Improvements Fund Act of 1909 established the Road Board, the first national authority for roads in England and Wales since the days of the Romans. This had powers to make grants and loans to local authorities for the improvements of roads. A petrol tax of 3d. a gallon and a motor licence duty provided the resources for this. A year later the Road Board appointed an advisory engineering committee.

It was an apt recognition of this laborious work that Rees Jeffreys, a leading figure in the agitation of the Road Improvement Association, should be appointed secretary of the Board and Colonel R. E. Crompton consulting engineer. Crompton, whom we have already met as an electrical engineer, had been Superintendent of the Government Steam Train Department in India, where he conducted experiments to test the possibilities of road as opposed to rail transport. Whilst in India he secured a road steamer made by R. W. Thomson (a pioneer of both solid and pneumatic rubber tyres) which was so successful that he was sent to England to superintend the construction of several more Thomson tractors. After winning a name as an electrical engineer, he had become a founder member of the Automobile Association.

In spite of Crompton and Jeffreys the Board itself was hampered by being virtually a sub-department of the Treasury. Yet some progress was made. The National Physical Laboratory (founded in 1902) at Teddington agreed to build, staff and operate a road laboratory to be sustained by the Road Improvement Fund, and Colonel Crompton designed a testing machine.

In 1913 Henry Maybury (1864–1943) became Chief Engineer with J. S. Killick as his assistant. Tar-spraying was generalized, road stone tested, and in the same year the British Standards Committee adopted a standard specification for road stones. Lord Cowdray formed a special company—Highways Construction Ltd.—which introduced Mexican bitumen.

The advent of the first world war gave the Road Board a great deal of work in camps and runways and increased the number of engineers with experience of asphalt and concrete construction. Henry Maybury was at first engaged on such work in England, then after 1916 in France. Over £5,000,000 was so spent.

F. R. Simms, who imported the first Daimler car into England and toured the English countryside inviting prosecution for infraction of the Act of 1865, took the first steps to form both the R.A.C. in 1897 and the Society of Motor Manufacturers and Traders in 1902. Simms, together with W. J. Leonard, a chemist, coined the name 'petrol' for the first refined product of petroleum for use in internal combustion engines.

The need for petroleum revolutionized the oil industry. Since 1859, when the first well was drilled at Titusville, Pennsylvania, kerosene (or paraffin) was the most important product for lamps, to replace whale and shale oil. The heavy oils resulting were used as lubricants and the gasoline (or petroleum as Leonard and Simms called it) was a drug on the market. The increasing number of internal combustion engines now transformed it into a primary product. Nearly half a million wells were completed in the U.S.A. up to 1917.

More attention was given to the cracking of other products of the oil to produce gasoline. In 1907 L. Edeleanu (1861–1941), a Rumanian, developed a process whereby extracts (low illuminants) and raffinates (high illuminants) were produced by mixing with liquid sulphur dioxide. Employed to treat gasoline, it was the beginning of a process of solvent extraction.

The increasing use of petroleum had, by 1914, become a matter of national concern. On 3 March 1914 the Institution of Petroleum Technologists was inaugurated in the Royal Society of Arts. Sir Boverton Redwood, its first chairman, remarked that its purpose was not to offer a new channel for the dissemination of knowledge relating to petroleum, or a 'special arena for the contests of mental gladiators over controversial questions' but to establish a hallmark of efficiency in connection with the profession. For the great development of the petroleum industry, and especially the multiplication of centres of activity, had made it increasingly difficult to procure the services of properly trained geologists, engineers and chemists, with the result that positions of responsibility were being given either to incompetent Britishers or to proficient men of other nationalities. Redwood spoke with feeling, for he had been secretary of the Petroleum Association as early as 1869, and was the author of the standard book on the subject. It was this body which was to nourish more chemical engineers in the generation to come.

THE INTERNAL COMBUSTION ENGINE TAKES TO THE AIR

German and American experimenters enabled the internal combustion engine to take the air. Otto Lilienthal (1848–96) began gliding experiments near Berlin in 1891. His gliders, built of cotton fabric stretched on willow ribs, enabled data concerning the behaviour of such machines in air currents to be collected. Both he and

227

his British follower, P. S. Pilcher (1866–99), a lecturer in engineering at the University of Glasgow, were killed before they could provide mechanical power to their gliders. Such experiments with steam engines as were being conducted at that time were vitiated by lack of good controls. Hiram Maxim (1840–1916) came to England as chief engineer of the United States Electric Lighting Company and in 1882 opened a workshop. His name is usually associated with a rapid-firing gun with completely automatic action adopted by the British Army in 1889 and by the Navy in 1892, but his experiments in mechanical flight, described in his book *Artificial and Natural Flight*, show that he was also interested in helicopter machines. In the same year as his gun was adopted by the army, Maxim began experimenting on the relative efficiency of aerofoils and airscrews driven by steam engines. His plane was prevented from rising more than two feet above the track. After three trials it succeeded in leaving the rails but, owing to a defect in the axles, crashed.

At the same time, there arrived in Paris the wealthy Brazilian inventor, Alberti Santos-Dumont (1873–1932). He began to experiment with automobile engines in airships, and his first ascent on 18 September 1898 was encouraging. In 1901 he won a prize of 125,000 francs for flying from St. Cloud round the Eiffel Tower and back again in thirty minutes. He subsequently flew from St. Cloud to his own house in the Champs Elysées. Santos-Dumont envisaged his airships would be able to observe submarines below the surface of the water and drop 'dynamite arrows'. He later worked on heavier-than-air machines, and in 1906 was the first in Europe to make a public flight in such a machine.

It was by two bicycle-makers, utilizing the scientific work of pioneers from Cayley to Langley, that the internal combustion engine was enabled to transcend the roads and take to the air. The Wright Brothers of Dayton, Ohio, had conducted a number of gliding experiments, and on 17 December 1903 made their first flight in an engine-powered glider at Kitty Hawk. After experimenting for four years with a wind tunnel and over 200 models, they solved the principle of simultaneous control of altitude in three axes of space. Following their successful flights, a number of similar biplanes were built, and the development of the internal combustion engine gave a fresh impetus to aircraft construction.

Engineers like J. C. H. Ellehammer, and motoring enthusiasts like Alberti Santos-Dumont pioneered in free flight in Europe. The British Army took an increasing interest in the new medium and S. F.

Cody, an American experimenter, was attached to the Balloon Section of the Royal Engineers for experimental purposes. The crossing of the channel by Blériot in 1909 and the introduction of flying meetings gave further fillips to the new machines.

Well-known names now began to emerge. Hugo Junkers (1859–1935) took out a patent in 1910 for an all-wing aeroplane. Horace and Eustace Short built an aeroplane factory at Shellbeach. Their planes, of Wright design, were flown by other motor engineers like the Hon. C. S. Rolls. A. V. Roe developed a biplane of his own design and later developed a tractor-type, whilst Edouard Nieuport in France began to make monoplanes. The military use of the airborne internal combustion engine was appreciated by the Italians in attacking Arab formations in Tripoli in 1911. In 1912, at the British Army manœuvres, the Royal Flying Corps provided two squadrons for reconnaissance purposes. In 1913, using a Nieuport machine, the Russian officer Nesterov first 'looped the loop' and initiated the aerobatics that were to be a prominent feature of the war to come.

THE REVOLUTION IN TRANSPORT

So, by 1914, man had broken through the 'oat barrier' and was emancipated from the horse. The car was no longer solely a laboratory curiosity nor an aristocrat's plaything. In the United Kingdom the total number of motor vehicles (including motor cycles) rose from 17,810 in 1904 to 265,182.

Motor-bus services grew slowly from the appearance of the first in 1898. London set the pace and the provinces followed. In 1904 there were little over 5,000. By 1914 there were 51,167. Many of them were just being developed as the Great War broke out and were immediately commandeered for military use. Before looking at that war it would be perhaps as well to see what effects the developments outlined in the previous four chapters had on Britain before it began.

CHAPTER 21

Social Adjustment in Britain, 1867-1914

INTIMATIONS OF MORTALITY

On 18 May 1866 Overend and Gurney, the great English banking house, closed its doors, Otto and Langen patented a 'free piston' gas engine, and Hoch was working on the first practical petroleum engine. The following year saw the practical self-exciting dynamo of Varley, Siemens and Wheatstone take shape and the first famous Babcock and Wilcox water-tube boiler: both of infinite portent when considered together. But no portents were as obvious as those revealed at the International Exhibition, held in Paris in 1867. There A. S. Hewitt was so impressed by the Martin open-hearth process that he returned to Trenton, New Jersey, to build the first open-hearth furnace in the United States of America. But to British observers the Exhibition was a shock. Lyon Playfair (1818–1898), one of the ablest chemists of his own generation and the next, left a comfortable teaching laboratory to point the painful moral that this country could now no longer rely on empiric skill and illiterate workmen to hold her own, much less excel her competitors.

Compared to the Germans, the British were slow in applying chemical techniques to other sciences which were still virtually empirical and mainly observational. Liebig, writing to Faraday of British geologists at the British Association in 1866, remarked 'in most of them, even the greatest, I found only an empiric knowledge of stones and rocks, of some petrifaction and a few plants, but no science. Without a thorough knowledge of physics and chemistry, even without mineralogy, a man may be a great geologist in England.' Edward Frankland, discoverer of the principle of valency, sorrowfully communicated the fact to *Nature* of 6 April 1871 that of the 97 chemists communicating papers to research journals from the United Kingdom, 'a considerable proportion were the work of chemists born and educated in Germany'. Perkin was even more explicit. The industry which he had founded, which depended on England's greatest asset, was by 1879 dominated by Germany. England had five colour works producing £450,000 worth of colours to Germany's seventeen producing £2,000,000 worth.

The British Government had appointed in 1870 a Royal Commission under the Duke of Devonshire to examine the whole question of technical instruction. In the same year the Cavendish Laboratory was established at Cambridge. Manufacturers began to see that money spent in their workmen's education was a legitimate widening of the field of investment: Joseph Whitworth's example in giving thirty scholarships for engineering students, though widely praised, was not followed as it should have been. The newly organized trades union congress agreed that improved educational facilities might afford a path to social peace, but they were thinking in terms of the new wage arbitration boards then being set up.

Technical competence began to tell as never before, since the élites had to manage and advise in highly specialized industries and efficient bureaucracies.

The Government, the municipalities and the joint stock companies needed specially trained technical staffs. Somewhere this staff had to be found. To provide it, new colleges were founded in the large towns, usually at the instigation and expense of local industrialists.

THE CIVIC UNIVERSITIES

The efflorescence (the architectural style justifies the image) of the new university colleges began in the industrial towns. They grew from, or were grafted to, the local mechanics' institutes, medical schools or literary and philosophic institutes. After 1870 their growth was accelerated by municipal pride and industrial munificence. Each college mirrored the techniques of its town. Newcastle Royal College of Science (1871) was specifically planted in the mining industry. Yorkshire College, Leeds (1874), grew in textile country and University College, Liverpool (1880), in a commercial setting. Both of them later emerged to full chartered status after a short period of affiliation to the University of Manchester. Firth College, Sheffield (1879), grew from the first people's college in Great Britain. The Mason Science College, Birmingham, was so practical that theology and mere literary study was excluded from its curriculum and even when they were admitted and Birmingham became a university in 1900, commerce was made a faculty and brewing set up as a department. The financial help of South Kensington science grants and the protracted stimulus of the University of London examinations helped other university colleges at Nottingham (1881), Reading (1892) and Exeter (1893) to emerge. These colleges did much to foster the teaching of science and engineering, and strangely enough were to cultivate an appetite for trained personnel in industry.

Yet private munificence could not, and did not, do much more. Whilst these efforts were going forward the eighty-nine city companies of London had drawn the limelight upon themselves by their extensive ownership of Irish land. A notion was current that their responsibility was for 'technical education', and one of the sheriffs of the City of London not only categorically said so, but added that they needed 'a little gentle persuasion' to make them discharge it. This 'little gentle persuasion' was applied by scientists like T. H. Huxley and radical politicians, and the City Companies began to assist the technical education which the times demanded. The Clothworkers helped the textile industry with endowments for Yorkshire College, Leeds and also tried to form a City and Guilds Industrial University in London. One of their members, Sir Sidney Waterlow, was aware of the rapid continental advances, having received reports from two commissioners especially sent for that purpose. With other companies, a committee was formed that took over the technological examinations which Donnelly had established in 1872 for the Society of Arts. By 1880 this committee was incorporated as the City and Guilds of London Institute for the Advancement of Technical Education.

The City and Guilds of London Institute founded the Finsbury Technical College in 1881, where W. E. Ayrton established a tradition of electrical engineering and H. E. Armstrong began to make his name as a teacher of chemistry.

THE NEED FOR CENTRAL STIMULUS

Three major industries suffered at this time from the lack of any central stimulus and encouragement. Road steam traction was penalized by the Red Flag Act of 1865. Steel production was hampered by the phosphorous content of the Cleveland ores and inadequate training of operatives. But both these, as will be seen later, were to some extent overcome. But a classic case of an industry which suffered from hesitant central stimulus was the telephone. Patented in 1876 by Alexander Graham Bell, an American citizen in Edinburgh, it was eagerly taken up a year later in Germany by von Stephan, who had just founded the Postal Union. Switzerland, too, welcomed it, and so did the Scandinavian countries. But in Britain, in spite of the fact that the post office urged that they should be nationalized, the Government allowed private companies to operate telephones in return for a royalty of 10 per cent on gross receipts. No competition developed, and by 1891 the National Telephone Company had acquired a monopoly. The Government then urged both the post office and the municipalities to set up separate services. This, too, was a failure: only seven licences were taken up and all but two were surrendered within a few years. The National Telephone Company was an indifferent provider, for, insecure in its tenure and coping with way-leaves, it had little confidence in its future. Not till 1912 did the Government decide to transfer the system to the post office. The result of this thirty-five years' paralysis was that the United Kingdom's ratio of telephones per head of population remained well below the European and American averages.

THE RISE OF NEW INDUSTRIAL STATES: JAPAN

This central stimulus was visible in several of the countries where an industrial revolution was beginning, like Japan, where, in 1868, the feudal system was abolished. This emancipated a labour force for industry and enabled the first cotton factory to be established by a Japanese landlord. Within twenty years the cotton industry was turning half a million spindles. The modified state capitalism which fostered this industrial revolution led in 1900 to the State establishing the first modern ironworks in Japan. A Japanese delegation visited England in 1872 and returned to recommend the establishment of an engineering college at Tokyo, which appointed as its principal Henry Dyer, a Whitworth scholar and a pupil of Rankine, and as one of the professors W. E. Ayrton. It was Norman Lockyer (1836–1920), who saw the significance of the establishment by the Japanese in 1872 of an Engineering College at Tokyo, where W. E. Ayrton established the first laboratory in the world for the teaching of applied electricity. As secretary of the Devonshire Commission Lockyer had a further opportunity of making his views heard. He knew that W. E. Ayrton was a pupil of Kelvin's and had worked in the Indian Government Telegraph Service. He saw that the Japanese technological university was becoming the largest institution of its kind in the world, and in *Nature* on 17 May 1877 he remarked: 'It is somewhat singular that this country, foremost as it has always been in matters of engineering enterprise, should be so behindhand in the systematic education of its engineers, there being no establishment in England devoted to that object which is recognised by the profession.' After commending Putney College (then defunct), Lockyer remarked: 'At the present time, with the exception of the technical classes at the Crystal Palace and at King's College, which, in a small way are doing good work, there is no institution in this country devoted to the education of engineers.'

THE TRAINING OF ENGINEERS

Lockyer's concern for training engineers to serve the new electrical and steel economy was shared by others. One eloquent advocate for training a cadre of engineers was John Scott Russell who, in 1869, had pointed out the virtues of the French system whereby the élite of the nation were selected for the *corps de génie maritime*, the *corps de génie militaire*, the *corps de génie civil*, or the *ponts et chaussées*, 'It is plain,' Scott Russell concluded, 'that we may judge of the wisdom of a nation by the foresight and forethought it bestows upon the rearing, training, and selection of this *corps d'élite* or *corps de génie*.' So he proposed the establishment of some fifteen technical colleges in the industrial towns.

The rapidly expanding technical press took up the cry. The *English Mechanic*, *Nature*, *Engineering* and the *Engineer* argued along the same lines. Lockyer, as the editor of *Nature*, was supported by W. H. Maw (1838–1926), the editor of *Engineering*. Maw (who had worked with Zerah Colburn on a book on the waterworks of London) had been founder of the Civil and Mechanical Engineers' Society in 1859. In 1870 with James Dredge he became editor of *Engineering*. Himself a consultant engineer of no mean standing, he gave himself to building up the engineering profession. Like William Pole, he arranged and collected data for the Institution of Civil Engineers on the education and status of civil engineers in the United Kingdom and foreign countries, he never lost an opportunity to press the case for engineers. He was to help, in later years, in the formation of the British Engineering Standards Association and was to serve on the Joint Board of the National Physical Laboratory.

The Institution of Civil Engineers made an inquiry in 1870 into the education and status of civil engineers in the United Kingdom and foreign countries, and found that there were only seven centres in Britain where an engineering education could be obtained, and that at five of these there was but one teacher and no laboratory.

So steps were taken to establish an engineering college at Cooper's Hill for cadets proceeding to India. The existing institutions, quite naturally, opposed this move, both in the press and by deputation to the India Office, arguing that the new engineering college, because it was protected from competition, would be prejudicial to the public service and narrow the field of recruitment. In March 1871 the House of Commons ventilated the whole question and the government went ahead. The managers of the Hartley Institution of Southampton decided to begin training engineering students for the Cooper's Hill College, and thus set a tradition which is still being fulfilled today in what is now the University of Southampton.

A second project was to begin engineering classes at the Crystal Palace at Sydenham, where a number of engineering models of great variety had been collected.

It was left to the private munificence of Bessemer, Whitworth and Charles Manby (the Secretary of the Civil Engineers) to initiate a third venture: a college of practical engineering at Muswell Hill, near London, which opened in September 1881, under John Bourne as principal.

THE GROWING CONCERN OF THE GOVERNMENT

Since 1859 the Science and Art Department had been making grants to every school or college which could satisfy its conditions. Slowly a complicated system of grants to teachers, managers and students in the British Isles had been built up. The intricate ramifications of these grants were exposed by the ten-part report of the Devonshire Commission on Scientific Instruction (1870–5).

The moving spirit (but not the head) of this government department was an army officer, Captain (later Major-General) Donnelly, a close friend of Huxley.

In 1871 he persuaded the Royal Society of Arts to establish the first technological examinations in the country, and three years later he became 'Director of Science' at South Kensington. His title, however, was more nominal than effective, because he was opposed by the very manufacturers who should have profited by his activity. Fearing that secret processes would be revealed, they refused to allow their workmen to discuss secrets of their own manufacturing process, and instead of assisting the scheme, they told him they would do everything they could to stop it. In vain Donnelly travelled through the industrial north and east to persuade them. He never secured any co-operation from them, and he complained bitterly to a subsequent Royal Commission on Technical Education, which sat from 1882–4 under Bernhard Samuelson, that fear of the revelation of trade secrets was the real stumbling block to any effective system of technical education. Yet in spite of such difficulties he earned Huxley's approval: 'The Science and Art Department', wrote Huxley in 1890, 'has done more during the last quarter of a century for the teaching of elementary science among the masses of the people than any other organization which exists either in this or in any other country. It has become veritably a people's university as far as physical science is concerned. . . . The University Extension Movement shows that our older learned corporations have discovered the propriety of following suit.'

Donnelly's problem lay in the wilderness of sheds and temporary buildings at South Kensington. Opposed by manufacturers outside, he also faced the opposition here of the mining enthusiasts who did not like the idea of establishing a college of science that might detract from the Royal School of Mines. In 1880 it was decided to accelerate the concentration of all existing science schools at South Kensington, where a school devoted to all branches of science applicable to industry could be formed. Also planned was a school to train science teachers. The mining enthusiasts were to be placated by retaining the title of 'the Royal School of Mines' to cover the final year at Jermyn Street.

The Treasury were not convinced that any serious increase of expenditure was necessary. Lord Spencer was very bitter about this and said later, 'In 1881–2 we were continually applying to the Treasury for the completion of the buildings (at South Kensington). We were always put off by excuses. They always said, "Well, we have got to build a new War Office or a new Admiralty," and there was always one excuse after another, and I fancy those excuses and those difficulties have not yet been overcome.'

Yet something was done. A Normal School of Science was established in October 1881 under Huxley. Officially it was a single organization, organized as a Royal School of Mines and a Normal School of Science. But, within the framework of each division, a vigorous growth began which led to the Royal College of Science and, in turn, to the Imperial College of Science and Technology in 1907.

THE SPIRIT OF EMULATION FOSTERED

As a manufacturer with interests in Britain and Germany, Mundella had repeatedly called attention to the superiority of the continent in matters of technical training. H. M. Felkin, his business representative in Chemnitz, made the first English translation of Herbart, and also published *Technical Education in a Saxon Town* in 1881. This description of efforts of the citizens of Chemnitz posed the question as to whether the citizens of England would be willing to make similar sacrifices. The head of the newly formed City and Guilds Institute was impressed, and so was Mundella.

So a Royal Commission under the ironmaster Bernhard Samuelson was appointed in 1882 to survey the whole panorama of foreign technical instruction and compare it with English efforts. As a leading chemist, Professor Henry Roscoe examined the research of foreign universities, Sir William Mather examined American practice and its bearing on industry, Philip Magnus visited and reported on the foreign schools and Swire Smith, a Keighley woollen manufacturer, reported on foreign methods of manufacture and the conditions under which they were conducted. Other commissioners had special briefs. The commissioners issued their final report in 1884. 'The best preparation for technical study', they reported, 'is a good modern secondary school . . . unfortunately our middle classes are at a very great disadvantage compared with those of the Continent for want of a sufficient number of such schools'. The commissioners 'looked for some public measure to supply this the greatest defect of our educational system'. That measure might have been taken had not the gigantic red herring of Ireland been drawn across the path the following year. The Commission also raised the pregnant issue of decentralization. They stressed the desirability of empowering 'important local bodies like the proposed County Boards and municipal corporations to originate and support secondary and technical schools in conformity with the public opinion for the time being of their constituents'. Here was the embryo of a new system.

As these proposals were being digested, another Royal Commission was appointed to examine the causes of the depression in trade. This Commission reported in 1886, and for the first time it was officially admitted that the industrial techniques of America and Germany were surpassing those of Britain. Voluntary efforts to remedy this state of affairs redoubled. In the large towns, industrialists (like F. T. Mappin, the railway spring manufacturer of Sheffield) gave generously to the nascent local colleges to enable them to develop technological studies.

THE BELL TOLLS AGAIN, 1887

The year of jubilee (1887) prompted T. H. Huxley to sound a serious note: 'We are entering', he wrote to *The Times* on 21 March 1887, 'indeed we have already entered, upon the most serious struggle for existence to which this country was ever committed. The latter years of the century promise to see us in an industrial war of far more serious import than the military wars of its opening years.' He indicated the menace both from the east (Germany) and from the west (America), and urged the organization of victory in this 'grave situation'.

Movement towards the certification of engineers was slow. Professor Osborn Reynolds had told the Manchester Literary and Philosophical Society in October 1876 that in his opinion representative bodies of engineers should constitute themselves into examining bodies like the College of Physicians and only admit such fresh members as could satisfy them. He added that the degrees and certificates conferred by the various colleges were not yet a sufficient recommendation. Reynolds' words fell on deaf ears.

Not until 1897 did the Institution of Civil Engineers set up its own examinations. Too many changes were needed in the educational system before a good basic training could be given to intending engineers and it is to these changes that we must now turn. Before doing so, it is worth recording the foundation of an Institute of Marine Engineers in 1889: a fitting pendant to the work of John Scott Russell, whose initial proposals evoked the foregoing remarks.

THE IRON AND STEEL INSTITUTE, 1869

It was the iron and steel trade that felt the complex impact of foreign advance, faced the prospect of coal exhaustion, feared the intensifying influence of chemistry and was puzzled by the increasing importance of electricity. The growing sophistication of mining techniques prompted the formation of the Iron and Steel Institute to diffuse technical information among members of the trade. Bernhard Samuelson, the great ironmaster of Middlesbrough, John Ramsbottom, the railway engineer, and the steel magnates like

Henry Bessemer, Edward Vickers and Mark Firth all subscribed and agreed.

One of its first collective actions was to set up a committee on mechanical puddling, and in 1871 it sent J. Snelus, the Dowlais works chemist, to Cincinnati to investigate a mechanical puddler. The president at both the provisional committee and the first general meeting of this body was Isaac Lowthian Bell (1816–1904) who had built up an iron works at Cleveland and a chemical works at Washington, near Gateshead. Bell represents the new type of industrialist. Trained in Germany, Denmark, Edinburgh, Paris, he remained all his life a student of engineering problems. In 1872 he worked out the theory of the blast furnace. His steel making was hampered by the phosphoric content of the Cleveland ores and he maintained that a larger 'blow' in the Bessemer convector would eliminate it. But it was left to a part-time student of a Science and Art Department class, Gilchrist Thomas, and his cousin, a South Wales iron-works chemist, to track down the real cause—the silica in the converter lining, which would not unite with the phosphorus. Thomas' remedy—perfected with the help of E. W. Richards, manager of the Bocklow-Vaughan works at Middlesbrough—was to add lime with the charge to absorb the phosphorus. Its first important application was in the basic open-hearth works at Frodingham, established by Maximilian Mannaberg.

The Iron and Steel Institute was the first real institutional response of the industry to the need for pure science. Thirteen years after it was founded, Sir William Siemens told a Royal Commission that 'a taste for science has been awakened among employers' and 'men who formerly ridiculed the idea of chemical analysis now speak of fractional percentages of phosphorus and sulphur with great respect'. The Institute pressed upon the 1851 Commissioners the advisability of applying its funds to give money to provincial science colleges and kindred organizations in preference to spending it on metropolitan bodies. Indeed, it took a leading part in establishing a provincial science college at Newcastle which subsequently became King's College in the University of Durham.

THE MINING ASSOCIATIONS, 1887–1889

The mining industry, heavily traditional, responded to the public need for training and certification. The National Association of Colliery Managers was founded in 1887, was a direct result of the

Coal Mines Regulation Act of that year which laid down that every coal mine employing more than thirty men was to be under the daily personal supervision of a certificated colliery manager.

The need for applying scientific techniques to the industry also led Lowthian Bell who, eighteen years earlier, had played a leading rôle in establishing the iron and steel institute, to work for the federation of the myriad professional groups concerned with the mining industry. Some of these groups (the South Staffordshire and East Worcestershire Institute of Mining Engineers, the North of England Institute of Mining and Mechanical Engineers (formed in 1851) and the Chesterfield and Midland County Institution of Engineers (formed 1871)) had long wished to do so, and had been discussing federation as early as 1868. Not until 1888 did their representatives, under Lowthian Bell's chairmanship, agree to establish the Federated Institution of Mining Engineers, for the advancement of coal and iron mining and the promotion of knowledge necessary for its effective prosecution. This was followed in 1892 by a third professional group specializing in minerals other than coal and metals other than iron: the Institution of Mining and Metallurgy. Both bodies were but a belated response to the technological improvements in mining that had begun taking place.

ASSOCIATION FOR THE PROMOTION OF TECHNICAL EDUCATION, 1887

The notion of an industrial army caught the public imagination, and the response to Huxley's appeal showed how little the parties were divided on this question.

Forty M.P.'s met in the Society of Arts (founded in 1754 and home of many earlier efforts to promote industrial education) together with delegates from school boards, trade unions, chambers of commerce, and other interested parties, to form a National Association for the Promotion of Technical Education. Familiar names appear among its officers: Sir Bernhard Samuelson, A. J. Mundella and Huxley himself. The secretaries Sir Henry Roscoe and A. H. D. Acland (the latter a young don, who had organized Oxford's first University Extension classes), supported by manufacturers like Sir William Mather, were active in Parliament.

Luckily, new county authorities were created in 1888 when the oligarchy of the quarter sessions had to give way to councils elected by the extra one and three-quarter million county voters created by

Werner Siemens (1816–1892) in 1880. This great German electrical engineer gave land for, and helped the physicist Helmholtz to organise, the Physikalisch-Teknische Reichsanstalt at Charlottenburg.

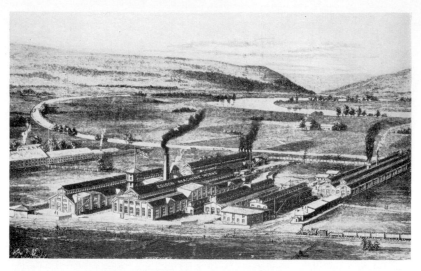

The Edison Machine works at Schenectady.

T. A. Edison (1847–1931) and C. P. Steinmetz (1865–1923) working on a problem. This was the embryo of the vast research organisation that, under E. F. W. Alexanderson was to discover, amongst other things, the electron microscope.

the Reform Bill of 1885. The idea that the county was the natural unit to undertake technical instruction was not new: increasingly it was appreciated that the devolution of returns, the institution of block grants and the donation of rate aid to voluntary and board schools alike would be facilitated. A. H. D. Acland saw in county control of secondary education a possible bridge between the board school and the university, a gap which was unknown in Prussia and Scotland.

And so in 1889 and 1890 the newly established county councils were granted powers (but unfortunately not obliged) to provide technical education. These powers were meagre in themselves: authority to levy a penny rate given in 1889 and, by a curious quirk of the Victorian conscience in 1890, a gift of part of the proceeds of a tax on whisky. This last was a political victory for A. H. D. Acland and Sir William Mather, the great Manchester electrical engineer, who was extremely active in education and politics during this decade. 'Whisky money' was a real stimulant to the construction of laboratories in technical colleges and secondary schools. Within five years 93 out of the 129 borough councils were spending the whole of their 'whisky money' on technical education. But only in 13 cases was a rate levied: a sad commentary on the permissive legislation of the day.

To supervise the expenditure of 'whisky money', their own rates and such other windfalls as came their way, the new county authorities set up Technical Education Committees. Many of these gave grants to existing grammar schools. The Association of Organizing Secretaries for Technical and Secondary Education (established in October 1891) agreed that help should be given both to technical and secondary education. One of Acland's friends, the organizing secretary to the county council of Surrey, argued that if his council were to be given a free hand over the endowed schools it would 'deal with secondary education in a much more comprehensive manner than we do at present'. The fruitful example of Wales, where in 1889 education committees (the majority of whose members were elected by the county councils) were empowered to control secondary education hitherto vested in the Charity Commission, was widely quoted by the 'Celtic fringe' then adorning so many departments of national life.

But the focus of events was to centre on London where, in the last decade of the century, the prime movers of events were to be those supple intellectual commandos, the Fabians.

THE FABIANS

Formed in 1883 and led by Sidney and Beatrice Webb, they took their name from a Roman general who wore away his opponents by attrition and persistent skirmishes rather than head-on collisions. They had no desire for precipitate or sudden action, but believed in 'the inevitability of gradualness'. Through wise administration they hoped to obtain equality of opportunity for every citizen. Their objectives, according to Beatrice Webb, were 'essentially collective ownership wherever practicable; collective regulation everywhere else; collective provision according to the need of all the impotent and sufferers; and collective taxation in proportion to wealth, especially surplus wealth'.

Administration of education, they believed, would secure some measure of equality of opportunity by means of a well-articulated educational ladder. This was Sidney Webb's aim when, on the motion of Alderman Quintin Hogg (creator of the Regent Street Polytechnic), the London County Council set up a Technical Education Board of thirty-five members and he became its chairman. In six years (1892–98) he made it the most important committee of the L.C.C. At its first meeting, thanks to his energy and industry, the fifteen co-opted members and the twenty London County Councillors worked as one body. Webb habitually wrote out the exact words of resolutions to be passed at each meeting, and personally signed every cheque to make sure that there was no danger of a surcharge. Yet with this minute industry went real largeness of vision. An extensive scholarship system was constructed, large in extent, elaborate in organization, and diversified in its ramifications, taking pupils from the lowest elementary school to the most distinguished institutions of university standing. The basis was the County Scholarship system which articulated elementary, higher grade, and endowed schools, and from them fed the higher institutions. Through this the T.E.B. (as it must be called to distinguish it from the London School Board) was brought into contact with all the schools in London, whereas the London School Board only controlled elementary and higher grade schools, with an occasional pupil-teacher centre. From the T.E.B.'s the languishing endowed schools received some much-needed help, and became aware of the existence of other members of the hierarchy. Technologically and pedagogically, London once more began to set the pace for the rest of England, and it was largely as a result of the famous *Cockerton Case*, when the

London School Board was surcharged for providing secondary education, that the county and county borough councils were in 1902 made responsible for primary, secondary and technical education within their areas.

COLLATERAL DEVELOPMENTS

A breath of the new state of things was blowing through the 'almighty walls' of some of the public schools. In 1892, the year in which Webb took over the chairmanship of the T.E.B., 35-year-old F. W. Sanderson was called in by the Grocers' Company to reorganize Oundle School. A graduate of Durham, he not only established science and engineering laboratories and workshops, but also introduced co-operative methods of study, and a wholly new approach to a synthesized curriculum. His activity impressed H. G. Wells, a Fabian who had suffered an education on the South Kensington model. Wells thought that Sanderson was 'beyond question, the greatest man I have ever known with any degree of intimacy'.

Relatively unappreciated at the time, another change was being initiated by Francis Galton. After a career of travel in Syria, Egypt and South Africa, he had returned to conduct experiments to establish the natural laws of heredity. At the famous Health Exhibition of 1884, he initiated an anthropometric laboratory, having begun his tests of ability. Five years later, in 1889, he published his book, *Natural Inheritance*. Galton may be called the father of objective mental testing, which those who believed in the collectivist idea of equality of educational opportunity were to find indispensable in their work. He also exercised an enormous influence on the development of statistical method.

AN EARLY OPERATIONALIST

As a Fabian, an admirer of Sanderson, and an imaginative exploiter of the work of Galton, Wells became increasingly critical of socialism. To him it lacked an 'analytic and experimental spirit' and languished in a state of 'exalted paralysis waiting for the world to come up to it while it marked time'. A social engineer was a far more attractive figure to Wells than a socialist, and in a steady stream of imaginative novels and books he said this many times over. He saw a vast new class emerging. which he described as 'a great inchoate mass of more or less capable people engaged more or less consciously in applying the growing body of scientific knowledge to the

243

general needs'. In *Anticipations* (1901) he prophesied that a war would enable this 'functional' class to seize power from 'the politicians' and organize themselves. The Fabians hit back. Shaw's *Man and Superman* (1902) contained in 'Enry Straker 'an intentional dramatic sketch of the contemporary embryo' class type whom Wells was limning and hymning. Stimulated and sharpened by these encounters Wells went on to sketch a Modern Utopia where engineers constituted the élite of a nation; a scientific samurai.

THE SERVICE GROUPS OF MUNICIPAL SOCIALISM

As it was realized that the socialization of public services was the price of safety, socialism, of the British variety, was nurtured in the cities. Its earliest protagonists were advocates of municipal control of technological developments in gas, water, light, air and transport, as well as education. Since only large capital investments could establish such undertakings to keep ratepayers healthy, specialized services developed around these functions and their scope may be seen in the latter part of the nineteenth century.

Habits of bathing and washing became commoner, thanks to Thomas Hawksley (1807–93) making a constant service water supply a standard feature of British towns and cities, building over 150 waterworks as well as a large number of gas works. After Downes and Blunt had established the bactericidal properties of light, urban reconstruction was accelerated, and town planners obtained another arguing point.

The 1875 Public Health Act, which set up local health boards consisting, in fact, of the town councils, with power to ensure proper sanitation, drainage and water supply, endowed the towns with yet another function. From 1835 onwards they had been empowered to supply water to householders. From 1848 they were empowered to arrange for the supply of gas. From 1870 they were also authorized to provide tramways and from 1882 and 1888 were permitted to supply electric light. All these activities required money, and from 1880 they had been allowed to float public loans in order to finance the schemes.

The specialized services of an increasingly socialized community led to the formation of five new professional service groups of engineers. The first we have already met: the Gas Engineers, founded in 1863. The second was the Institution of Municipal Engineers, founded in 1873. This aimed at the promotion of high professional competence amongst engineers and surveyors employed by public authorities, and provided a focus for such problems as town planning which were to develop as large-scale industries began to need thousands of employees near the works.

The close of the century saw three more professional groups take shape: the Institute of Sanitary Engineers (1895), the Waterworks Institute (1896) and the Institution of Heating and Ventilation Engineers (1897). The Institute of Sanitary Engineers was incorporated in 1946 and changed its name in 1953 to the Institute of Public Health Engineers. It issues a quarterly journal, conducts examinations in public health and holds meetings to discuss professional problems. It has defined a public health engineer as 'a person who is engaged as an engineer in connection with public health, and who designs, or controls, or undertakes, or advises upon constructional works or other like matter affecting the health of the community'. Its affiliation with the Federation of Sewage and Industrial Wastes Association of the United States ensures that members are kept informed of current progress.

The Waterworks Institute changed its name on incorporation in 1911 to that of the Institution of Water Engineers, which apart from its issue of an annual volume of *Transactions* (since 1947 replaced by a *Journal* issued seven times a year) and *Year Book*, made its most notable contribution by publishing a *Manual of British Water Supply Practice* (1950). Its study groups and official representatives on public bodies, have served to promulgate a code of practice recognized by the Government. Its *Manual* is the most comprehensive work on the subject to be attempted in the world.

The Institution of Heating and Ventilation Engineers, established in 1897 to discuss problems of heating, ventilating, hot-water and air conditioning, and to give 'an impulse to inventions likely to be useful to members of the Association and to the community in general', has recently expanded its scope and membership. Problems of humidity, extracting poisonous dusts from factories, and even aviation matters now come within its scope, and a national college was established in 1948 for advanced teaching and research in these problems.

THE NEW PROFESSIONALS

The application of electric power can be traced on various fields of national life as new associations emerge to cope, by exchange of information and techniques, with the problems so created. Two dealt with the technology of food: the Institute of Refrigeration (founded in 1900) and the Institute of Brewing (founded in 1903). A third, also founded in 1903, was the Faraday Society. Taking root in the new university centres, this aimed at promoting the study of the sciences lying between chemistry, physics and biology.

New foundry problems led to the establishment of the *Foundry Trade Journal* (founded in 1902). This carried accounts of the American Foundrymen's Association (established in 1896), and stimulated the formation of the British Foundrymen's Association, the fifth professional group of the new century. To supplement the annual convention (or conference as it was called after 1914), branch meetings were organized, not only in England but also in the Commonwealth. A Royal Charter was obtained in 1912 and, by a curious error, the name Institute was conferred upon it at the same time. In its early years Professor Thomas Turner, of Birmingham, who had in 1885 carried out classic work on the influence of silicon on cast iron, gave it great help. By 1954, with a membership of 5,300, it was probably the largest metallurgical association in Europe. Today it is a major organization concerned with progress in foundry technology, and, as such, has promoted the City and Guilds of London Institute Examinations in pattern-making and foundry practice.

The new light alloys produced by electricity were the concern of a sixth group, the Institute of Metals, formed in 1908 to do for these what the 40-year-old Iron and Steel Institute had been doing for the ferrous metals. The Institute of Metals did much to educate engineers of all kinds in the multifarious uses of aluminium alloys, and played a part in persuading aeroplane engineers to adopt them in place of wood.

Electricity generating stations, and hydro-electric projects and steel-framed buildings especially needed concrete. So, in 1908, a seventh group, the Concrete Institute, was established in Britain. Its value was soon recognized by Parliament, which a year later accepted it as one of the bodies to be consulted by local authorities when regulations affecting the stability of buildings were being considered. The Royal Institute of British Architects consulted it on preparing a report on reinforced concrete. Within four years, the scope of the institute had widened to cover all branches of structural engineering, and by 1922 it became the Institution of Structural Engineers. By 1924 its journal, too, was renamed the *Structural Engineer*. Ten years later, in 1934, it obtained a royal charter. By 1944 a Chair of Concrete Technology was established at Imperial College, London. This has been followed by the establishment of Chairs of Structural Engineering at the University of Witwatersrand (1952), and at the Manchester College of Science and Technology (1957), and a Chair of Engineering Structures at the City and Guilds College, London (1958).

As one of the basic uses of electricity is illumination, an eighth group, the Illuminating Engineering Society, was formed in 1909. Here the pioneer was Leon Gaster, a consulting engineer, who persuaded electrical and gas engineers to unite. It has been represented on all Departmental Committees on Factory Lighting since the first was set up in 1913.

Four smaller groups took shape around the new technologies. The Association of Supervising Electrical Engineers, formed in March 1914, began to issue *The Electrical Supervisor*. Today it has representatives on committees of the Institution of Electrical Engineers and other bodies. The Junior Institution of Locomotive Engineers was formed in 1911. Small as it is in comparison to the major engineering societies, it has linked together a large proportion of the mechanical and electrical engineers. An Institution of Railway Signal Engineers was incorporated in 1912 and held its first meeting on 25 February 1913. Its aim, the advancement of the Science of Railways Signalling, was achieved by close liaison with the Railway Signal Association of America and the publication of *Proceedings*.

THE BRITISH STANDARDS INSTITUTION

To provide a focal point for scientific and technological standards, the veteran Institution of Civil Engineers took the lead by establishing the Engineering Standards Committee in 1901. Incorporated seventeen years later as the British Engineering Standards Association, it became the British Standards Institution in 1930 and received its charter in 1931. It comprises several divisions—Building, Chemical, Engineering and Textiles—each with its committees and subcommittees—and is governed by a general council. It promulgates British Standard terms, definitions, codes of practice and specifications for materials and methods of testing.

The Institution of Civil Engineers was also very active in promoting engineering education. In 1903 the council appointed a committee for that purpose, under Sir William White, which reported in 1906. Five years later they took a leading part in promoting a conference on engineering education, whilst in 1914 W. C. Unwin once more reported on the practical training of engineers. It was an uphill task.

THE CRITICS WITHIN

By the twentieth century an orchestra of critics was playing upon the 'fatal torpor' (as E. E. Williams called it) of the British. In *Made in Germany* (1896) Williams argued that the deceleration of industrial growth could be checked by state action, especially if better technical training was provided. Others, like W. T. Stead, pointed out that there were three 'American secrets' which should be adopted by Britain: education, increased incentives to production, and a democratic way of life. The penalty for not doing so, he argued, was 'our ultimate reduction to the status of an English-speaking Belgium'. Stead's *Americanisation of the World* (1901) was followed by the even more alarming book, *The American Invaders: Their Plans, Tactics and Progress* (1902) by F. A. Mackenzie, which expressed the pious hope that American penetration of British markets might prove a blessing in disguise if it wakened the British to a sense of purpose. The President of the British Association caught the mood, and in a moving address entitled 'The Influence of Brain Power on History', pointed to Germany and America as two countries where science was encouraged.

GOVERNMENT PATRONAGE OF ENGINEERING, 1900–1914

In the first decade of the twentieth century the state took five major steps to foster science and technology. The first was the establishment in 1902 of the National Physical Laboratory at Teddington. With a grant of land and an annual subvention it was modelled on the Physikalisch-technische Reichsanstalt at Charlottenburg. The second was the institution of a larger government grant to the maturing civic universities—then emerging from their infancy as university colleges. In this, as in so many similar activities, the Government were stirred by the energetic personality of R. B. Haldane, who chaired the Departmental Committee set up to consider the future of the Royal College of Science, and recommended its fusion with the School of Mines and the City and Guilds of London Technical Institute. From this came the third step: the establishment of the Imperial College of Science and Technology in 1907. The fourth was the establishment of a subsidiary committee on aeronautics in 1909 which, under Lord Rayleigh, advised on work at the Royal Aircraft Factory and, in conjunction with an aerodynamics department at the National Physical Laboratory, made possible British air supremacy in the first world war. The fifth was taken in 1910, when Development Commissioners were appointed under the Development and Road Improvement Funds Acts to recommend grants for the improvement of agriculture, rural industries and fisheries in Great Britain. This, in embryo, was the first fillip to be provided for agricultural and microbiological engineering.

The first of these five steps was the most decisive. The National Physical Laboratory, with its nine main divisions, conducted tests of all kinds. To its work can be attributed the great use made of light aluminium and magnesium alloys in every department of British engineering, and especially the encouragement of aeronautical research in the war that began in 1914.

AN ENGINEERING PRESSURE GROUP

Perhaps the best comment on the tangled and delayed response of an old industrial economy to the new challenge of the age was the formation, in September 1911, of the British Engineers' Association. Douglas Vickers, M.P., was mainly responsible for its first convention. Beginning with twenty, it soon embraced over six hundred of the leading engineering firms in the country, aiming—as the *Memorandum of Association* put it—at providing 'a central national organization in the engineering industry for the promotion and protection of the interests of British Manufacturing Engineers and British Engineering.'

Its first director, D. A. Bremner, who held office from 1918 to 1943, was to be leading spirit in the establishment of the World Power Conference. Its president was the representative of the engineering industry on the advisory council of the Board of Trade. Its observations in 1912 struck a note that is ominous even today:

'Britain's two strongest competitors, Germany and the United States, are making a very serious bid for the education of the Chinese engineer. The Germans, with their model dockyard at Tsingtao, have properly organized engineering works where young Chinese of the right class are encouraged to attend and learn thoroughly not only the German language, but engineering and dockyard practice, and large sums of money are expended in bringing to Germany the right class of Chinese for their engineering training. . . .

'Then again, German manufacturing engineers on their own account have an Association for dealing with their overseas interests, and have subscribed a large sum for the purpose of creating three purely German engineering schools in China. These schools are to be effective nurseries for the German machinery trade. German diplomacy, too, has succeeded in introducing the German professional element very largely into Chinese native schools.

'The Americans have gone a step further. They waived a portion of their war indemnity against China after the Boxer rebellion on the understanding that that money should be spent on sending over the right class of Chinese to the United States for the purpose of theoretical and practical education.'

This technique was to be followed, on an even larger scale, by Russia after the second world war.

CHAPTER 22

The First World War and after, 1914-1935

The first world war was a chemists' war. The savage senseless-
ness of stalemate in the trenches was only circumvented by
technological imagination, by poison gas, tanks, machine-
guns, and massed artillery. One of the great disasters of the war,
Passchendaele, where there were 448,614 British casualties, was
partly due to Haig's destruction of the drainage system of the land
by a barrage of four and a quarter million shells. The Allied offensive
was stopped in a sea of mud. All the participants had to gear their
economies to the production of gunpowder, motor engines, guns,
ships and drugs.

D.S.I.R. AND THE RESEARCH ASSOCIATIONS

The khaki dye needed for British uniforms came from Germany.
So did the glass lenses for range-finders, the cameras and the very
aspirins needed by the harassed staff officers.

When the first world war broke out, the Board of Trade set up a
committee under Haldane to 'consider and advise as to the best
means of obtaining for the use of British Industries, sufficient
supplies of chemical products, colours and dyestuffs of kinds
hitherto largely imported from countries with which we are at present
at war'. The various professional societies urged the Government to
establish a National Chemistry Advisory Committee. The Government responded by establishing an Advisory Council for Scientific
and Industrial Research and, in 1916, the Department of Scientific
and Industrial Research—a far more comprehensive body.

The Department of Scientific and Industrial Research (or D.S.I.R.)
with its own office, establishment and a purposely indefinite rôle,
together with a committee of the Privy Council, directed the application of parliamentary funds for developing scientific and industrial

research, and established a number of co-operative research associations. Co-operation and co-ordination were the essence of its policy, designed to reinvigorate the industries of the country by new ideas and new methods. To encourage industry to contribute, the incomes of the associations were to be free from income tax.

Twenty-one of these research associations were established in the years after the first and before the second world war. Nine of them were in London: Scientific Instruments (1918), Rubber (1919), Non-Ferrous Metals (1919), Confectionery (1919), Laundering (1920), Leather (1920), Food Manufacturing (1926), Iron and Steel (1930), and Printing and Allied Trades (1930). Others were concerned with trades near London, like Electrical and Allied Industries (established in 1920 at Perivale), Flour Milling (established at St. Albans in 1923), Paint, Colours and Varnishes (established at Teddington in 1926), Automobiles (established at Brentford in 1931), or Coal Utilization (established at Kingston in 1938). Provincial industries were not neglected. The Yorkshire Woollen Industries obtained a research association at Leeds before the end of the world war in 1918. The Midland Boot, Shoe and Allied Trades obtained a Research Association at Kettering in 1919. The Lancashire Cotton, Rayon and Silk industry followed with one at Didsbury, now the B.C.I.R.S. at the Shirley Institute, and the Linen Industry of Northern Ireland with one at Lambeg; both in the same year. Staffordshire obtained two: a Refractories Research Association at Stoke-on-Trent in 1920 and a Pottery Research Association in 1931. The Black Country's Cast Iron Research Association was established at Birmingham in 1921.

In the second world war the number of these research associations was raised to 25 by the addition of Gas Research (1941), Internal Combustion Engines (1943) and Shipbuilding (1944). By 1950 they had increased to 40. These were the backbone of a vigorous offensive operation by industry to overcome inertia generated by past successes.

EXPLOSIVES AND CHEMICAL ENGINEERS

Further advances in the three major explosives, nitroglycerine, picric acid and T.N.T., all called for still further supplies of nitrogen. It was a pupil of Hofmann's, Sir William Crookes, who called attention to the possibility of a nitrogen famine. The nitrogen cycle in agriculture had by now become so important that deposits of guano from Chile and Peru were being consumed very quickly. It was the success of the German Haber process in fixing nitrogen from the air that enabled them to fight the first world war—secure in their reserves of ammunition and fertilizer.

During the first world war Lord Moulton (1844–1921) was chairman of the committee responsible for the production of explosives and propellants for the British forces. By 1918 these were being produced at the rate of 1000 tons of high explosive a day. In all 612,697 tons of the explosives and 450,487 tons of propellants were produced during their operations. Because trinitrotoluene (which had replaced lyddite) required a great deal of coal (600 tons of coal produced one ton of toluene, which in turn produced only two-thirds of a ton of trinitrotoluene), Moulton's committee augmented it by ammonium nitrate, thus producing amatol. Experience in organizing the gasworks, coke ovens, fat and oil supplies of the country led to his crusade for the proper training of chemical engineers. A team of chemical engineers under K. B. Quinan realized that these enormous amounts could not be produced by expanding existing factories and new ones had to be built. For this the collection of data by H.M. Factories Branch was necessary, and a supply of chemical engineers even more so. Quinan found that 'either intentionally or through negligence, it was customary at many places to keep the chemists in complete ignorance, not only of the costs of their plants, but also even of their efficiencies'. Ignorance of such facts Quinan remarked has often prevented 'important alterations and improvements'. The explosives industry created such a demand for phenol that it stimulated the post-war production of synthetic resins, and so of man-made fibres.

Four New Chemical Pressure Groups, 1916–1922

Four chemical pressure groups emerged from the war. In 1916 the Association of British Chemical Manufacturers was formed to act as a co-ordinating body for the industry to deal with the government and as a pool of ideas for promoting efficiency. In 1917 the British Association of Chemists was formed to promote the interests of the chemists in Industry and their efficiency. In 1920 the Chemical Plant Manufacturers Association was formed to co-operate with the above bodies and to act as the trade association for firms engaged in the design and manufacture of plant for the conduct of unit operations and unit processes of chemical engineering.

But the fourth was the most significant. At a meeting called by Professor J. W. Hinchley at the Society of Chemical Industry on 29 July 1918 some 70 members, under the chairmanship of Professor G. T. Morgan, F.R.S., decided to form a subject-group of Chemical Engineering. This came into existence on 20 October 1918 and by 1920, at their fourth conference, this group debated whether or no a separate Institution of Chemical Engineers should be formed. The advisability was further discussed in the *Chemical Trade Journal, The Chemical Age*, and other papers. As a result of these discussions Hinchley called another meeting on 9 November 1921 at the Engineers' Club under Sir Arthur Duckham. After studying the constitution of the existing Institutions and Societies of Engineers, and hearing from G. K. Davis of attempts to form a similar institution in 1880 and 1890, the Provisional Committee held the inaugural meeting of the Institution of Chemical Engineers on 2 May 1922. A Chemical Engineer, as defined by the Institution, was 'a professional man experienced in the design, construction, and operation of plant in which materials undergo chemical or physical change'. Formally incorporated on 21 December 1922, it held its first annual meeting at the Engineers' Club, Manchester, on 8 June 1923.

THE DEVELOPMENT OF RAYON AND THE RISE OF SYNTHETIC FIBRES

These chemical engineers were to play their part in another spurt of development in rayon production, which by 1918 had dropped to 3,000,000 tons: less than half that produced in 1913. But in the years that followed output rose to 38·7 million lb. in 1927, and 115·2 million lb. in 1939. World production over the same period rose from 25·9 million lb. through 295·1 to 1,149·6 million lb. respectively.

The expiry of the basic patents, the continuous improvement of the product and the high profits earned in the industry stimulated a number of new firms to enter the industry. The imposition of a custom duty in the 1925 Budget also helped. And when the 'shine' was taken from rayon, the attraction increased. By 1928 nearly thirty firms had entered the trade as compared to the two that existed in 1920. Courtaulds at the same time agreed to exchange shares and technical information with Glanztoff of Germany, with Sia Viscosa of Italy, and with Enka of Holland.

The figures cited in the above paragraph include rayon staple and acetate rayon. Rayon staple, a product made of low-grade wool pulp, grew rapidly in popularity. Production rose from 2 million lb. in 1929 to 58 millions in 1939. Acetate rayon, which in Chapter 21 we saw emerging in the hands of the British Cellulose and Manufacturing Company (which became British Celanese in 1923), tempted Courtaulds to enter the field too. Lengthy legal action lasting till 1935, ended in favour of Courtaulds. For acetate rayon was cheaper in the finer deniers than rayon and was much in demand for weaving. Whereas in 1927 it contributed only 10 per cent to the total of 38·7 million lb. of the British rayon produced, by 1939 it provided 26 per cent of the total of 115 million lb. produced.

THE AEROPLANE

The aeroplanes, for whose cotton-covered wooden wings the cellulose acetate factory at Spondon, Derbyshire, was built during the first world war, themselves became agents of innovation. From reconnaissance instruments, intermittently and ineffectively used for long-range bombing, they developed into integral parts of the war machine. At the Battle of Amiens (August 1918) aircraft were employed in close co-operation with ground troops and the Royal Flying Corps was bombing lines of communication behind the enemy front. Britain's independent air force was used as a garrison police force in the deserts of the Middle East.

Collaterally, civil aviation developed. Germany began first with a local service before the first world war, but in 1920 three British and two French companies were flying modified military aircraft. These companies did not have much success, so in 1924, with Government help, Imperial Airways was formed into which the British civil airline companies were amalgamated. This, under Woods-Humphreys and Mayo, formed a cadre of engineers, technicians and pilots. It was just at this time that metal began to be generally used as a structural material for aeroplanes, especially in Germany where duralumin was very highly regarded.

So another branch of engineering developed. In addition to the Royal Aeronautical Society (founded 1866), the centres of aircraft engineering and research in 1914 were Farnborough, Eastchurch, Teddington and Brooklands, operated by the Army, Navy, National Physical Laboratory and private firms respectively. At Farnborough the Royal Aircraft Factory under Colonel O'Gorman had a team of designers, engineers, physicists and mechanics, but just towards the end of the war a parliamentary inquiry led to its reconstruction as the Royal Aircraft Establishment, and many of the designers, like Geoffrey de Havilland, Folland, F. M. Green and others, went into private industry. As a result of the enormous advances in aircraft manufacture during the war, the Society of British Aircraft Constructors was formed in 1917.

Beginning the war with about four squadrons of fifty aircraft, Great Britain went on to produce over 55,000 aircraft in the four years that followed. The Americans, during their short incursion into hostilities, produced some 29,000. Money was spent on research equipment and wind tunnels and the habit survived the war. One of the by-products of such advance was the improvement in fuels and lubrication. The numerous engine failures of the combat planes led to research into the 'knocking' of engines. To overcome this, Midgley and Boyd, working in the laboratory of General Motors, discovered tetraethyl lead.

The Air Ministry invited the Spanish engineer, Juan de la Cierva, to come to England and pursue his investigations into the autogiro. A company was formed in the following year, and Germany also began building them under licence. In 1926 Dr. A. A. Griffith of the Royal Aeronautical Establishment suggested that the gas turbine was a feasible proposition, and an air force officer, Frank Whittle, demonstrated that it was by patenting in 1930 a gas turbine with a centrifugal compressor as a source of a high-velocity propulsion jet. The Air Ministry were cautious and when in 1936 power jets were organized it was with private money. The Germans took jet engines seriously, and by 2 August 1939 had in the air the He178. Coterminously, the British Government financed the construction of two of the largest airships yet built, the R100 and R101. When the latter was completed, it was tried out on a maiden voyage to India early in October 1930 with disastrous results, for it crashed, killing the Secretary of State for Air and all the passengers and crew. This disaster when compared to numerous flights of endurance undertaken by various people in heavier-than-air machines, hardened opinion in favour of the heavier-than-air plane. Yet before dismissing the airship as a cul-de-sac of engineering development one should remember that the vast hangars it called for elicited the structural ingenuity of Freyssinet, whose corrugated parabolic shells of concrete were the first of this kind in the world.

An early sorter and counter of Herman Hollerith (1860–1929), one-time instructor at the Massachusetts Institute of Technology, who designed the machine for compiling the U.S. Census returns in 1890 and later organised a company that is now (with others) the International Business Machines Corporation.

Wallace Hume Carothers (1896–1937), inventor of nylon and the first organic chemist working in industry to be elected to the National Academy of Sciences in the United States of America.

THE RISE OF RADIO

The great advances which radio engineering made in Britain during the war, coupled with the establishment of the British Broadcasting Company in 1922, led to the formation of a British Radio Association in the same year. This was forced to admit members who were not actually radio engineers, and the professional elements in the association ventilated, through the technical journals, the need for a body which would cater for the rapidly growing science of electronics and radio. So, in October 1925, after numerous debates about the title, the Institute of Wireless Engineers was formed, and soon amended its title to the Institute of Wireless Technology, which began to hold examinations in November 1926 and to issue *Proceedings*. Six years later it was registered as a corporate body.

The radio engineers hived off in 1933 from the Institute of Wireless Technology to form a separate Institution of Radio Engineers which, four years later, added the prefix British to their title. This existed as a separate entity until 1940, when it fused once more with the parent body. With the conjoint drive of both bodies now coupled in one organization, together with the great wartime need for radar operators, a City and Guilds Certificate in Radio Engineering was established in 1942.

THE GENESIS OF THE NATIONAL CERTIFICATE SYSTEM

In 1919 a new step in the training of engineers was taken on the initiative of H. S. Hele-Shaw (1854–1941), a former Whitworth Scholar at the University of Bristol, who after holding the chairs of engineering at Bristol (1881–5) and Liverpool (1885–1904) had become professor and principal of the Transvaal Technical Institute in South Africa. Hele-Shaw was not only a founder member of the R.A.C. and a judge of innumerable early road trials, but invented the friction clutch that goes by his name. In experimenting with this at Liverpool, he was prosecuted under the 1865 Act, for the clutch failed and he careered downhill, backwards, at more than the statutory limit of 4 m.p.h. His interest in the production of motor cars was only equalled by his enthusiasm for the training of engineers, and it was he, as much as anyone, who helped to organize the National Certificate System after the first world war.

Under this scheme the Institution of Mechanical Engineers arranged with the Board of Education for the certification of

R   257

mechanical engineers who had passed a qualifying examination after three years (for the Ordinary National Certificate) or five years (for the Higher National Certificate) at a local technical college. Their example was followed by the Institute of Chemistry (1922), the Institution of Electrical Engineers (1924), the Institute of Gas Engineers (1924), the Institute of Naval Architects (1925), the Institute of Builders (1930), and the Textile Institute (1931). In addition Higher National Diplomas were instituted on similar conditions, but by 1938 only 37 were awarded.

When Shaw retired in 1937 from the chairmanship of the Joint Committee of the Board of Education and the Institution of Mechanical Engineers that administered the National Certificate Scheme, over 3,327 students were working for it in 139 technical colleges in England and Wales. In 1959 altogether 37,745 students sat the Ordinary National Certificate examination and 19,654 passed; while 15,454 sat the Higher National Certificate and 10,546 passed.

THE CHANGED POSITION OF BRITISH INDUSTRY, 1919–1935

After the war countries accustomed to obtain manufactured goods from Great Britain were making them for themselves. Sometimes, like India and Japan, they made textiles which seriously threatened the Lancashire Cotton industry. War shortages also led to an intensification of agriculture in the 'larder lands' not yet industrialized. As a result of this, improved techniques of growing and preserving food led to lower agricultural prices in those countries which further complicated the position of British farmers.

Lastly, the loss of overseas investments, especially in America, led to an increasing dependence on the American stock market. In general the heavier industries suffered from the war and the lighter industries grew from it. Perhaps the most typical engineering and industrial advance was connected with the rise of the automobile and electrical industries.

THE RISE OF THE MOTOR INDUSTRY

After the war 86 makes of car were introduced. The 96 firms in business in 1922 declined to 41 in 1929 and 33 in 1939 as amalgamations and federations into ever-larger groups (a phenomenon common enough at this time) prevailed.

Of the post-war manufacturers W. R. Morris soon out-distanced his competitors by lowering costs. Morris was originally a cycle repairer at Oxford who, before the war, rented the old Military Training College at Cowley and because the number of engines he wanted was quite outside the range of his supplier's experience, he placed his orders in America. The first world war, however, intervened and he was soon making hand grenades and then machine cases for the bombs fired from trench mortars. From these he went on to assemble mine sinkers in great numbers for which he also designed the jigs and tools. His success was such that 2,000 a week of these mine sinkers were soon being produced.

At the end of the war he registered Morris Motors to take over W.R. Motors, his original company, and turned to car manufacture again, using independent suppliers like Smith, Lucas, Dunlop, Ransome and Marles. Since the American Continental Manufacturing Company which had previously supplied his engines decided not to seek further orders, Morris invited the Hotchkiss Company of France (which had been manufacturing machine-guns in England) to make engines for him. In 1923 he bought this firm and Morris Engines was formed. Three years later he formed Morris Motors Limited. It was said that Morris Cars threatened Ford's position in foreign markets and helped persuade him drop the model T for the model A.

A second large organization was built up by Herbert Austin, who began making cars in 1905, but only became a world figure through the manufacture of the famous Austin Seven in 1922. This car was produced up to 1934 and was being made in France, Germany and the United States: a total of 352,000 being sold.

By 1920 there were 650,148 motor vehicles on the roads in Britain, nearly forty times as many as in 1904. This number was doubled in five years and by the beginning of the second world war had more than doubled again. By 1938 444,877 motor vehicles were being produced annually in Great Britain, which was second only to the United States, which was producing 2,489,085 a year. Germany, which was third, produced 352,369.

It is worth mentioning that Lord Nuffield also long wanted to bridge the gap between the theoretical and the practical engineer. Having given the University of Oxford two million pounds for medical research, he offered one million pounds to found a College where engineers could garner information about the theoretical aspects of their profession, and modern business techniques. But the University thought otherwise. The money, it was suggested, should be used for a college for post-graduate study in the social sciences. Perhaps they realized that the gap between theory and practice in this field is wider than in engineering! They also suggested that some of the endowment might be utilized for research in the basic sciences, especially physical chemistry. Lord Nuffield fell in with these proposals in 1937. The college was finished as this book was being written, and the writer, for one, regrets that Lord Nuffield's original intention was not carried out in the spirit in which the other senior university has seen fit to found Churchill College.

The Social Impact of the Motor Car

The two million private cars registered in Great Britain by 1939 ran through the old social boundaries of the towns and, with their variants, the bus and the lorry, helped to evacuate the cities. Suburbias, verging to subtopias, mushroomed. Whether they were large municipal housing estates fed by motor buses (like Kingstanding, near Birmingham), or garden cities fed by the car (like Welwyn, founded in 1920), or ribbon development along the motor roads out of the town, they represented the new civilization of the time and excited the compassionate criticism or contempt of John Betjeman, George Orwell and J. B. Priestley. Civilization seemed suddenly to become less brassy and gilded and more polychromatic and plastic; less metallic and more variegated.

Here too the motor car helped in stimulating the development of plastics. The production of plastics had spread so extensively by 1931 that the Plastics Institute was formed. In 1933 it was estimated that there were thirty firms in Birmingham making plastics of one kind or another. In the same year the Research Department of I.C.I., experimenting on the effect of high pressures on chemical reactions, manufactured a minute amount of a new plastic of remarkable properties: polythene. This owed much to W. H. Carothers of Du Pont in America, as well as to P. Morgan in England. Polythene proved one of the major materials during the second world war and its availability transformed the design of radar. Its high dielectric strength, its low loss factor and its ease of moulding led it to be chosen as a basic material in a great deal of electrical equipment. By 1957 plans for production of no less than 760,000 tons were afoot.

Moreover, motor cars, by their very utilization, helped to stimulate food processing industries. Picnics, week-ends in the country, long journeys drew women from the kitchen. Chocolate bars and ice cream stimulated old firms like Mather and Platt of Manchester to establish a works at Radcliffe in 1932 to make food processing machinery. New industries like artificial silk, electrical goods, motor cars increased in output by an average of over 10 per cent between 1923 and 1935.

THE PROBLEM OF THE ROADS

The motor car and coach demanded better roads. The old Road Board was superseded by a Ministry of Transport in 1919 and in the following year a 'road fund' was established, under the Ministry's control, from the new motor taxes and licence fees. This 'road fund' rose from £9½ million in 1921 to £25 million in 1928. Expenditure increased fourfold from £5 million to £21 million in the same period. Roads were classified into A and B classes with numbered qualifications. A roads were aided by a 50 per cent grant and B by 25 per cent.

Traffic grew faster than improvements. There was 1 car per 136 persons in 1921, but 1 per 57 in 1927. Schemes like the Mersey Tunnel (1925), the Watford and Kingston bypasses (1927), the Great West Road (1925) were possible in large areas, but the general inadequacy of the rural local governmental units led to their responsibilities being transferred to County Councils in 1929. In the following year the Road Traffic Act introduced compulsory third-party insurance and established traffic commissioners to control licences for public service vehicles.

By 1935 traffic hazards had increased to such an extent that a speed limit was imposed on all vehicles in built-up areas. This increase of traffic led to the Road Fund accumulating so much that by 1937 the yield was applied 'for general purposes'. Unfortunately these 'general purposes' did not include the provision of fast safe motorways on a scale sufficient to satisfy current needs.

Sir Henry Maybury (as he now was) was made Director General of Roads until 1928, and later became Consulting Engineer and Adviser to the Ministry of Transport on road and traffic questions until 1932. Whilst Director-General of Roads he set up an experimental station at Harmondsworth which, in 1933–4, was transferred to the Department of Scientific and Industrial Research.

Under no less than fifteen consecutive Ministers of Transport, official road policy was at best intermittent: a marked contrast to the railways, which persistently opposed the construction of a virtually free permanent way for their competitors. It took the trading estate promoters, the town planners and the burgeoning commercial interests time to form a coherent view of their own about roads, and in this way they were undoubtedly helped by the genesis of a professional group concerned with motorways: the highway engineers.

The International Association of Road Congresses, which resulted from the Paris Conference in 1908, exerted an increasingly powerful stimulus on the profession of highway engineering. This was especially true of the fifth (held at Milan in 1926), the sixth (held at Washington in 1930) and the seventh (held at Munich in 1934): all countries where bold innovations were being made. The Italian autostrada in Tripoli, where Mussolini opened a 1,000-mile road, played a major role in the campaigns of the second world war. American development covered a practically roadless country with what Rees Jeffreys called 'the finest system of roads in the world', yet, as he ruefully commented, it 'made no impression upon the public mind or the governmental authorities of this country'. Rees Jeffreys himself attended the Munich Conference in 1934 with Professor R. G. H. Clements. Clements inspected German autobahnen again three years later and was vastly impressed by their safety, cost and aesthetic appeal, and his report put new life into the new roads movement.

Though the Americans did not adopt the device of taxing petroleum to sustain the mounting costs of road maintenance till 1923, their energetic engineers soon laced the country with highways on which fast-flowing traffic could travel. Similar enthusiasm in Germany and Italy led to attractive careers being offered to engineers as road builders. Major achievements not only provided America with a national asset, but also trained a large number of road engineers, whose skills were invisible exports in the same way as British railways engineering skills were in the corresponding decades of the nineteenth century.

Some realization of this prompted the Imperial College of Science and Technology to establish a Chair of Highway Engineering in 1928. It began to function in October 1929 under Professor R. G. H. Clements. In the following year the activities of those civil engineers who specialized in highway engineering were such that it was decided to form a special institution. Its founder was H. Slipper, and the first president was Colonel R. E. Crompton, who had been the first consulting engineer to the Road Board. No less than £60,000,000 were being spent on roads at this time and it was felt that the promotion of research and other work in highway engineering was eminently desirable. And so, in 1930 the Institution of Highway

Engineers was formed. Unfortunately it had no funds with which to sustain a campaign for professional status, and until the second world war, it had to share an executive secretary with the Institute of Water Engineers.

THE GREAT COMBINES

The combination so evident in the motor industry was evident in other branches of industry. The marriage of each fertile scientific discovery and human need produced a host of products and problems that, by their interrelation, demanded close co-operation. Each new discovery had a reciprocal effect on existing techniques.

In Britain this led to an increasing combination of steel and chemical firms respectively. Steel firms began extending forwards and backwards. Stewarts and Lloyds (1902) and the United Steel Companies (1918) were two early examples. They were followed by the English Steel Corporation (1927)—a merger of Armstrong Whitworths, Vickers and Cammell Lairds—who were joined two years later by John Brown and Bolchow, Vaughan and Armstrong. In addition the National Federation of Iron and Steel Manufacturers (established in 1918) and the International Steel Cartel (established in 1926) showed how co-operation on both the national and international level was growing.

Also in 1926 the United Alkali Limited united with Brunner Monds to form Imperial Chemical Industries, the largest industrial undertaking in the British Empire and the largest manufacturing company in Great Britain, which now produces 12,000 different products, from explosives to slide fasteners. The other companies in this great merger show how chemical engineering necessitates combination: the British Dyestuffs Corporation (which itself had incorporated Perkin's original company), Nobels and Kynochs. From these five companies the present firm of I.C.I. has built some eighty factories which straddle the United Kingdom. In 1937 they acquired the Salt Union, itself an amalgamation effected in 1888 of the salt producers of Cheshire, Worcestershire, Durham and Northern Ireland. Of the million tons of salt produced in England, industry claims a major share for dyeing processes, soap-making, metallurgy and refining.

A second great combine was formed in 1929 when Lever Brothers, based originally on soap from vegetable oils, then on food processing, and later built up by the acquisition of chains of grocery stores (Liptons, Maypole, Home and Colonial) became Unilever.

The need for administrative coherence and the standardization of units, services and repairs also led to the growth of corporations, and many of these corporations tended to be publicly owned. The model was in the older port and harbour trusts, which in turn were preceded by public local trusts for turnpikes, canals and sewers. The Belfast Harbour Commissioners (1847), the Tyne Improvement Commissioners (1850), the Mersey Docks and Harbour Board and the Clyde Navigation Trust (1858), and the Port of London Authority (1909) offer cases in point.

The first of the inter-war public corporations was the B.B.C. Originally a private company, it became a public service as a result of the recommendations of the Crawford Committee in 1926.

The second was the Central Electricity Board, also created in 1926 to ensure uniform frequencies and voltages and standardized equipment. A high-tension grid of power lines was built connecting a number of new generating stations. The production of electricity increased to 22,877 kwh by 1937, supplying million consumers. In the same year over a million electric cookers were in use. Refrigerators, wireless sets, irons, vacuum cleaners and of course the electrification of railways created a growing power hunger.

The railways themselves provide a third example of inter-war public corporation. The idea of the London Passenger Transport Board stemmed from Herbert Morrison, whose scheme was adopted by a Conservative Government in 1933. All the tube, underground railway, train and omnibus services were taken over by the Board and nationalized. Fifteen years later, though remaining as an essential unit, it was in turn taken over by the British Transport Commission.

NEW PROFESSIONAL GROUPS

The registration of each advance can be seen in the formation of new groups. The Institution of Production Engineers (1921) and the Institution of Chemical Engineers (1922) were founded directly as a result of experiences in war production. The importance of human relations in industry was recognized by the formation of the Industrial Welfare Society in 1918, and three years later, by the foundation of the National Institute of Industrial Psychology. Easier, safer and healthier, and more interesting work was now a goal to which a number of organizations were working.

Since the welfare and safety of the whole nation was seen to be largely dependent on the prosperity of the engineering industries, it was felt that they should have a larger share in the national councils. A Joint Council was therefore formed in 1923 by the Institutions of Civil, Mechanical and Electrical Engineers and Naval Architects to foster engineering interests and to be ready to take action in any national emergency. It was anticipated that others would desire representation, either as constituent or affiliated institutions: such representation only being granted on the basis of a large corporate membership and the high standard of examinations. In May 1923 the four main engineering associations adopted a constitution. The scientific and technological advances of the war and its aftermath stimulated the formation of yet more institutes.

The Institutes of Physics (1918), Fire Engineers (1918), Transport (1920), Welding (1923), Fuel (1926), and Highway Engineers (1932) all sprang into vigorous life at this time, together with the Oil and Colour Chemists Association (1918), the Institution of Engineering Inspection (1919), the Society of Consulting Marine Engineers (1920), the Television Society (1927), and even the British Interplanetary Society (1933). The ridicule which greeted the last-named body has been much reduced by the launching of the sputniks and satellites. This phase of institute-formation culminated in the Engineers' Guild founded in 1938 to promote 'the unity, public usefulness and interests of the profession.' The Guild (which was incorporated in 1951) is not, like those listed above, concerned with the advancement of research, but with the status of engineers and their professional security, and, through a joint consultative committee, it sustains a close liaison with the three senior professional institutions.

The decline of manual skills and of heavy physical work, accelerated by the educational ladder created after 1902, enabled women to enter the 'white overall' as well as the 'white collar' class. Their extensive employment in both capacities during the first world war led to the formation, in 1920, of the Women's Engineering Society.

CHAPTER 23

The American Oasis and the Twentieth Century

THE NEW FUEL AND ITS TECHNOLOGY

If Britain is a piece of coal surrounded by fish, the United States is an oily oasis surrounded by thirsty travellers. President Grant had prophesied in 1871 that by utilizing her vast oil deposits 'America may soon become one of the great industrial nations of the world'. He was right: oil needs no stokers, leaves no ash, and possesses a high energy content to weight. A ship carrying an equivalent amount of oil can go three times as far as one carrying coal. In 1900, of the total world consumption of 150,000,000 barrels (a barrel = 42 U.S. gallons = 34·97 imperial gallons = 159 litres), the United States provided 64 per cent, and when by 1937 world consumption had increased by two hundred and eighty times to the staggering total of 4,273,000,000 barrels a year, the U.S. still provided 63 per cent of it.

Not only was America providing well over half the oil needed by the thirsty world, but its own rate of consumption per head of population was ten times greater than that of any other country in the world. This new fuel was the lifeblood of the American economy. From it the vast Rockefeller fortune was rolled up: a fortune which has been applied to more socially useful ends than perhaps any other in history. The Rockefeller Foundation, amongst other things, helped to keep research in British universities going in the years between the two world wars, as well as eliminating hookworm in South America. It was the first of many examples of international help in the emergent Atlantic economy.

THE AUTOMOBILE AND THE ASSEMBLY LINE

As British coal fuelled the steam engine, so American oil fuelled the American car. The first motor-car exhibition, held at Madison

266

Square Gardens in 1900, symbolized the advent of this great American engineering achievement that was to become its greatest national industry. By 1913 most of the nearly two million automobiles in the world were in the United States. This twenty-years' expansion, fired as it was with the new fuel, reacted on the whole American economy. From the gas buggy of Charles E. Duryea, first built in 1892, there had been a rapid kindling of interest. Ransom E. Olds, of Lansing, Michigan, had built several steam cars before this. His Oldsmobile, manufactured in 1897, encouraged him to set up a factory in Detroit, which by 1901 had produced 425 cars. Others paced him. The Haynes Winton, Stanley, White, Locomobile, Packard and Cadillac were becoming known, if not so well known as the Benz Daimler, De Dion and Panhard. By 1904 there were 121 manufacturers assembling cars of their own, organized in an Association of Licensed Automobile Manufacturers. The licence in the title referred to the fact that they all were paying royalties to William C. Whitney's Electric Vehicle Company of Hartford, which owned a comprehensive patent for a self-propelled vehicle driven by an internal combustion engine. This comprehensive patent, filed in 1897 by G. B. Selden, had been upheld by the courts in a test case in 1903.

There was one manufacturer, however, who refused to pay royalties: Henry Ford. An enthusiastic mechanic who had built his own steam engine at the age of 15, and at 33 had become the chief engineer of the Edison Illuminating Company of the United States. Ford built his own car in 1896 on four bicycle wheels. Encouraged by its success he formed a company in 1899 which collapsed, the only positive result being that the maker of his engines organized the Cadillac Automobile Company. In 1901 his second company collapsed. He then secured a bicycle racer to drive his famous '999', which won public acclaim and enabled him to form on 16 June 1903 the Ford Motor Company of Michigan, financed by a Detroit coal dealer. It produced the famous model 'A'.

The model 'A' Ford immediately provoked a court case for infringement of patent which continued for eight years. Ford's profits mounted, and ultimately he successfully contested the claim. In the meantime he had bought out the coal dealer and several other stockholders. By 1907 he was making some $5,000,000 in sales.

Perhaps the decisive technological invention that transformed the car from a luxury to a necessity was the electrical self-starter, the work of Vincent Bendix and Charles F. Kettering in 1911. By eliminating cranking, it enabled a car to be driven by any adult. Henry Ford took full advantage of it.

At this stage Ford introduced vanadium steel as the basis of his famous model 'T'. To cope with the enormous demand his cars had stimulated, Ford set up, in 1913, a moving assembly line which enabled a chassis to be produced in 1 hour 33 minutes as opposed to 12 hours 28 minutes. In the following year he was producing 264,972 cars a year as opposed to 181,951 in 1913 and 76,150 in 1912. Moreover, he laid out the huge River Rouge Plant for the manufacture of all the complex constituents of his cars. This last act provoked the Dodge brothers, who wanted a more generous distribution of dividends. Ford bought them out in 1919 for $25,000,000 By 1923 his policy of integrating production had paid rich dividends: over 2,000,000 cars a year were being produced by his company. The model 'T' with its standardized parts assembled by conveyor-belt techniques had by then become a classic example of the third American contribution to Europe: production engineering.

Parallels to the story of the Ford can be found in Chrysler, Nash, Hudson, Studebaker and White cars. Equally extraordinary was the transformation of the Buick factory at Flint into the General Motors Company in 1908 which absorbed Olds, Cadillac and nine other companies. This company was itself to be absorbed by the Chevrolet Motor Company and both in turn were to emerge as General Motors in 1916.

One index of this development was that the U.S. bought 83 per cent of the total number of cars produced in the world during the year 1920 to 1924, 77 per cent during the years 1925 to 1929, just over 60 per cent during 1930 to 1934, and nearly 70 per cent in the period 1935 to 1939. This colossal dominance stimulated the production of steel, petrol, rubber, plate glass and plastics.

THE AEROPLANE INDUSTRY

Pacing the development of the motor-car industry and initiating further social revolutions of its own, was the contemporary development of the aeroplane. Glenn Curtis, a motor-cycle racer, began to build his planes and engines under the joint stimulus of the Aeronautic Experiment Association and the *Scientific American* and won a prize at the first international aviation exhibition at Rheims. 'Stunt' flying publicized the new move and by the beginning of the first world war it was being used by the U.S. post office.

This war stimulated the creation of acetate, non-inflammable dopes for the linen-covered wings, the mass production of the famous 'Liberty' engine to the tune of 150 a day, the improvization of an air-ground radio, and after the war led to the rise of a new industry. The first Atlantic flight in 1919, the establishment of a trans-continental service in 1921, and Lindbergh's historic flight in 1927 showed that aviation had a great future. By 1929 the U.S.A. had created a Bureau of Aeronautics under the Civil Aviation Act of that year. By 1938 a further Civil Aeronautical Act set up a Civil Aeronautics Board to administer the growing transport.

Big new companies like Pan American, government interest in the new weapon (expressed through the National Advisory Committee on Aeronautics, set up in 1915) and flourishing schools of aeronautical engineering in the universities led to great advances being made in aerodynamics till by 1939 the United States had the largest air force in the world.

The necessities of the aeroplane elicited improved techniques. Precision engineering, the machining of light alloys, powder metallurgy leap to mind. As it came to dominate the production economies of belligerents in the second world war, aeroplane materials, assembly techniques, fuels and lubricants, no less than their power units, initiated changes in all branches of the engineering industry. In materials, the high tensile light alloys, high tensile steels, and the research on fatigue of airframes also made possible the successful construction of gas turbines. Duralumin, beryllium, titanium, iridium, fibre glass and plastic adhesives, are now as commonplace outside the aeroplane industry as in it. The forging, pressing, skin milling and other techniques of assembling airframes have themselves evolved special forms of testing by optical and ultra-sonic methods. The search for better fuels led, as we shall see, to the development of catalytic cracking in the second world war. It has even affected one of the most traditional forms of engineering, mining, in that the manufacturers of hydraulic undercarriages have built the Dowty pit prop, which is now being manufactured in quantity all over the world.

It is not too much to say that aviation has had a more comprehensive and far-reaching effect on all fields of material endeavour embarked upon by humans than has any other previous single development. Beginning with surveys in the first world war, it was later used for crop-spraying, irrigation works, railways, plans for towns and for checking progress generally. Spectacular results were obtained in underdeveloped areas where land transport could not penetrate.

Aeroplanes demanded better fuel than was obtainable by normal distillation processes and here a further revolution was initiated in one of America's greatest national assets: petroleum.

THE CRACKING OF THE NEW LARDER: PETROLEUM

Up to the first world war much of petroleum was almost a waste product. After it, the process of thermal cracking, producing smaller hydrocarbon molecules, introduced by William Burton in 1913, was improved. Tetraethyl lead was introduced in 1920 to stop the 'pinking' of aero engines, and continuous distillation began. E. J. Houdry introduced the fluidised bed catalytic process. But the most valuable figure was W. H. Carothers (1896–1937), who affords an interesting example of the extrapolation of ideas from one scientific field to another. Interested in the application of electronic concepts of valency to organic molecules, he went on to work on the catalytic hydrogenation of aldehydes over a platinum catalyst. Early in 1928 he became director of a new experimental station of E. I. du Pont de Nemours and Company, and working on the discoveries of the Rev. Julius A. Nieuwland, he developed a synthetic rubber known as neoprene, made from acetylene, salt and sulphuric acid. It was used for insulation gaskets and power transmission.

His investigation of the synthesis of polymers of high molecular weight by means of esterification and amide formation led to the commercial production in 1939 of nylon fibre: the first synthetic fibre to compete with natural fibres. Unfortunately he did not live to see its popularity.

The loss of Malaya during the second world war accentuated research in synthetic rubbers. In June 1940 the B.F. Gooderich Company produced rubber tyres from such startling materials as potatoes, grain sugar, petroleum and coal tar. Production expanded till by 1942 400,000 tons were being produced and by 1955 the annual production was 1,000,000 tons: 250,000 tons more than the total of natural and reclaimed rubber produced in that year.

Though the Massachusetts Institute of Technology had been offering lectures on chemical engineering since 1888 and an American Association of Chemical Engineers had existed since 1908, it was the enormous demand for explosives and chemicals between 1914 and 1917 that really put the chemical industry on its feet in

America. In those three years the value of explosives exported abroad increased from $6,000,000 to $802,000,000, and chemicals and dyes from $22,000,000 to $181,000,000. Concomitantly iron production increased by 34,000,000 long tons (from 41,000,000). The Allies liquidated $2,000,000,000 of their American securities to buy American goods and, when that money ran out, raised loans in the United States, making it the creditor national of the west.

During the second world war the Allied armies used fourteen times as much petrol in a day as the entire Allied armies had used in the whole of the first world war. Aeroplanes needed 100-octane fuel, explosives called for toluene, the loss of the Malayan rubber resources stimulated the production of butadiene, airfields needed asphalt, radio sets needed plastics and resins, whilst the enormous demand for medicines created a sub-industry of petromedicaments. Anaesthetics like cyclopropane, disinfectants, shock drugs like allonal, vaso-constrictor drugs, diuretics for kidney troubles: all flowed from the catalytic cracking plants in great abundance. Indeed, it has been calculated that half a million compounds can be developed from petroleum if uses can be found for them. Synthetics like Paracril for hoses and gaskets, inner tubes, cattle sprays, hormones, alcohol, ketones, esters, all come from tailored and remodelled molecules which chemical engineers can use in petroleum. Even the waxes that were used to pack other products were obtained from petroleum.

The Sinews of Hegemony

To the motor car and petroleum, two other factors in American industrial hegemony in the early twentieth century must be added: electric power and production engineering. All four interrelated and coupled with concomitant stimuli in the three wars of 1914–18, 1939–45 and 1950–1, built up the sinews of the subcontinent. The whole world was to feel the impact of developments there. One example (of many) must suffice.

The development of South Africa (where the British were involved in a wearing war with the Boers from 1899 to 1902) was materially assisted by American engineering skill. Cecil Rhodes, the great British imperialist, employed as his chief consultant the American J. H. Hammond, for deep-level gold-mining operations on the Witwatersrand, whilst in Kimberley his diamond mines were indebted to another American engineer, Gardner Williams—who was succeeded by his son. In England, American railway engines were being purchased by the leading companies: 'Atlantics' by the Great Northern and 'Pacifics' by the Great Western. By 1914 the Great Eastern Railway had even got an American manager, in H. W. Thornton. The correspondence columns of the *Engineer* reverberated with praise of American machine tools and one early twentieth century editorial remarked: 'In truth, it seems that we must adopt American mechanical engineering and metallurgical methods in their entirety . . . and this means little short of social revolution.'

This social revolution meant learning American skills. Here the clearest insight and initiative lay with an Englishman who saw American engineering skills in South Africa at first hand and was so impressed that he paid the expenses of two delegations to go to the United States and find out how it was done.

The first delegation of trade unionists included, amongst many others, the secretary of the Amalgamated Society of Engineers, the second, of educationists, included two university professors of engineering. Both delegations were impressed with what they saw. Their findings, usually named the Mosely Reports, after the Englishman who promoted them, influenced the changes which took place in the early years of this century to incorporate engineering more securely into the fabric of the nation.

THE DEVELOPMENT OF ELECTRIC POWER

By 1900 no less than 200,000 horse power was being exerted by industrial electric plant in America. The factories where it was employed dispensed with shafting, pulleys and belts and were sited without regard to geographical limitations. Domestic appliances like irons, fans and heating emancipated the American housewife. Electricity was applied by L. H. Bailey of Cornell to grow plants and in Alabama to process cotton. It enabled the skyscrapers to be built since it powered some 3,000 elevators by 1900. In that year $5,000,000 worth of electrical equipment was being exported.

The policy of instituting research sections in the electrical industry and departments of electrical engineering in American universities paid rich dividends. Charles Proteus Steinmetz (1865–1923), who became the consultant engineer to the General Electric Company at Schenectady (an Indian name which aptly means 'Gateway to the West') and professor of electrical engineering at the University there, did amongst other things valuable work on hysteresis (a problem of alternating magnetism) and on man-made lightning. W. A. Antony (1835–1908), who gave at Cornell one of the earliest courses on electrical engineering, was a leading spirit in the formation of the American Institution of Electrical Engineers in 1884. This so accelerated the establishment of similar courses that by 1896 Senator Hoar could remark 'every technical school in the country has established, or is seeking to establish, a department of electrical engineering'. And one of the great sources of strength was the free movement of 'academic' and 'business' electricians into their respective fields. Professor Elihu Thompson gave his services to the Thompson-Houston Company and later to the General Electrical Company, which, as we have just seen, established the first research laboratory in this field in the world.

The research laboratories in the electrical, chemical, automobile and other industries multiplied enormously. Before 1915 there were less than 100 employing less than 1,500 people. By 1950 there were 2,700 of them, employing a total personnel of 175,000, of whom 70,000 were scientists. Some of them were to win names as incubators of fundamental as well as applied research.

Americans romped around this new power medium. In 1900 the professor of electrical engineering at Pittsburgh, R. A. Fessenden, transmitted speech by radio, and in the following year took out the first patent for voice transmission. With the help of Charles Steinmetz of the General Electric Laboratory, in 1906, he made his first broadcast of music and speech. A year later Lee de Forest patented the three-electrode vacuum tube that made it possible not only to rectify but also to amplify the signals. The United States Navy adopted the new system. Another electrical engineer, E. H. Armstrong, developed the positive feed-back, or regenerative circuit, in 1914, the superheterodyne system, and later the frequency modulation system. Armstrong's teacher, M. I. Pupin (1858–1935) was perhaps the classic example of the internationalisation of electrical knowledge. Himself a Central European exile, he studied at Columbia, went to England to study under Clerk Maxwell, and Germany to study under Helmholtz. Following the theoretical investigations of the Englishman Heaviside, Pupin became the first to use the X-Ray tube in the U.S.A. He pioneered the use of distributed inductance coils to perfect long-distance telephone calls, and lectured to business men as a professor of electro-mechanics at Columbia University.

NEW MEDIA

The opening of the Nickelodeon at Pittsburgh in 1905 brought to a commercial head the various experiments in projective photography that had culminated in the Edison-Dickson kinetoscope. This industry stimulated a new art form when D. W. Griffith made his film *Birth of a Nation* in 1915. When sound was added by General Electric in 1926, it became the most formidable instrument for spreading the American way of life, setting standards of dress, behaviour, speech and habits throughout the world.

Television was another laboratory-born entertainment. Stemming from the British cathode-ray tube (pioneered by Crookes, Campbell-Swinton and Baird) and the German scanner (invented by Paul Nipkow), the basic techniques of modern television were worked out by the Americans H. E. Ives (of Bell Telephone), V. K. Zworykin (of Westinghouse and the Radio Corporation of America) and P. T. Farnsworth. This has now become the most influential medium of communication yet given by science to society.

The new media themselves profoundly affected technology. The T.V. camera has become a new machine tool in the service not only of expression but of discovery, whilst the electron microscope is currently an eagerly coveted item of basic research equipment. It is indispensable in atomic energy development as a means of remote control. Indeed we seem to be but on the threshold of its manifold uses.

ELECTRICAL TECHNIQUES

In 1909 the Heroult electric arc furnace was introduced from France and within six years 323 were in operation, more than a third of the total number then operating in the world. Machinery in the steel industry was also electrically operated, especially at Gary, Indiana. Such techniques, both industrial and educational, became hot gospels throughout the western world. The glass industry too was revolutionized by the need for lamp bulbs, and the Owens method of mass manufacture of these spread to bottles. Electricity made possible the more efficient refining of copper and zinc, and was later to help in the creation of a virtually new non-ferrous metallurgy based on the recovery of metals from the sea and the sand, like magnesium and zirconium. Zirconium was discovered by a German refugee in America, Dr. William Kroll, who presented his discovery to the United States Bureau of Mines. Now it is used as a material in the construction of nuclear reactors.

Aluminium (electrically made from bauxite), the third most abundant metal on the earth's surface, was supplemented by titanium (electrically made from the ore ilmenite)—the fourth most abundant metal on the earth's surface. And here, too, Kroll was the pioneer. Beryllium (from beryl in Argentina) was another element even lighter than aluminium. Germanium was discovered to have electrical properties which led to the making of transistors, which in turn made possible great advances in electronics.

All these complex metallurgical advances were reciprocally connected with the development, on a vast scale, of the hydro-electric resources of the United States.

SCIENCE AND SOCIETY: MULTI-PURPOSE HYDRO-ELECTRIC SCHEMES

America realized that science was the dominant force in modern life. Thorstein B. Veblen (1857–1929), in *The Place of Science in Modern Civilisation* (1906), observed 'in our modern culture, industrial processes and products have progressively gained upon humanity until these creations of man's ingenuity have latterly come to take a dominant place in the cultural scheme. They have become the chief force in shaping man's habits and thoughts.'

The social position of the scientist in American life was illustrated by the convention in 1908 of a conservation conference at the White House. This drew up the first inventory of American national resources, and within two years thirty-eight of the states had set up bodies to consider projects for soil conservation, irrigation and afforestation. The states began to think increasingly of planning rather than plundering their natural resources, especially after 1929 when an engineer, Herbert Hoover, became president of the United States. Hoover had obtained experience as the promoter of a report entitled *Waste in Industry* (1921). After becoming president he appointed another committee, and its chart of *Recent Social Trends* (published in 1932) drew attention to the unsynchronized developments in technology which were so profoundly affecting society. Hoover's successor, F. D. Roosevelt, continued this line of thought, and his National Resources Committee's report on *Technological Trends and National Policy* (1937) was an attempt to forecast long-range changes in American life. Roosevelt's administration was, by

275

necessity, marked by a great experiment in social engineering: the Tennessee Valley experiment. T.V.A. techniques were in turn to be imitated on an equivalent scale in the U.S.S.R.

With the passage of the Reclamation Act in 1902 the United States Government began to exploit the work on turbines and generators by building large storage dams like the Roosevelt Dam in Arizona and the Pathfinder Dam in Wyoming to provide power. Spurred by a power hunger, others followed. The Shoshone (now Buffalo Bill) Dam, also in Wyoming, was when finished in 1910 the largest in the world. By 1920 40 per cent of American electricity was hydraulically generated.

The drainage basin of the Tennessee River comprises an area half that of Great Britain—about 41,000 square miles. Its recurrent floods and difficult navigation had tantalized engineers since the Civil War. In 1918 the nitrogen fixation plants and hydro-electric power dam built by the federal government at Muscle Shoals, Alabama (the latter not finished till 1925) had to be disposed of to private industry.

The advent of President Roosevelt to power and his determination to cope with the greatest industrial depression America had yet experienced, led to the whole undertaking becoming public property. A multiple purpose corporation known as the Tennessee Valley Authority was established in 1933. This set about building sixteen major dams and modifying five others already existing to unify the development of the river. At a cost of well over £200,000,000 a 9-foot navigable channel linked the inland waterway system of the Mississippi Valley with the seaborne commerce of the Gulf of Mexico. Flood control, power supply, soil conservation, fertilizers, public health and reafforestation are all part of the development of the natural resources of the region. It has reversed the natural erosion of the land, restoring some 3,000,000 acres to cultivation. It is, without question, the biggest effort in social engineering America has made. Other projects were also put in hand. By 1930 the Boulder Canyon Project began, with Hoover Dam as its central feature. It was followed by the Columbia River Basin Project, in the Northwest, which in the Grand Coulee, produced the largest concrete dam in the world; the Central Valley Project in California, and the Missouri River Basin Project with some twenty multi-purpose dams.

The T.V.A. was the largest single construction programme ever undertaken in the United States. It supplied 15 per cent of the national hydro-electric capacity and 5 per cent of the entire national generating supply for public use. It was equivalent to eight Boulder Dams or three Grand Coulees. This power has been distributed over 80,000 square miles, and with its system of high navigation dams from Paducah in Kentucky, to Knoxville, Tennessee, it has reduced floods, promoted irrigation, has nine major hydro electro-plants and sells energy to forty-five municipalities and rural power corporations. Since 1933 a million acres have been terraced and another two million improved.

Its success has led to a number of proposals for the establishment of similar autonomous development corporations in other major regions of the U.S.A. and encouraged English socialists, like the late Professor H. J. Laski, to hope for similar benefits from state socialism. In the second world war it provided electrical energy to produce aluminium (for aeroplanes) and for the gigantic atomic energy plant at Oak Ridge.

In the north-west, the Columbia River, the greatest single source of power in the United States, was harnessed by two dams, Bonneville and Grand Coulee, the latter when completed in 1941 being the largest hydro-electric station in the United States. These powered the Kaiser shipyards at Vancouver, the Boeing aircraft plants at Renton and Seattle, the light metals industry of the west, and the plutonium plant at Hanford, near Washington, that made possible the atomic bomb.

The example was contagious. After the war continental countries hastened to construct similar hydro-electric projects. Switzerland had the Grand Dixence (912 feet) and the Mauvois (745 feet) which, when finished, will be the highest and second largest dams in the world. France built, or is building, forty post-war dams and Italy five. Scotland has the Errochty, Pitlochry, Loch Sloy and Cluny dams; India the Maganadi Valley project and the Bhalera Dam in the front range of the Himalayas, India's greatest. Australia's Snowy Mountains project is another of the world's largest. 'White coal' has apparently come to stay.

The major integrated piece of international land surgery of post-war years, however, has been the creation of a seaway to any port on the shores of the Great Lakes. Begun in 1954 after a tussle with a combination of vested interests, ranging from Atlantic ports to private producers of electric power, it now enables power to be generated for New York State as well as Ontario, and gives access to Labrador iron ore for mid-western steelworkers and outlets for mid-western wheat to Europe. It has converted the great lakes into an eighth sea or a North American Mediterranean. Not the least significant feature of its opening is the construction of an aluminium plant at Massena.

PRODUCTION ENGINEERING

These massive projects, involving problems of efficiency in men and materials as well as organizational gifts of a high order, rested on another great American contribution to the twentieth century—production engineering. Here the pioneer was H. R. Towne (1844–1924). H. R. Towne had superintended the erection and installation of the machinery in the monitors by John Ericsson. In 1868 he realized that Linius Yale's newly invented pin tumbler lock could be made by quantity production (except the tumbler cylinder). A company was formed to do this of which, owing to Yale's death. he was left in charge. He extended it to manufacture electric hoists, testing machines and, above all, cranes, of which he was a pioneer in the U.S.A. By 1883 the Yale and Towne Manufacturing Company (as it was then called) was really specializing in production engineering. His experience in this field was embodied in a paper he read in 1886 to the American Society of Mechanical Engineers on 'The Engineer as Economist': an eloquent plea for the marriage of production and management. From this time onwards various engineering societies encouraged the management movement, justifying the title 'production engineering' that is now usually applied to this important activity.

'Efficiency' has been hitherto applied to prime movers and machines. Now it was applied to men. Here the pioneer was Frank W. Taylor (1856–1915), an American who had abandoned the study of law because of bad eyesight. Entering the Enterprise Hydraulic Works—a pump manufacturing company in Philadelphia—he became a pattern-maker and machinist. In 1878 he joined the Midvale Steel Company, Philadelphia, as a labourer. By studying at night at the Stevens Institute of Technology at Hoboken, New Jersey, he became a Master of Engineering in 1883. He also found time to win the tennis doubles championship of the United States in 1881. By 1884 he was chief engineer at the Midvale Steel Company. He designed and built the largest steam hammer ever built in the United States in 1890 and went on to become general manager of the Manufacturing Investment Company, Philadelphia, which were converting the forests of Maine and Wisconsin into paper. In 1893 he set up in Philadelphia as a consultant on efficiency engineering, and two years later his paper, *A Piece Rate System* (1895) initiated a number of studies.

Taylor believed the capabilities both of men and machines could be ascertained, and that by so doing the antagonisms between employer and employee could be eliminated. By 'job-analysis' in various works he proved his point so well that the great Bethlehem Steel Company of Pennsylvania retained him to undertake a study of the treatment of hot steel. As a result of his collaboration with J. Maunsel White, the Taylor-White process of heat treatment of tool steel resulted in increased cutting capacities of 200 to 300 per cent.

So enthusiastic an evangelist for the 'Science of Management' did Taylor become that he resigned from the Bethlehem Steel Company to promote it by lecture and free consultation. His efforts resulted in the formation of a society with that name in 1911. In the same year his *Principles of Scientific Management* was published, and in a paper written for the American Society of Mechanical Engineers he synthesized his creed: 'We can see our forests vanishing, our water power going to waste, our soil being carried by floods into the sea; and the end of our coal and iron is in sight. But our larger wastes of human effort, which go on every day . . . are less visible, less tangible.'

Psychologists and psychoanalysts reinforced his findings, notably by Hugo Muensterberg's *Psychology and Industrial Development* (1913). Current developments in cost accountancy and the work of others emphasised it still more. Henry L. Gantt utilized graphical systems and charts for disclosing variations in sales and production, Carl G. Barth developed a production slide-rule, Harrington Emerson applied job analysis to the working of the Santa Fe Railroad, while Morris Cooke applied similar principles to city management.

Yet another development in 'Scientific Management' was that of Time and Motion Study. Here the pioneers were Frank B. Gilbreth and his wife. Gilbreth, a vice-president of the Society for Promoting Engineering Education (founded in 1893) used cinematograph films to study work motions. Among her other work, his wife's application of these techniques has done much to make the modern American kitchen the labour-saving machine that it is today. The Gilbreths had, like Taylor, just published a book, *Motion Study* (1911). Frank Gilbreth and F. W. Taylor had, however, met and come into contact thirteen years before when Gilbreth, then the 30-year-old owner of a successful building contractor's firm, found one of Taylor's assistants timing his bricklayers with a stop-watch. Gilbreth disapproved of time studies before actual analysis of the motions had

279

been made, and the rest of his life was spent in showing how motion study should go forward.

He was well qualified to do this. After leaving high school at 17, he began work as a building contractor's apprentice and within ten years had not only became a foreman and superintendent, but had established his own business. This was in 1895. Three years later, having studied the motions of his men, he increased the output from 1,000 to 2,700 bricks a day. In 1904 he married. In 1912 he began to practise as a consultant with his wife Lillian Moller Briggs. By using the cinematograph and cyclegraph they eliminated unnecessary motions. He classified motions into basic types, called 'therbligs'— an anagram of 'Gilbreths'.

Taylorism spread in early twentieth-century America like a prairie fire. It was given great publicity when at the hearings of the Interstate Commerce Commission in 1910 Louis D. Brandeis presented industrial engineers who reported that railways could save a million dollars a day by such means. Two years later the Efficiency Society was organized in New York, and in the same year a society to promote the Science of Management was organized, which later became the Taylor Society.

The suspicions of the trade unions were immediately aroused. They suspected that efficiency was another word for speeding; they feared that skill and craftsmanship would be superseded by specialized routine dictated by job cards, and they were convinced that some engineers were bent on eliminating collective bargaining—the very basis of their existence. At their instance, the House of Representatives appointed a special committee to study the Taylor system in Government arsenals; this resulted in Congress prohibiting time study on employees there. Then in 1914–15 Professor R. F. Hoxie was appointed to investigate further claims, and his *Scientific Management and Labour* (1915) led to a further retreat from time studies. If Taylor's system was not popular in his own country at first, it certainly spread abroad.

TAYLORISM IN EUROPE

Coincidentally two British industrialists were thinking along the same lines. J. Slater Lewis, the General Manager of the Salford Rolling Mills, published his *Commercial Organisation of Factories* in 1896, and A. J. Liversedge was incubating *Commercial Engineering— By a General Manager*. In France Henry Fayol, general manager of one of the mining and engineering combines, addressed the Jubilee Congress of the Société de l'Industrie Minérale in 1908 on 'L'Administration Industrielle et Générale'.

During the first world war mass production techniques accelerated the study in Britain of what was called Engineering Production, and a journal of that name was published in which H. E. Honer suggested in 1920 that the time had arrived for specialized institutions to cater for engineers engaged in manufacture. A meeting was held on 26 February 1921 and the Institution of Production Engineers was founded.

In France Georges Clemenceau issued a circular on 26 February 1918 pointing out the necessity of all plants under the Ministry of War employing 'in every kind of work . . . the minimum of labour through scientific research with the most advantageous methods of procedure in each individual case'. Planning Departments were established in each works and the heads of these were recommended to study certain specific texts by Taylor, notably an article in the *Revue Metallurgie*, XII (1915), and two books by Vincent Cambon and Henry le Chatelier respectively. The Michelin Foundation sent young engineers to America to study management methods, and sought to promote Taylorism in the technical colleges. In Vienna a periodical was started called *Taylor Zeitschrift*.

Taylorism's most distinguished convert was to apply it to building, not a business, but a state. On 28 April 1918 Lenin, in an article in *Pravda* on 'The Urgent Problems of Soviet Rule', urged 'We should try out every scientific and progressive suggestion of the Taylor system . . . we must introduce in Russia the study and teaching of the new Taylor System and its systematic trial and adaptation.' It is to Russia we must now turn.

CHAPTER 24

The Awakening of the Russian Giant, 1920-1946

At the end of the nineteenth century Russia was the most rural of the European countries, with only 12·8 per cent of its population in towns as opposed to France and Germany's 40 per cent and England's 70 per cent. From 1722–1897 its population had increased from 14 to 129 millions, without the concomitant industrial expansion of Western Europe. The retardation of Russian life was a complex affair. Tartar invasion, serfdom, the village commune, a restrictive political atmosphere, based on autocratic government and large estate administration, all contributed; but the heart of the matter was poor communication and altogether inadequate exploitation of power. Military defeats by Japan (1904–5) and Germany (1914–17) helped detonate the Russian Revolution of 1917.

When Lenin took power he found that three-quarters of the Russian cornfields were lost: the Ukraine, Poland, Finland, Lithuania, Estonia, Latvia, and part of Caucasia. Under his leadership a new society was planned by a committee of politicians, geologists and engineers. Known as Gosplan, its outlines hardened after 1924, when the Ukraine, White Russia, Trans-Caucasia and the Central Asian Republics had been brought, with Russia, into the Union of Socialist Soviet Republics. The first and subsequent five-year plans transformed the U.S.S.R. into the most scientific economy in Europe. Under extreme economic pressure and national insecurity the Communists were trying ruthlessly to achieve in a decade what it had taken the Western powers over a century to accomplish.

Heavy industry was moved eastwards and labour camps set up, first in the Urals, and later in the Arctic Circle (Yagoda) and Kazakhstan (Karaganda). Consumer goods were limited and piece work was introduced. A ferocity unknown in Russia since the time of Peter the Great, forced the 25,000,000 peasants, very much against their will, into vast collective farms, to provide food for this gigantic operation.

These labour camps grew. Operated by a body known as Gulag they came to number about 40 comprising some 14,000,000 forced labourers and were run (after 1946) by the N.V.D., or the Ministry of Internal Affairs. The development (after 1930) of the Ural–Kuznetsk complex of mining and metallurgical enterprises in the Urals and Siberia, the very heart of Russia, provided the sinews for driving back the German invasion of 1941.

The shift in the centre of gravity of the Soviet Union to Siberia was accelerated during and after the second world war. At Magnitogorsk in the Urals, and further east at Kuznetsk and Krasnoyarsk the industrial skeleton of a new country equivalent in size to Europe is taking flesh. There in forty years, 560 new cities and 1,000 new urban communities have been founded.

THE PETRINE PRECEDENTS

In many such matters, the Russians acknowledge Peter the Great (1672–1725) as a great forerunner. He encouraged a school of navigation in Moscow, for which British engineers and mathematicians were recruited. He conceived the idea of constructing a Volga–Don Canal to enable ships from the Volga to join ships from the Don in attacking Azov. He asked the English engineer, Captain John Perry, to build it together with docks and other projects. Perry, during his thirteen years in Russia, found such procrastination and incapacity amongst Russian workmen that he was ultimately obliged to put himself under the protection of the British Ambassador and return to England, where in the intervals of draining and embanking, he wrote an account of his experiences: *The State of Russia under the Present Tsar* (1716).

Peter the Great also established the Imperial Small Arms Factory at Tula (which still exists) under an Englishman called Trewheller. Ironworks at Neviansk and Kamensk were built in 1700–3, and worked by conscript artisans. Demidov and his son Aklinfi developed blast furnaces on their own, whilst Tatischev and De Hennin developed others for the state. By 1740 Russia was producing 31,975 metric tons of iron to England's 20,017, France's 25,979 and Germany's 17,691. By 1745 there were 49 metal works, and by 1762 97. The market for the iron they produced was England, which by 1781 was importing 50,000 tons of pig and bar. England ceased to take Russian iron when Henry Cort perfected his puddling process in 1787.

To Peter, as to most eighteenth-century rulers, 'the arts' meant the *Masterstva*—mathematics, mechanics, surgery, architecture, anatomy, botany and hydraulics. So when he 'cut a window through on Europe' by founding St. Petersburg (now Leningrad) he also cut windows through on Europe by building in addition to his school of mathematics and navigation (1701), a translators' school (1703), a medical school (1707), an engineering and gunnery school (1712), and a number of mining and navigation schools (1721). Just after his death the Imperial Academy of Sciences was founded in 1725. Plans for this were modelled on the French Academy of Sciences of which Peter was a member.

FOREIGN SKILLS AND RUSSIAN DEVELOPMENT UP TO THE REVOLUTION

Russians either had to go abroad to acquire scientific knowledge or foreign scientists had to come to Russia. M. V. Lomonosov (c. 1711–1765) was sent to the University of Marburg to study metallurgy, where he imbibed such a great love of literature that, on returning to Russia as professor of chemistry, he literally reformed the Russian language by synthesizing the complicated ecclesiastical Slavic tongue with vernacular speech. Lomonosov was the Russian Leonardo, and his course of practical chemistry given from 1782–6 is the first for which any details have survived.

England was a Mecca for others. Roman Dmitriev was sent to study under Smeaton from 1777 to 1779. He was followed by Fedor Borsoi, who spent four years studying 'fire and other machines, mechanics and architecture', and returned with a model steam engine and boiler. Lev Sabakin of Tver translated Ferguson's *Mechanics* in 1787 and he published it, supplemented by his own lectures on fire engines. He was under no illusions about his English hosts and wrote: 'They preferred to treat me and to take me to gardens rather than their works and factories; and when we did happen to go there, I suspect they forwarned all these places by letter to conceal this or that or not to lead me to this or that place at all.' In 1794 Kokushkin was sent to England to obtain knowledge of fire engines, but he too does not seem to have organized an engineering cadre.

Amongst English engineers who went to Russia, John Robison and Samuel Bentham stand out. Robison taught there and Bentham built ships. James Watt was invited to become 'Master Founder of Iron Ordnance to Her Imperial Majesty', Catherine the Great, but declined—his friend Erasmus Darwin warning him against 'the embraces of the Russian Bear'. Smeaton made an engine for Kronstadt, but the earliest English steam engine in Russia seems to have been built at the Olnetz Ironworks at St. Petersburg. In 1787 Charles Gascoigne was induced to leave the Carron Ironworks for Russia, and took several workmen with him. James Stuart built a small foundry at St. Petersburg on Ghutuevsk Island. There seems also to have been a native steam engine devised by Barnaul, roughly comparable to that of James Watt, put into action by Polyzunov.

WHY DID RUSSIA FALL BEHIND EUROPE?

The industrial renaissance of Russia in the eighteenth century did not last. Though the export of iron stimulated the iron industry of the southern Urals round what is now Magnitogorsk, yet there were no convenient deposits of coking coal to expand the economy as there were in England. Moreover, the Urals were over a thousand miles away from a centre of population, and Moscow was only a brown coal region. So, whereas the production of iron in England during the period 1750–1825 increased by thirty times, in Russia it only doubled. Poor communication and transport lay at the heart of the matter.

The first railway was only a short line from St. Petersburg to the summer residence of the Czar (Tsarkoé-sélo). John Cockerill travelled to Russia to persuade the Czar to undertake the construction of a network of railways, but failed. The journey was his last, and he died in 1840. Three years later the Czar undertook the lines from Warsaw to the Austrian frontier and from St. Petersburg to Moscow. The Russians found it cheaper to import coal from the Ruhr than to haul it from the Donetz basin. The modern cotton-spinning industry in Russia was started by Ludwig Knoop, who was entrusted by C. W. Morosoff with the task of establishing a mill on the English model after arriving in Russia in 1839. By the end of his life Knoop had established 122 spinning mills in Russia.

If industrial skills came from England, chemical techniques came from Germany. Indeed, two of Russia's outstanding chemists were Liebig's pupils: N. N. Zinin, who first described the reduction of nitrobenzene to aniline in 1843, and A. M. Butlerov, the eminent organic chemist. There were other Russians who pioneered in agricultural engineering, like Dimitry Pryanishnikov.

The Academy of Sciences, which during the eighteenth century had been a growing point of science, was by Statute of 1836, allowed to become a virtual ivory tower. In 1841 it was merged with the Imperial Russian Academy (founded in 1783) and began to operate in three divisions: Mathematical Physics, Russian Language and Literature and Historico-Philological Studies. Its hostility to new ideas, whether in chemistry or politics, was illustrated by its rejection of D. I. Mendeleyev (1834–1907), the Russian chemist whose formulation of the periodic table of atomic weights enabled him to predict the properties of several new elements like gallium, scandium and germanium, which were actually discovered a few years later. Ironically enough, his work was to be invaluable in the discovery of new rocket fuels.

Though there are ample signs of other industrial activity before the 1918 revolution, indeed a Commission for the study of Natural Productive Resources of Russia was set up in 1915, yet it never got under way before the series of national five-year plans initiated by the Bolsheviks provided the drive and direction.

In 1918 foreign capital owned 90 per cent of the joint stock of the mining industry, 50 per cent of the chemical industry, 42 per cent of iron, steel and engineering, 37 per cent of the timber trade and 28 per cent of the textiles. Of the total foreign capital one-third was French and between one-fifth and one-quarter was British. There were no 'cartels' or 'trusts' of their own on which the Soviets could build.

THE PLANNED EXPLOITATION OF POWER

To Lenin, Russia was to be regenerated by 'socialism plus electrification'. His successors have carried out this policy with vigour and imagination, so though coal production increased by 1,500 per cent since 1913, electricity has increased by 10,000 per cent. The Volkhov hydroelectric station was the first step. Constructed between 1921 and 1926 to supply electric power to Leningrad, it was the fore-runner of innumerable projects that rapidly opened up the really vast resources of what Sir Halford Mackinder called 'the heartland'.

Russia's energy resources amount to 23 per cent of the world's known types, whereas those of all other countries in Europe amount to 10 per cent. With one-half of the world's total resources of iron ore and over one-half of the petroleum resources, and one-sixteenth of the ice-free land surface of the globe, Russia's policy of massive industrialization almost inevitably was based on electricity, helped (as we saw) by modifications of the Taylorian system of production engineering.

Along the Volga a glowing necklace of large industrial towns is strung: Kalinin (railway carriages), through Yaroslav (mixed), Vladimir (tractors), Invanovo (textiles), Gorky (heavy industry), Kazan (wrist watches) to Kuibyshev. At Kuibyshev the waters of the Volga are harnessed for one of the largest power stations in the U.S.S.R., and the harnessing continues further down at Saratov and Stalingrad, where the largest mass of man-made water in the world will be collected.

The greatest Russian hydro-electric project, the Kuibyshev barrage across the Volga, involved more sheer moving of earth than did the Panama Canal. Opening it in August 1958 Nikita Kruschev expressed dissatisfaction at the length of time—seven years—involved in building generating stations of this kind, and opted for the more swiftly built thermal power stations. In pursuance of this policy the Siberian projects at Saratov and Krasnotarsk were to be abandoned. So proficient had the Russians become in building dams that it was suggested by Vasili Zakharchenko (in *U.S.S.R.* December 1956) that the Behring Straits, the narrow water gap between Siberia and Alaska, should be bridged to block the Arctic waters from penetrating the Pacific and so prevent the far-Eastern ports from becoming icebound.

THE BUILT-IN ACCELERATORS

The built-in accelerators of the Soviet economy are education and research. Through a complex of institutions and organizations the whole outlook of the youth of Russia is oriented towards the mastery of their environment. This orientation becomes sometimes such a lust that at times the bald elaboration of a Soviet scientific theory (e.g. Lysenko's) reads like a political extrapolation into genetics. Organically, education is Russia's breeding heart. It drives the state. Instead of allowing the administrative machine (as in France) or the universities (as in England) to exercise a determinative voice in curricula, the whole Russian system is designed to provide the technological revolution with the engineers and scientists to service and improve it. Lenin began by requiring certain polytechnic institutions to shed their elementary teaching loads and develop as specialist technological institutes. He initiated this revolution within the revolution and from 1928 the emphasis in all the technicums (for technicians) and technological institutes (for technologists) was the overriding need for scientific manpower in every field, especially in engineering.

In 1927 20,200 engineers were at the disposal of the Soviet and another 25,200 were needed for the first five-year plan. To produce these was itself a major operation, and was considered as such. By 1933 7,900 engineers were graduating a year. By 1937 that figure had quadrupled to 34,700, and a quantitative and qualitative analysis showed that a total strength of 250,000 engineers and architects and 810,000 trained technical personnel, were available. These became a virtual technocracy. Well they might be, for by 1938 coal production had quadrupled, so almost had oil. In metals steel production had quadrupled, that of iron had increased by a third and pig iron by over a third. Each Russian engineer's training is supplemented by at least three industrial practice assignments totalling 25 weeks and reaches its climax with a thesis. Respect for 'science' is obligatory, and for applied science especially. Engineers are third from the top in the salary scale, only outranked by élite groups and by research scientists. They certainly rank higher than physicians, farmers or teachers.

Incentives played as large a part as discipline in the progress of Russian industrialization. When industry was reorganized in 1929, a determined effort was made to capture the loyalties of engineers of the pre-revolution regime. As Stalin told a conference of managerial workers on 23 June 1931, 'It would be silly and unwise to see an uncaught criminal in practically every specialist or engineer of the old school. . . . Hence, the task is to change our attitude toward the engineers and technicians of the old school, to show them greater attention and solicitude, to display more boldness in enlisting their co-operation.' So in 1935 the Stakhanov movement was launched. On 15 May that year at a conference of the managers, engineers and technical personnel, P. Ordzhonikidze demanded further improvement in production and on 30 August, in a prepared session, Stakhanov—a miner—hewed out 102 tons of coal in six hours. This led to an even more intensive study of production engineering by the intelligent workers. Special courses for masters of socialist labour were organized for them. The second five-year plan, confirmed in 1934, approved an increase of 19·5 per cent in the engineering and metal-working industries to give an overall increase of 207 per cent.

From 1928 to 1953 the number of persons who completed higher professional education in the U.S.S.R. increased from 280,000 to

1,920,000, of which 630,000 were engineers representing 25 per cent of all the graduates. This should be set against comparable numbers in the U.S.A. where, in the same period, though twice as many students graduated, only 480,000, 9 per cent of the total, were engineers.

In 1928 the U.S.S.R. employed 40,000 engineers. By 1953 it was employing 510,000 engineers, most of whom had a sound engineering training. To this number must be added another 20 per cent of 'practicals', i.e. those with little formal education but plenty of on-the-job experience, giving a total force of 610,000 engineers.

So many scientists and engineers are now being produced that there is a virtual surplus for export. These new technological jesuits are the market-winners for the Russia of tomorrow.

TABLE OF THE PROPORTION OF ENGINEERING TECHNICAL PERSONNEL
TO PRODUCTION WORKERS

in all branches of Soviet industry (1930–50)
(per 1,000 production workers)

	Higher Education	Technicians	Practicals	Total
1930	7	8	20·53	35·5
1936	13	17	40·5	70·5
1940	19·7	23·3	67	110·0
1950	28	37·0	65	130·0

The 'liberal' arts were brought in to provide a hagiography for this new technical aristocracy. Literature and historical inquiry were focused on the dynamics of industrialism and the factory. A long and increasing list of Russian 'firsts' in the field of technology began to appear. The steam engine, it was alleged, was invented by L. I. Polzunovin in 1766, the electro-magnetic telegraph by L. L. Schilling, electric lights by A. B. Lodygin in 1875, the aeroplane by A. F. Mozhaisky in 1882, the radio transmitter and receiver by A. S. Popov in 1895, the television tube by Boris Rozing in 1907, 'agricultural machinery' by V. P. Goryachin, and basic theoretical work for the multi-stage space-rocket by K. E. Tsiolkovsky in 1911. Such zeal and imagination was displayed in rewriting the history of Russian engineering that Stalin himself was credited with the remark that a writer is 'an engineer of the human soul'. As far as Russian literature is concerned, his remark has a grim relevance. L. L. Sedov, who is generally accredited with the successful launching of the Sputnik, has had another distinguished Russian predecessor, N. I. Kibal-tchitch in St. Petersburg, who visualized in 1892 rocket propelled airship motors fed with charges of solid propellant. To these 'firsts', the Russians have recently added one that is incontestable: that of launching the first rocket on the moon. In this major technological triumph, great credit undoubtedly accrues to A. V. Topchiev (1907–) a graduate of the Mendeleyev Institute of Chemical Technology, and head of the Institute of Petroleum and Gas Chemistry, whose work on hydrocarbon propellants generally is of prime importance.

THE ORGANIZATION OF RESEARCH

The plutonium heart of the whole apparatus is the Soviet Academy of Sciences: today the third most important body in Russia after the Presidium and the Council of Ministers. Symbolically it was purged and reorganized in 1929 when the first five-year plan began. Institutes of special technical studies were established. Engineers and 'heroes of practical work' were appointed to membership, to marry up industry and science. Two of these indicate the scope of the Academy's duties from now on. The first, I. P. Bardin (1883–), a graduate of the Kiev Polytechnic and now vice-president of the Academy, was a chief metallurgical engineer in the Donbas and subsequently a powerful agent in the exploitation of the mineral resources of the Urals before and after the war. The second, A. V. Vinter (1878–), also a graduate of Kiev and a distinguished power engineer, built the first peat-fuelled electric power plant and was responsible for directing the construction of the Dnieper Industrial Combine. His subsequent work on the Kuibyshev and Stalingrad hydro-electric power plants showed how research was planned in accordance with the actual needs of socialist construction.

'Individualists' were discouraged and seminars instituted to entangle those who remained with current problems. Even the Academy itself had to submit a plan of its research projects to the Council of Ministers.

Much more important, the Academy was obliged to develop its own scientific personnel. It became a virtual ministry of science and technological development, subordinated to the Council of Ministers: the centre of a nucleus of research organizations which in time became academies of sciences on their own. As a symbol of its changed status it moved from Leningrad to Moscow in 1934. By 1938 its Technical Department (established in 1935) had been supplemented by another five to encompass all fields of intellectual endeavour. The change was such that whereas in 1917 it had 45 regular members and 220 scientific and technical associates, by 1941 it had 5 honorary members, 119 regular members, 182 corresponding members and a total personnel of 10,000.

Its work swelled to include not only the study of practical problems, but the preparation of compendia, encyclopaedias and national surveys. It incubated a number of scientific research institutes: 18 in physics and mathematics, 16 in chemistry, 20 in geology and geography, 32 in biology, 32 in technical sciences, 10 in history, 3 in economics, politics and law, and 4 in legislature and language. In addition, 16 branches of the Academy straddled the whole of the Soviet Union, from Sverdlovsk (founded in 1932) to Ufa (founded in 1951). More than half these branches have been formed after the second world war.

In addition to these branches of the Academy of Sciences at Moscow, others have developed as Academies of Sciences in individual Union Republics. Kiev (in the Ukrainian S.R.) was founded in 1919 and Minsk (in the Byelorussian S.R.) in 1929. The war stimulated the foundation of Kaunas (in the Lithuanian S.R.) in 1941, Tashkent (once a branch in the Uzbek S.R.), and Nakhichevan (once a branch in the Armenian S.R.) in 1943, Baku (once a branch in the Azerbaidzhanian S.R.) was founded in 1945, and Alma Ata (once a branch in the Kazakh S.R.) in 1946. In the new satellite countries Tallin (in the Estonian S.R.) and Riga (in the Latvian S.R.) were founded in 1946. In the same year Tbilsi (once a branch in the Georgian S.R.) was founded. Stalinbad (once a branch in the Tadzhik S.R.) and Ashkabad (once a branch in the Turkmenian S.R.) reached full Academy status in 1951.

Today the Academy of Sciences with less than 200 full members directs the work of a hundred different research institutions and over a quarter of a million Russian scientists, with a staff of 37,000 on a budget six hundred times greater than it enjoyed in Czarist Russia. Unlike the Royal Society, it recognizes the social as well as the exact sciences and excludes only art, music and medicine which have their own academies. Its eight panels cover every subject of human study and endeavour. It is at once pansophist, philosophic and polytechnic.

In addition to publishing nearly 30,000 papers a year it has founded in 1953 an All-Union Institute of Scientific and Technical Information which publishes, with the help of 1,800 translators, four times a month, abstracts from all the major journals. Its president, the biochemist Alexandr Nesmeyanov (1899–) established the Metallo-organic Compounds Laboratory at Moscow University. His methods of obtaining organic compounds of mercury, tin, lead, thallium, antimony and bismuth are known as 'Nesmeyanov's diazo-method'. Contemporary academicians include L. D. Landau (1908–), a graduate of Leningrad State University, who since 1937 has worked at the Vavilov Institute of Physical Problems of Moscow on the so-called phase transitions of second order in solid bodies elucidating their profound connection with qualitative changes in symmetry of the body during transition. Another is V. I. Veksler (1907–), director of the Electrophysics Laboratory of the U.S.S.R. Academy of Science and director of the High Energies Laboratory at the Dubno United Institute of Nuclear Research. He made possible the production of new types of accelerator-sychrotrons, phasotrons and microtrons. With D. V. Yevfremov and A. L. Mints he directed the construction of the ten billion electro-volt sychrophasotrons in the Academy of Sciences Electro-physics Laboratory.

A third is Y. G. Mamedaliev (1906–), whose extensive research on chemical methods of petroleum refining began with the development of the first industrial installation for the synthesis of high-octane compounds. He is now rector of Azerbaidzhan State University at Kirov: having been head of the Organic Synthesis Laboratory of the Azerbaidzhan Petroleum Research Institute. A fourth, D. M. Chizhikov (1895–), a graduate of the Moscow Mining Academy, directed the planning and construction of the zinc distillation factory in Donbas, and now directs the newly organized Research Institute of Non-Ferrous Metals. A fifth, P. P. Budnikov (1885–), a pre-war graduate of the Riga Polytechnic, is the outstanding Soviet authority on the chemistry and technology of cements and silicates. His hydraulic and anhydrate cements are widely used throughout the U.S.S.R. A sixth, A. N. Tupolev (1888–), a graduate of the Moscow Higher Technical School, is a leading aircraft designer, with over 100 types to his credit, the latest being one which carries 220 passengers, has two elevators, a telephone system and a restaurant for 48 persons and is considered the world's cheapest plane, the first really to challenge rail transport as a cheap mode of travel.

These six leaders in the fields of metallurgy, applied physics and power symbolise the results of forty years of education, ingenuity, toughness and discipline. They and others like them are beginning to tame the Siberian sub-continent, make habitable its frozen lands and shrink them into a manageable whole. Symbolically the gateway to it, Chelyabinsk, has the second largest tractor plant in the Union, and nearby Magnitogorsk has the largest iron foundry and steel mill in the world. Further east, the Kuzbas, with its coal reserves of nearly a million million tons, feeds the steel foundries of Stalinsk, the chemical plants of Kemerovo and the zinc works of Belovo. And further east still are the rising towns of Komsomolsk and Vladivostock, the latter a nursery of I.C.B.M.'s.

The I.C.B.M., or intercontinental ballistic missile is today's major national armament. When the U.S.S.R. launched the first artificial satellite to circle the earth (humorously called Sputnik, or Fellow Traveller) on 4/5 October 1957, another a month later on 3 November and a third on 15 May 1958, the general public obtained an insight into the vast engineering and technological skill that lay behind their rocketry. This insight was sharpened when on 12/14 September 1959 a Soviet rocket hit the moon near, of all places, the Sea of Serenity, and another put into lunar orbit the first instrument to record the other side of the moon.

These and other developments in Russia have in turn set off elaborate programmes in other countries. Chains of radar warning stations, and electronic scanning systems show that Russian rocketry is respected throughout the world.

THE RISE OF CHINA

An equally relevant, if less spectacular ascent has been that of China to potential world power status. Boosted by Russian precedents and skills its second (1958–62) and third (1962–7) five year plans aim (among other things) at increasing output from Sinkiang, an area bordering on the Soviet Republic of Kazakhstan. Here rich deposits of uranium, pretroleum, oil shale, vanadium, titanium, manganese and mica have been tapped since 1953. As Chou En-lai said in January 1956, 'By the end of the Third Five Year Plan (1967) our most vital scientific departments will approach the world's most advanced levels'.

To implement this programme, enrolments at Chinese universities and technical institutes are to be increased from 360,000 (1956–7) to 850,000 by 1962. This means as G. A. Modelski has pointed out, that by that time China 'may already be ahead of Great Britain in the number of engineering enrolments' and graduating classes in engineering 'may approach the level of the U.S.A.' The percentage of engineers produced (38 per cent of the student enrolment) is at present even greater than that of the U.S.S.R. (30 per cent).

CHAPTER 25

Purpose and Planning in Britain, 1935-1945

'The ideas of economists and political philosophers, both when they are right and when they are wrong, are more powerful than is commonly understood. Indeed, the world is ruled by little else.'

So J. M. Keynes, in the *General Theory of Employment, Interest and Money* (1936), began to deploy his argument that employment rests on the twin propensities to consume and invest, and that these propensities should be stimulated by state action. In the place of an evangel of 'doom' then currently affecting western thinkers, he outlined a blueprint for 'boom'. Such state action, he insisted, whether in the form of low interest rates, public works or equalization of incomes, was necessary social therapy for economic depressions. This became almost a doctrine, furtively acknowledged by the pre-war governments to remedy unemployment. Coupled to the growth of new industries noted in Chapter 22, and increasingly pressurized by an armaments programme that began in 1935, the British industrial machine began to work faster. This acceleration fell into two clearly marked phases, the first a slow overture from 1935 to 1939, and the second an intensification leading to a crescendo until 1945. Taken together, these two phases are virtually equivalent to two five-year plans. 'Planning' as such was not particularly popular in Britain at this time. For ironically enough, a planning group formed in 1931 under the name P.E.P. (Political and Economic Planning), had lost one of its members—Aldous Huxley—who wrote a widely read novel, *Brave New World*, that was broadly interpreted as a warning against it, and indeed, was.

From 1935 to 1939 British industry enjoyed a nutrient in the increasing state expenditure on armaments. From an annual average of £23,000,000 a year, this expenditure rose till it reached £273,000,000 by 1939. It began with Hitler's re-militarization of the Rhineland, and until 1939 was relatively small compared to the subsequent five years. In this phase the Royal Air Force was developed, the navy renovated and the army re-equipped with anti-aircraft armament.

Allied to the development of aircraft was that of radar. In the early thirties H. E. Wimperis, Director of Scientific Research at the Air Ministry, suggested that the detection of aircraft at a distance should be investigated. Knowledge of the reflecting properties of radio waves had, of course, been known to Marconi, but it was the Radio Division of the National Physical Laboratory, under Sir Robert Watson-Watt, which, in 1935, when the concern about our air defence was most expressed, suggested the scheme of transmitters and receivers that ultimately became the radar defence system of Great Britain by the time of the second world war. Both the United States and Germany had made progress along similar lines. When the war began, the Telecommunication Research Establishment was set up to improve the basic principle.

Aircraft production was paramount. Since Brigadier-General William Mitchell strikingly confirmed the importance of air attack by reducing an obsolete anchored battleship to a smoking hulk by aerial bombs in 1921, this had dominated British planning. The use, by Mussolini, of air power in the Abyssinian Wars, and of dive-bombers by Germany in the Spanish Civil War, showed the potential of the new weapon. So, by 1939, the main preparations of this country were against air attack. When the second world war did break out the British were soon forced to send supplies to armies in the Middle East round by the Cape of Good Hope because German air power controlled the Mediterranean. Conversely the Germans failed to invade Britain because they lost air supremacy. The Japanese were able to launch a surprise attack on America by means of a strong, swift, naval air force at Pearl Harbour, and later, thanks to air surprise, sank the great British battleships *Prince of Wales* and *Repulse* in the Pacific.

The astounding acceleration of the aircraft industry in Britain is best illustrated by the fact that not more than 50,000 people were employed in it in the year 1935. By 1943 this figure had increased to 1,500,000, with a further 300,000 employed on balloons, bombs and other requirements. Airframes and engines absorbed £93,000,000 and £117,000,000 respectively of Government expenditure.

To supplement plant existing in 1935, seven motor-car engine firms were brought into service, manufacturing and assembling components. In 1939 another group of four firms was brought in and by 1942 both groups were producing 1,500 engines a month. At the peak of engine production there were 20 major factories in operation. Then in turn there were carburettors, magnetos, sparking blocks, radio and radar equipment, light-metal industries and various kinds of hydraulic apparatus to be made as well.

Next to aircraft came motorized equipment: lorries, cars, tanks. This was especially brought home in 1939–40 when German motorized divisions consisting of tanks, motor cycles and trucks overran first Poland then Western Europe. In 1941 the Germans' advance in Russia was only materially halted when the cold froze the oil in their engines. When the United States Government entered the war they geared the motor industry to produce, by 1945, a million vehicles a year and had a ratio in France of one motor vehicle to every four men. To carry them across water, new bridging devices, tank-landing craft, and other novelties were devised.

BRITAIN, THE ISLAND ARSENAL

In its first year of deployment on the Continent in 1939–40 the British Army had to leave behind in France equipment of from eight to ten divisions, including 45,000 motor cars and lorries. In its second year Britain undertook to supply Russia with industrial machinery, medical supplies, raw materials, foodstuffs, as well as armaments and aeroplanes.

A new and even more intensive phase began in March 1941 with the planning of a final operation in which armoured formations were to play a dominant role. Labour was mobilized under Mr. Bevin, priorities and allocations were channelled, and for the first time in its history Britain was an island arsenal. By the middle of 1942 1,250,000 people were employed in the engineering industry alone, and a manpower shortage was barely solved by a virtual conscription of women. From February 1942 till the end of the war, Mr. Oliver Lyttleton gave his office as Minister of Production a far greater substance and definition than it originally possessed. The Joint War Production Staff not only satisfied industrial demands for labour but supplied tools and mobilized industrial capacity. New problems like 'Operation Bolero' (the reception and maintenance of the American forces in Britain), 'Pluto' (a pipe-line under the ocean—hence its name—to convey oil to France) and 'Mulberry' (a floating dock to enable troops to land abroad) were minor burdens to the major roles, producing 1,300 landing craft and 2,000 aircraft a month between

1942 and 1944, and a million gross tons of shipping a year. The army, operating in a number of theatres, required 600, 700 and 800 per cent more warlike stores from 1942 to 1944 than they did in the first three months of the war. The production of radio and radar equipment increased from £5,000,000 to £123,000,000 worth per annum.

As M. M. Postan (p. 356) has said 'the most remarkable achievement outside the conventional range of weapons was undoubtedly the production of medical stores'. Mepacrine (in 1939 only in experimental production), scrub typhus vaccine, sulphonamides, and penicillin were four drugs brought into mass production as a result of the war.

EMPLOYMENT (in thousands)

	June 1939	June 1943	Increase
1. Shipbuilding and repair	145	272	127
2. Marine engineering	59	99	40
3. Aircraft, motor vehicles and cycles	473	1122	648
4. Mechanical and general engineering	742	1503	761
5. Electrical engineering	139	189	50
6. Electrical cables, apparatus, etc.	196	291	95
7. Scientific instruments, watches, clocks, etc.	87	99	12
8. Non-ferrous metal manufacture	56	114	68
9. Bolts, nuts, screws, hand tools, cutlery, brass and metal ware	378	448	70
10. Railway and other carriage construction and repair	66	59	−7
11. General iron founding and heating apparatus	117	86	−31

(William Hornby, *Factories and Plant*, p. 385)

For all this a greatly improved and enlarged programme of machine tool equipment was needed. Indeed, from 1940 to 1943, 230,000 tons of machine tools were imported, three times as many as were imported during the whole of the first world war. As William Hornby, in his history of *Factories and Plant*, pointed out (p. 327) 'without supplies of these from overseas many sections of war production would have had to be re-planned at a very much reduced level of output and many programmes abandoned'. Never before were so many employed in the basic engineering industries. The total increase was some 70 per cent over June 1939 and this was mainly concentrated in the engineering, shipbuilding and chemical industries.

Mass expansion, planning, phasing—all the operations involved in making decisions on highly technical matters—led to the emergence of operational research. Of course, Archimedes conducted a kind of operational research for Hiero of Syracuse, but the emergence of the activity as a specialized field for scientists began in the first world war when F. W. Lanchester and Edison wrote memoranda on military and naval matters. The decisive step was taken by General Pile of Anti-Aircraft Command in the second world war when he secured the appointment of Professor P. M. S. Blackett to organize an operational research group for him. Blackett subsequently organized other research units in coastal command and the navy. It was from his efforts and those of Professor J. D. Bernal and Professor (now Sir) Solly Zuckerman that J. B. Conant was to learn so much in 1940. In 1941, the U.S.A. began exploring this new field with characteristic vigour. After the war operational research was adopted by large industrial firms in both countries.

The Development of Reaction Propulsion

A particular result of the war was the development of reaction propulsion. As with most inventions this was not new. John Barber of Nuneaton had, in 1791, patented a device which gave constant torque to a turbine. In 1872 the Stoltze gas turbine was designed but it was not constructed until the first decade of the twentieth century, and then not successfully. The Société des Turbomoteurs in Paris had been also active in promoting experiments in this field: René Armengaud and Charles Lemale constructing an engine in 1894 which worked at 5,000 r.p.m., producing 500 b.h.p. In Germany H. H. Holtzwarth began to experiment in 1905 on the constant-volume gas turbine. In Hungary G. Jendrassik gave it much thought. In the United States C. G. Curtis obtained a patent for a gas turbine in 1895.

A desire to utilize the energy of the exhaust gases of the internal combustion engine inspired Dr. Alfred Buchi of Switzerland and the Brown Boveri Company to experiment with Diesel engines, and Professor A. Rateau to use these gases in aircraft engines flying at high altitudes in the first world war. Rateau's investigations excited the interest of Dr. S. A. Moss of Cornell University. The General Electric Company financed research work and Elihu Thomson and R. H. Rice constructed several successful exhaust-driven superchargers for aeroplane engines flying at high altitudes.

The credit for developing the first jet aircraft to fly—the Gloster E 28/29—is given to Air-Commodore (now Sir Frank) Whittle who before the war had been selected for an engineering course at Cambridge and so united in his person both pure theory and practical application. From this developed the Meteor fighter. Once crossed, the barrier between a new idea and its practical application seems to disappear.

But even more startling developments were inspired by Hermann Oberth's *Die Rakete zu den Planetenraümen* (1923). A Society for Space Travel was founded in Germany in 1927 to experiment with Oberth's ideas, and three engineers, Rudolf Nebel (1897–), Klaus Riedel (1910–44) and Wernher von Braun, built a motor that worked on petrol and liquid oxygen. In four years, using liquid methane and liquid oxygen, they dispatched a rocket from a parade ground at Dessau. An American Interplanetary Society was founded in 1930 and it too began to experiment through a firm called Reaction Motion. A British Interplanetary Society was formed in 1933.

It was the German rockets, however, which received government backing when the German army appointed Walter Dornberger (1895–) to command an experimental laboratory at Kummersdorf in 1932. In five years Braun had produced a 25-foot rocket which had a thrust of 3,300 lb., and the Institute was transferred to Peenemünde on the Isle of Usedom in the Baltic. On 3 October 1942 an experimental rocket was completely successful. Hitler would not grant priorities at first and it was not until 8 September 1944 that they were used against London: 1,300 V2's, as they were called, were subsequently launched against England up till 27 March 1945.

The Peenemünde Institute, together with the parts of nearly 100 rockets, was captured by the American Army and transported bodily to the United States of America where, at Whitesands in Mexico, preliminary explorations of the upper atmosphere were undertaken. The firing of these V2's enabled the United States to develop their own Viking, Corporal and other rockets. New propellants like nitric acid and aniline were devised.

The adaptation and application of the rocket principle to aircraft, and to the exploration of space by means of satellites, is part of current history and it is significant that in 1946 the American Rocket Society became affiliated to the American Society of Mechanical Engineers. Not so well known is the fact that the United States used 30,000 rockets to capture Okinawa, and that by the end of the war the U.S. rocket bill alone was 100 million dollars a month.

A New Source of Energy

Perhaps the greatest result of the war was the ripening of a new source of energy. At the first world power conference, held at London in 1924, Sir A. P. M. Fleming, one of the 'Holy Forty', prophesied on 10 July 1924: 'In the future, the greatest economic advances in engineering will probably come through improvements in materials for the development of which increased knowledge of the constitution and nature of matter is required. All power developments, whether from solar energy, energy of the atom, or combustion of fuel, are based on the application of laws of physics and chemistry. The future of engineering requires for its success therefore a very close alliance with modern physics, and this can only be brought about by the training of a group of physicist engineers.'

The 'physicist engineers' which Sir Arthur Fleming described were taking shape in Manchester where, in 1907, Ernest Rutherford gathered round him a distinguished team of workers at the University working on radio-activity, a phenomenon discovered by H. Becquerel eleven years earlier. Helped by Hans Geiger, Rutherford devised a method of counting the alpha-particles emitted by radium. They found that 136,000 were ejected each second from 1/1000 of a gramme. Among the brilliant group of research students were Boltwood, James Chadwick and Andrade. In 1911 Rutherford was able to put forward the theory that the atom contained a nucleus bearing a positive charge—around which revolved a number of negatively charged electrons. Extensions to his laboratory opened by S. Z. de Ferranti, then President of the Institution of Electrical Engineers, enabled other research students like H. G. J. Moseley and Niels Bohr to work with him. In 1912 Bohr put forward a theory of the structure of the atom far more detailed than Rutherford's, utilizing the 'Quantum Theory' of Max Planck. This enabled Bohr to suppose that the electron's orbit round the hydrogen atom was determined by its 'quantum' of energy, i.e. that if it gained a 'quantum' it would move from an inner to an outer orbit. The Rutherford–Bohr theory of the atom was to hold the field. By 1919 Rutherford used alpha-particles from radium to bombard nitrogen nuclei, which ejected high-speed particles: the artificial disintegration of an element had been demonstrated for the first time. This unlocked a storehouse of potential energy. Fourteen years before, Einstein's theory of relativity had shown that matter was congealed energy. If one ounce of it was completely converted into energy, enough would be released to turn a million tons of water into steam. His equation $E = MC^2$ (where M is the mass of matter, C the velocity of light and E the energy) dangled like an inn sign at the doors of scientific laboratories all over the world.

From 1919 onwards, the investigation of nature's storehouse of energy, matter, went on in every country in the world where physics was taken seriously. In 1931 F. W. Aston (1877–1946) detected the line of uranium 235 by spectograph. In the following year H. C. Urey (1893–) announced his discovery of an isotope of hydrogen, known as deuterium or 'heavy hydrogen' and G. Hertz separated it from hydrogen. The whole structure of the atomic nucleus was considerably simplified by Chadwick's discovery of neutrons (or uncharged particles) which were found to be most valuable missiles for bombarding nuclei since they were not deflected by electrical forces in the atom. By 1935 A. J. Dempster had detected uranium 235. Fresh light on the possibilities of the atom was shed by H. A. Bethe, who suggested in 1938 that the sun itself converted hydrogen into helium at high temperatures and pressures.

Around the possibility of separating uranium 235 many scientists were working in 1939. Theoretical physics was advancing even more rapidly than it was possible to verify in practice, and the possibility of a bomb based on nuclear fission had been explored by, amongst others, F. Joliot in France, Peierls and Frisch in England and numerous others in America. In June 1941 the Thomson Committee reported that an atom bomb could probably be made before the end of the war. In June 1942, after the United States entered the war, British scientists went over to co-operate in its fabrication at Los Alamos. The results of their activity were not only to destroy Hiroshima and Nagasaki, but many of the cherished concepts which engineers had held for thousands of years.

The potential value of this new source is best illustrated by the fact that the complete fission of all the atoms in 1 lb. of U 235 is equal to the energy liberated by the complete combustion of 3,000,000 lb. of coal.

THE THERAPEUTIC EFFECTS OF THE WAR IN BRITAIN

Other results of the war were manifold. A first was the increasing intensification and popularization of mass production techniques in the metal, vehicle, shipbuilding and electrical industries.

A second was the tilting of the balance of the national work force towards the engineering industries. An increase of some 35 per cent in electricity, chemicals, instrument-making, marine and general engineering labour took place, whilst in non-ferrous metal manufacturing and shipbuilding the increase was nearer 50 per cent.

A third was the building up of a substantial amount of new industrial capital in the shape of large new factories (capable of being transformed to peacetime production) and also through the involvement of small works in contracts of one kind and another. This involvement enabled them to obtain insights into the use of materials and techniques of production not otherwise accessible to them. It was as if the 14,000 engineering works that were engaging on aircraft work at the peak of production had undergone a collective training course.

A fourth was a major revolution in British agriculture, where the mechanical power used on the land increased enormously by some 158 per cent. This did not include the increasing number of motor vehicles used by farmers in the form of lorries, vans and cars. Brawn was supplemented in its last stronghold by ingenuity when Harry Ferguson and David Brown developed a mobile hydraulic lift for farm use. This lift was tractor operated, and between 1939 and 1946 the number of tractors nearly quadrupled from 56,200 to 203,400, representing an increase of 2,850,000 horse power. During this time the number of horses declined from 649,000 to 520,000. Sir Keith Murray in retrospect estimated that in the six years after the war the use of mechanical draught power increased a further 70 per cent. Arable land was increased by 6,000,000 acres, and the operation called for ploughs, harrows, fertilizer drills and other equipment that left British farming, at the end of the war, as the most highly mechanized in the world. For this, some credit must go to the Agricultural Machinery Development Board, which in 1945 was supervising the work of a Central Institute of Agricultural Engineering. Four years after the war ended the supervision of the National Institute of Agricultural Engineering was returned to the Agricultural Research Council and a separate Agricultural Machinery Advisory Committee was set up. Concomitantly, the Institution of British Agricultural Engineers, founded just before the war in 1938, was followed by the establishment of the National Institute of Agricultural Engineering in Scotland. Both played a part in the production of agricultural machinery, so that Britain became the second largest agricultural machinery exporter in the world. In 1951 a National Diploma in Agricultural Engineering was established and eight years later the Ministry of Education decided to establish a National College of Agricultural Engineering.

ROADS AND MOTOR-CARS

A fifth result of the war is now taking place. The construction of airfields, the heavy reinforcement of British roads to carry heavy military transporters from abroad, the engineering improvisations consequent upon coiling up the striking forces for the invasion of north-west Europe, all intensified the efforts of all professional engineers to secure an expansion of roads at the end of the war.

Other professional bodies joined them in the campaign. The County Surveyors' Society, and the Institution of Municipal and County Engineers endorsed in wartime the need for fast motorways in the years of peace. *Post-war Planning and Reconstruction* (1942), issued by the Institution of Municipal and County Engineers, was followed by a *Report on Post-war Development of Highways* (1943), issued by the Institution of Highway Engineers. The latter excited great interest throughout the country and evoked a leading article in *The Times* of 9 April 1943 entitled 'High Speed Roads—Engineers on Motor Ways of the Future'. The then President, J. S. Killick (who had been Engineer of the Road Board from 1916 to 1918, and Chief Road Engineer to the Ministry of Transport from 1919 to 1921), was instrumental in initiating the Road Traffic Engineering Conference of various organizations concerned with traffic matters, and the training, status and scope of traffic engineers. In 1944 some 200 Road Transport Engineers met in London and resolved to establish an Institute of Road Transport Engineers which would act as an educational and examining body and provide facilities for the exchange of ideas and knowledge amongst members of the profession. In ten years it acquired a membership of 2,000, who are mainly concerned with the maintenance and repair of fleets of road vehicles. In 1951 the first Traffic Engineering Conference was held at Harmondsworth.

A sixth result of the war was the gearing of the motor engineering industry with state needs. Following the establishment in 1943 of the Joint Committee of Vehicle Manufacturers and Retailers, an autonomous body, another organization—the National Advisory Council for the Motor Manufacturing Industry—was set up in 1946. This was to enable 'the location of industry, exports, imports, research, design and technical developments, production methods and the general progress of the industry' to be discussed in a national context. Undoubtedly it has been a major factor in standardization of components amongst the six major firms in the industry.

This national context is most important. By 1948 for the first time since perhaps the earliest phase of the industry Britain sold more cars abroad than the U.S.A. It was a dramatic rise. In 1920 the total value of British exports of motor vehicles and parts was £8·4 millions in 1929 (after a decline) £10·5 millions, in 1938 (the highest inter-war year) £15·9 millions. But by 1946 it was £48·2 millions, in 1947 £74·6 millions, and in 1948 £115·5 millions. Even allowing for the changing value of the pound this was a significant advance during those years. For a world in which the possession of cars afforded indices of a country's level of prosperity, British exporters had a seller's market. The U.S.A. here led the field with 1 car to every 3·5 of the population, New Zealand second with 1 to every 6, Canada was third with 1 to every 6·6, and Australia fourth with 1 to every 9. Britain, with 1 to every 18, was obliged to sell abroad. The British exports of motor-vehicle products in 1948 (£115,500,000) were just a little short of the total payments for imported wheat and flour and this was important in the post-war world, as we shall now see.

PETRO-CHEMICALS

A seventh and almost consequential result of all the foregoing was the vastly enhanced importance of the chemical engineering industry.

Man-made materials benefited enormously from the cross-fertilization of ideas between the engineers, chemists and industrialists effected during the war. Three factors accelerated what Dr. Ronald Holroyd F.R.S. called the post-war 'explosion' of the petro-chemical industry in Britain. Firstly the large demand for synthetic rubber in the second world war called forth and improved techniques like catalytic cracking, polymerization, alkylation, platforming, and isomerizing of petroleum for the mass manufacture of styrene and other acyclic (straight-chain or non-aromatic) compounds. Secondly, the equally large demand for coke at the same time produced a great deal of tar

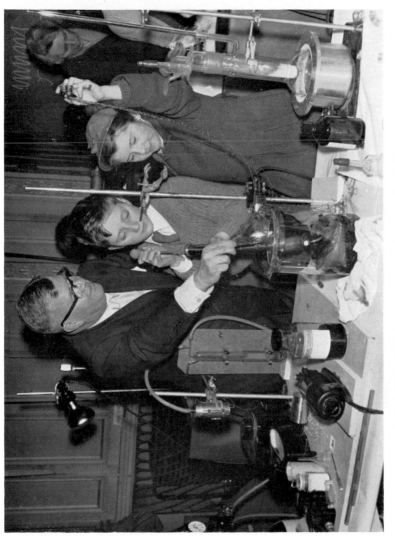

G. J. Whinfield demonstrates terylene to schoolchildren.

Calder Hall in May 1958.

which, by fractional distillation or hydrogenation, produced the cyclic aromatics used in plastics, insecticides, soaps and solvents. A small proportion of the benzene so produced was also used for nylon. Thirdly, the extensive and effective use of nylon for two ropes and parachutes opened men's eyes to its possibilities. As a result they now challenge the cellulosic fibres. Whereas the cellulosic fibres had made only moderate invasions into the clothing, textile and industrial (e.g. tyre cord) fields, nylon, terylene and the tailored molecule synthetics are spreading in all directions, from hosiery, raincoats, drip-dry shirts, household utensils, to industrial belting, and cordage, gear wheels, bearings, lenses and glasses.

A brief table gives the picture of the growth of this virtually new industry.

	CELLULOSIC FIBRES (in million lb.)			SYNTHETIC FIBRES (in million lb.)		
	U.K.	*U.S.A.*	*Japan*	*U.K.*	*U.S.A.*	*Japan*
1939	173	380	540	—	—	—
1951	374	1294	369	14	171	7
1955	434	1261	732	38	379	35
1956	429	1148	917	54	400	63
1957	425	1139	967	70	516	94

From this it will be seen that the rate of growth of the synthetic fibres has been greater than that of the cellulosic fibres.

Being resistant to mildew and biological attack, later synthetic fibres are being increasingly used for fish nets, marine ropes, and filter cloths. Having a slow rate of water absorption they are easily washed and have come to be used for making drip-dry clothes, sheets and socks, and evening clothes. Being lighter and more durable, they reduce replacement costs (e.g. in raincoats). Perhaps their greatest asset is that they save transport charges on raw material, and prevent reliance on cotton-growing countries. Portable, expendable, cheap, adaptable, colourful, durable, and (above all) available, these new products most literally lighten man's lot on earth. They have made possible the democratization of the motor car, the radio and television set, and toys. They are even making books more attractive to possess.

CHAPTER 26

The Sinews of a New State, 1945-1960

THE NEW LOOK

The profile displayed by the economy of the United Kingdom at the beginning of the sixth decade of the twentieth century can be best appreciated in the vital statistics of three basic activities. The generation of electricity has quadrupled, the consumption of petroleum products has trebled and the production of steel has nearly doubled in twenty years since 1938. The figures tell their own story.

	1938	1958
Electricity generated (in million k.w.h.)	24	98
Petroleum: main products (in million tons)	8	29
Steel: U.K. production (in million tons)	10	19

A glance at the more familiar, if less appealing segments of the British economy, coal and cotton, is equally revealing.

	1938	1958
Coal produced in the U.K. (in million tons)	226	215 (of which 14 millions was open-cast)
Cotton: U.K. consumption (in million pounds)	1,310	712

The second world war accelerated the process of transforming Britain into an engineering state. The purchase of munitions and armaments cruelly dissipated the vast hoard of overseas investment capital accumulated in the golden days of the nineteenth century. £428,000,000 in the U.S.A. and £571,000,000 elsewhere were thus lost, together with the invisible exports that enabled the dollar

306

and the pound to be mutually convertible. As if this were not enough, new external debts of £3,000,000,000 were incurred, upon which interest had to be paid. The price of raw materials had risen, and the gold and dollar reserves were only half of their pre-war level. And so, instead of enjoying a relaxed peace, Britain had to reorganize and re-deploy its industrial forces and intensify production efforts to win a battle for subsistence and survival. In this battle all the resources of accumulated engineering skill and scientific knowledge had to be exploited to the full, others developed, and a more collectivist approach than ever before adopted in social policy.

Since the operation was of such grave national importance, the State took over the main power resources of the community. The coal industry, which had been under control since 1942, was nationalized in 1946 and an extensive plan for its modernisation, costing some £635,000,000 was initiated. Two research establishments were set up at Cheltenham and Isleworth and a central engineering establishment at Bretby in Derbyshire. By 1956 87 per cent of the coal obtained from deep mines was being mechanically cut and 93 per cent being mechanically conveyed, and in the same year a pilot plant for the underground gasification of coal was initiated with a second state body, the British Electricity Authority.

This second body, the British Electricity Authority, was established in 1947, and took over in 1948 the assets of municipal and private electricity supply stations in England and Wales and part of Scotland. Its name was later changed to the Central Electricity Authority.

The third was the gas industry, which in 1949 became by an act of the previous year, publicly owned, when it absorbed 991 under-takings into 12 Area Boards, administered by a Gas Council and advised by a Consultative Council in each Board area.

The three nationalized industries are responsible to the Minister of Power on policy issues. This Ministry is one of two created during the war which have now found a permanent status in the English governmental hierarchy. The Ministry of Fuel and Power, as it was called on establishment in 1942, grew from the Mines and Petroleum Departments of the Board of Trade. It was made permanent in 1945 and its name altered to the Ministry of Power in January 1957. Its rôle is to co-ordinate and develop fuel and power supplies and to exercise a supervisory eye over the petroleum industry. It has, since 1957, been responsible for atomic energy as a motive power for industry.

The petroleum industry was literally revolutionized by the war.

Before it, 75 per cent of the British supply was refined overseas; now that percentage has substantially diminished. Its reliance on overseas sources of supply (for native sources in Britain yield little more than half a million tons a year) has led to the whole economy of Britain reverberating with each Middle Eastern growing pain. Abadan, Suez and Jordan are all manifestations of a fundamental disharmony between communities living on the margin of technological development and those living at the centre.

Home refineries, like the Stanlow platformer in Cheshire and the Fawley refinery at Southampton, based on the latest electronic techniques, have led to a new industry growing up in Britain, the manufacture of equipment for the petroleum industry, that is worth ten times more than that of the steelworks plant equipment.

With another eight at Shellhaven, Partington, Spondon, Grangemouth, Wilton, Billingham, Canvey Island and Milford Haven, they represent another revolution: the massive supplementation and supplanting of the traditional sources of industrial chemicals, like molasses (which yielded ethyl alcohol), wood (which yielded acetone) and coal (which yielded benzene). Coal and sugar are particularly vulnerable because it was only the demand for their main products that made their by-products cheap to obtain. On the other hand ethylene, the raw material from the cracking operation, had a diverse range of application. Reacted with sulphuric acid it gives ethyl alcohol. With oxygen it gives a fumigant and with water it gives ethylene glycol anti-freeze and in turn this gives a new range of products. With benzene it gives styrene, one of the components of synthetic rubber, latex paints and plastics. Reacted with itself it gives polyethylene, an increasing source of material for household utensils and containers. Another valuable raw material from petroleum is propylene.

But perhaps one of the most dramatic products of the molecular engineers has been polytetrafluorethylene (Teflon) in which fluorine atoms are substituted for hydrogen atoms. It is rigid, insoluble, acid and heat resistant, non-inflammable and has the lowest coefficient of friction known. Naturally, it is becoming increasingly popular in the whole range of engineering activity.

RESEARCH: THE KEY TO INNOVATION

The traditional research associations connected with D.S.I.R. were reorganized and increased. They rose from 21 to 38 between

1939 to 1948 and from 1948 to 1957 to 46. At the same time their grant was increased, first to an aggregate of £2,500,000, then to £4,000,000. The new research associations, together with those in the nationalized industries, were in 1956 placed under the new Executive Council for Scientific and Industrial Research, which replaced the Advisory Council set up forty years before.

One is the Shirley Institute of the British Cotton Industry Research Association. Founded in 1919 it has extended its province to include man-made fibres.

The national importance of science on one hand and invention on the other, and the importance of bridging the gap between a new idea and its application in industry, led to the establishment, in 1948, of the National Research Development Corporation (or N.R.D.C. as it is generally known). Though under the Board of Trade, this has a wide field in which to develop in the public interest inventions resulting from research carried out by government departments and other public bodies. By a further act of Parliament in 1954, N.R.D.C. was allowed to initiate research as well as develop it. To it were assigned the rights in the Bailey Bridge and the 'Denny-Brown' ship stabilisers, both devised in the service departments during the war. Other service inventions which are in the Corporation portfolio as a source of revenue are the light-alloy Cutting Electric Arc, the Air Position Indicator, the Course Setting Dial Adjustment for Aircraft, the M.O.S. Electrical Ignition Systems for jet engines, the R.A.E. Parachute Harness, the Backward Travelling Wave Tube, and the T.R. Switch (enabling the same aerial to be used for reception and transmission). In 1949–50 its annual income was £3,029 and its expenditure £19,595. By 1959 its annual income was £285,908 and its annual expenditure was £567,521.

In ten years the National Research Development Corporation could report that it had sponsored eight substantial projects which derived from the efforts of professional engineers or scientists. They were the Packman Potato Harvester (1949–50), the Ricardo Light Steam Engine (1950–1), the Thompson Map Plotter, the Menton method of making Diffraction Gratings, and the Hallpike Aural Microscope and Chick-Sexer (1951–2), the Clerk 'Gyreacta' Regenerative Braking System (1952–3), the Spottiswoode Stereoscopic Cine-Camera (1953–4) and printed electrical circuits (1954–5). This last was a production technique of great importance in the light current electrical industry. The Corporation served a most useful purpose as a transmitter of knowledge from one front to another. Thus, when

309

the Medical Research Council discovered that waste sisal juice was a relatively copious source of hecogenin (a basis for synthesizing Cortisone), the N.R.D.C. was able to commission a plant in Kenya and in Tanganyika for hecogenin, before the pharmaceutical industry had even made up its mind.

Three of its most recent projects have been concerned with transport. The first is known as the Dracone D.1, a 100-foot long tube of nylon and rubber in which oil and other liquid cargoes can be transported on water. These multi-capacity Dracones were the brainchild of a group of Cambridge scientists, Professor W. R. Hawthorne, Mr. J. C. S. Shaw and Sir Geoffrey Taylor, at the time of the Suez crisis. As 'sea trains', towed by small motor boats, they may well be a solution to the problem of transporting fuels and gases.

Another is the Hovercraft, conceived by C. S. Cockerell, a British electronics engineer, who began experiments in 1953 to prove its feasibility. The Corporation took it up and through a subsidiary company, placed an order with Sanders-Roe, Ltd., of Cowes. On 7 June 1959 it made a successful first flight and on 25 July it crossed the Channel. The Hovercraft principle of suspension on an air-cushion is being further applied, under the Corporation's auspices, to the building of a 40-ton Channel ferry. The third is even more fundamental, a new power unit. Electricity is generated by hydrogen and oxygen gases, circulating in porous nickel tubes packed with caustic soda, and this, pioneered by Francis Bacon at Cambridge, has been sponsored by the Corporation, who have licensed an American firm to use the invention.

Its largest single project was concerned with digital computers, and this will increase. A number have been built for the Corporation, and they have declared that 'we intend to give this inventive field our full support for some time to come'.

ELECTRONIC COMPUTERS

Both atomic and rocket research stimulated another activity, that of mechanical computers. In 1935 a young English mathematicisn, A. W. Turing, investigated the possibility of a machine being constructed to analyse logical problems. At about this time a machine was actually built in the United States at the Massachusetts Institute of Technology by the International Business Machine Corporation on mechanical principles. Christened Bessie (after the German mathematician, F. W. Bessel), it was capable of adding 15,000 numbers, and worked through punched cards. It proved a great help to the United States for various missile and bombing calculations. But its greatest accomplishment was to calculate the mathematical problems concerned with nuclear fission. This took 103 hours to accomplish.

The development of electronic computers began when two electrical engineers, Vannevar Bush (1890–) and O. H. Caldwell (1888–) joined forces. Caldwell was editor of the American journal *Electronics*. Bush had been with the General Electric Company, a teacher of mathematics at Tufts College and specialized in submarine detecting devices in the first world war, becoming professor of electric power transmission at the Massachusetts Institute of Technology. In 1929 he published a book, *Operational Circuit Analysis*, and built a differential analyser—a machine for solving differential equations.

The next stage was the construction of an Electronic Numeral Integrator and Computer (E.N.I.A.C.), the result of further needs in ballistic calculation, begun in 1942 at Harvard under Professor Howard Aiken by J. W. Mauchly and J. P. Eckert. Howard Aiken (1900–) was formerly an engineer with the Madison Gas and Electric Company and the Westinghouse Electrical Manufacturing Company before joining the staff at Harvard in 1939. By 1946 he was director of its computation laboratory. J. P. Eckert (1919–) went on to design B.I.N.A.C. in 1946–9 (the Binary Automatic Computer which has two brain valves which check each other and is portable), and U.N.I.V.A.C. in 1948–51 (the Universal Automatic Computer built for the American Statistical State Bureau for the evaluation of the 1950 census). He and Mauchly set up the Computer Corporation in 1947 which was taken over by Remington Rand and by the Sperry Rand Corporation.

One of those engaged on the construction of E.N.I.A.C. was an English physicist, D. R. Hartree, who after anti-aircraft research in the first world war, had been Beyer Professor of Applied Mathematics at Manchester from 1929–37 and later Professor of Theoretical Physics. During the second world war he was acting Chief of the Institute of Numerical Analysis, U.S. Bureau of Standards. He built the first machine at the N.P.L. laboratories at Teddington, known as A.C.E. (the Automatic Computing Engine), which was a considerable advance on E.N.I.A.C. with 4,500 valves. The University of Manchester followed this up under F. C. Williams, F.R.S., and A. W. Turing. Professor Williams was, for his work with Ferrantis and on this, awarded the first Franklin medal in 1958. The stage was set for even more rapid development under the aegis of the National Research Development Corporation.

Apart from the impact of computers on industrial production, one of the interesting consequences of the growth of electronic computing techniques was the entry of the famous catering firm of J. Lyons and Co. into co-operation with the engineering industry. This firm had established an Organization and Methods Office as early as 1922, and when the E.D.S.A.C. computer was built at Cambridge, Lyons began, in 1951, a two-years period of trial and experiment to utilize a similar computer for their weekly pay-roll, and for information on which to base their buying and pricing. Their computer did the work of 200 to 400 clerks and occupied only 1,150 square feet, and time on it was soon hired out to other firms and government departments.

It was so successful that a new company was subsequently formed to manufacture electronic office equipment: Leo Computers Ltd.

AUTOMATION

The rapid development of analogue machines and computers, coupled with the wartime advances in servo-mechanisms, made it possible to visualize an automatic factory. These two trends, grafted on to the existing tendency to run assembly lines in factories, led to the convention in June 1955 by the Institution of Production Engineers of a conference on 'The Automatic Factory—What Does it Mean?' They concluded that there was no reason to fear technological unemployment. The Department of Scientific and Industrial Research regarded automation in a report issued in the following year as an expansion of the scope of mechanization and materials-handling. The Amalgamated Engineering Union, however, adopted a different attitude. They suspected unemployment and resolved that the worker had a right to keep his post if affected by automation. The Trade Union Congress, however, have adopted a co-operative and constructive attitude which augurs well for the future.

Automatic controls made it possible to produce motor cars, chain cables, steel tubes and printed circuits with a rapidity and exactness hitherto unequalled.

But the prime utilization of automation was in the new nuclear power stations. Hinkley Point was designed at a capital cost of £105 per kw. as opposed to the £120–£140 per kw. for Berkeley, Bradwell and Hunterston. Analogue computers can now simulate the performance of a complete nuclear power station and enable the automatic control systems to be planned.

Yearly, almost monthly, improvements in the electronics that underlie automation are being recorded. Cryotrons are replacing vacuum tubes, transistors are making computors lighter and simpler, automatic readers are facilitating communication, automatic chemical analysers are enabling chemical engineers to obtain quick visual checks of their processes. The full history of this development will make exciting reading.

ATOMIC POWER

After the war an Atomic Energy Research Station was established at Harwell in Berkshire under Sir John Cockroft, and a centre of applied knowledge was established at Risley in Lancashire under Sir Christopher Hinton. Cockroft had been one of Rutherford's lieutenants. Hinton, an engineer who had organized ordnance factories during the war, directed the establishment of atomic piles at isolated spots in the country. The first was in Cumberland. D.S.I.R. controlled atomic research and development until 1946, when it was taken over by the Ministry of Supply. This arrangement lasted till 1954, when the new President of the Council took control. From 1 August 1957, the United Kingdom Atomic Energy Authority took control. By 1957 the Prime Minister became responsible for the development of atomic energy. In February 1957 the Government established a National Institute for Research in Nuclear Science. Having built in Calder Hall the first large-scale nuclear power station in the world which supplied public electricity, it decided on 5 March 1957 in view of the rapid engineering advances and the Suez crisis (which emphasized our undue dependence on external supplies of oil), to raise the programme for nuclear energy stations to 5,000 to 6,000 mw., instead of from 1,500 to 2,000 mw. This revolutionary policy was to cost £3,350,000,000. Further stations at Bradwell in Essex (300 mw.); Berkeley, Gloucestershire (280 mw..); Hunterston, Ayrshire (320 mw.); Hinkley Point, Somerset (500 mw.); Trawsfynnyd, Merionethshire (550 mw.), and Dungeness, Kent (550 mw.) were under construction in 1957. Sir John Cockroft expected these to provide 25 per cent of the total power in the United Kingdom by 1966.

There were also hopes that electricity might be directly generated in a fusion or thermo-nuclear reactor without the need for boilers, turbines, or any form of heat engines if the temperature was raised

313

to 1,000 million Centigrade for about a second. Sir George Thomson, outlining this at the British Association in 1957 was prepared to suggest that this would become a feasible proposition by 1972. The experimental device ZETA had given reasonably encouraging results.

THE SPECTRUM WIDENS

The unitary concept of engineering, once envisaged as subduing nature by methods almost crudely surgical, has now been broken up by the prism of history into a spectrum band of professions, each qualified by an adjective. This spectrum is widening at either end.

One end, the crudely surgical, has expanded enormously since the war. We have seen, in Chapter 25, how the Institute of Road Transport Engineers began in 1945. Related skills involving the use of dozers, scrapers, rippers, rooters, tractors, and miscellaneous earth-moving machines devised or imported to accelerate the construction of airfields, roads, camps and parks in wartime stimulated, by the war's end, the creation of another group: The Incorporated Plant Engineers, which arose from the need for exchange of information and spare parts on the part of civil engineering plant managers responsible for the large American excavators imported during the second world war for working open-cast coal. The benefits obtained made these engineers want to continue this service in the form of an Institution, and Incorporated Plant Engineers was incorporated on 18 September 1946. Facilities have now been available for some time in the larger technical colleges for the two-year part-time course (Plant Engineering, No. 187) organized by the City and Guilds of London Institute in collaboration with the committee of Incorporated Plant Engineers.

The manufacture of such equipment has become a major engineering industry in Britain. By 1956 £76,000,000 worth were being produced, of which over 50 per cent was exported. And this itself has not been without its influence both on open-cast coal mining and also in the new highways policy that is currently engaging national attention.

Its use in overseas constructional work in the African territories—like Uganda's hydro-electric scheme at Owen's Falls, Ghana's harbour works at Takoradi, Pakistan's dry docks at Karachi, Iraq's airfields, and other public works—is increasing yearly.

At the other end of the spectrum the delicacy of harnessing elements in nature, like the platinum catalyst to raise the refining of motor gasoline from fuel oil, has been extended to organisms. The biological (as opposed to the chemical) engineer is stirring and striving to use the enzyme to produce foods. Micro-organisms may yet enable man to keep the nitrogen cycle going, and biochemical engineering is looming up on the academic horizon.

Claims for yet another primary technology—control engineering —are now being made. The rapid coalescence of interest in satellites, long-range ballistic missiles and automatic controlled continuous process industries led Sir Harold Hartley to urge that it should receive recognition as a primary technology. Like many other technologies of the age, it is dynamic, co-operative and dependent on the matching and mating of a number of disparate engineering skills.

Thus the three senior professional engineering associations, together with the Institute of Physics and the Institution of Chemical Engineers, founded in 1955 the British Nuclear Energy Conference to provide a forum for the discussion of problems of nuclear energy. Joined by the Institution of Structural Engineers, the Institution of Metals and the Institute of Fuel, the Conference publishes a journal which offers the most valuable picture of current technical development.

Nor should the Society of Instrument Technology (founded in 1945) be forgotten, as instrumentation is a singularly important feature of automation and atomic energy.

THE PROBLEM OF STRESS AND STRAIN

The increasing mental tensions consequent on the intensification of the productive process have led engineers to be far more sympathetic towards the social sciences. The foundation of the National Institute of Industrial Psychology in 1919 marked one significant step forward in the realization that in works discipline it was not enough to secure the co-operation of the employee. The second step forward was taken when the Tavistock Institute of Human Relations was founded in 1946. This seeks to bring the resources of psychology and the social sciences to bear on the solution of practical human problems which arise in industry, new satellite towns and the big new housing estates and blocks of flats.

The work of psychiatrists, too, is being increasingly enlisted in the selection of key personnel, and the group techniques of Elton Mayo, Kurt Lewin and Dorwin Cartwright in America have been utilized in this country, notably by Dr. Elliot Jacques, whose *Changing*

Culture in a Factory has become a classic in this field. Psychological techniques have been more and more employed in the vestibules of the engineering industry and in schools, especially since the 1944 Education Act.

SCIENTIFIC MANPOWER

Indeed, the nation was concerned, as never before, with the education and training of scientists and technologists for research, production and management. Russian and American examples were frequently quoted. The progressive sophistication of manufacturing processes imposed a stern timetable on governmental action in this, perhaps the most important, of the productive process: an increased supply of scientific manpower. In 1945 the Percy Committee estimated that technical colleges should produce 1,500 qualified engineers a year for the next ten years to keep the total output at 3,000 a year. In 1946 the Barlow Committee recommended that the number of students in university faculties of technology should be doubled in the same time. Both committees underestimated the need, for by 1955 technical colleges were producing not 1,500 but 4,000 qualified engineers, whilst the universities had doubled their production of technologists in two years. The success of the technical colleges was partly due to the institution of new Higher National Certificates in Metallurgy (1949), Applied Physics (1945), Applied Chemistry (1947), Chemical Engineering (1951) and Mining (1952), and the rise in the number of successful candidates in these examinations generally, during the ten-year period. For whereas in 1945 4,629 gained the certificate at Ordinary Level and 1,723 at Higher Level, by 1955 13,922 gained the certificate at Ordinary Level and 7,507 at Higher Level. And, of the 7,507, nearly 50 per cent were continuing their studies to obtain admission to professional institutions. These figures did not include some 1,400 technical college students who obtained degrees from the Universities.

There was during these ten years a protracted dispute as to the best ways of expanding the supply of technologists. As *The Times* remarked on 21 September 1951, 'the controversy about how to speed the study of technology is so strong that it impedes progress'. *Cmd. 8357* recommended the establishment of a Royal College of Technologists: a proposal that was coldly received. Lord Adrian remarked on 30 December 1951 that even 'a utility version' of one of the great American institutions would be an advantage.

It was left to the Advisory Committee on Scientific Policy (set up in 1947) and the National Advisory Council on Education for Industry and Commerce (set up in 1948) which ultimately initiated another great forward movement. Apart from those engaged on national service and postgraduate research at universities (some 7,000) there were in 1956 135,000 scientists and engineers who held degrees, diplomas or membership of some recognized professional institution. The annual recruitment to this small, select body was some 9,500 a year, a deplorably low figure, which the Advisory Council on Scientific Policy recommended should be increased to 16,900 by 1966, and to 20,000 as soon as possible afterwards. The National Advisory Council on Education for Industry and Commerce had, from 1950 onwards, drawn attention to the virtues of 'sandwich courses' (alternative periods of study and industrial practice, equivalent to a three- or four-years' University course, and practised in the Royal Technical College, Glasgow, as early as 1880). These, thanks to increased grants by the Ministry of Education grew to number 100 by 1955, enrolling 2,300 students. Full-time advanced courses grew over the same period to some 616.

In the meantime, following a recommendation of the Percy Committee, a fruitful financial partnership between the Ministry of Education and Industry set up six new national colleges for Horology, Foundry Work, Rubber Technology, Heating Ventilating, and Refrigeration Engineering, Food Technology and Leather Industry for industries too small to provide more than enough students for one college. Full-time one- or two-year courses are run in these six colleges which are expanding. By 1958 they were catering for 489 students.

THE COLLEGES OF ADVANCED TECHNOLOGY

A new step was taken in 1955 when, on the recommendation of the National Council on Education for Industry and Commerce, the Government established a new national award; the Diploma in Technology. This, comparable with a University honours degree, was to be accredited by an independent National Council for Technological Awards established in the same year under the chairmanship of Lord Hives. Lord Hives, assisted by Sir Walter Puckey (Chairman of its Board of Studies in Engineering) and Geoffrey Loasby (Chairman of its Board of Studies in Technologies other than engineering), considered courses submitted by colleges, and the National Council announced that a 'Dip. Tech. (Eng.)' for engineering and a 'Dip. Tech.' for other technologies were to be established with first and second class honours. The entry standard would be tenure of either ordinary National Certificates or passes in the G.C.E. at Advanced Level. Courses for the diploma were to be either on a full-time or on a sandwich basis, and it was suggested that three years for full-time and four years for sandwich courses (i.e., 26 weeks full-time study) should be standard practice.

With these recommendations in mind, the Government announced in February 1956 its intention to spend £95,000,000 within the subsequent five years upon expanding and improving the country's technical colleges, to raise the number of technologists from 9,500 to 15,000 and doubling the number of apprentice technicians and craftsmen released by their employees for part-time courses from 350,000 to 700,000. It had previously announced measures for expanding the universities.

Such a declaration, published as a White Paper on *Technical Education* (*Cmd.* 9703) marked, in every sense of the phrase, a turning point. For it virtually created a new type of institution: the College of Advanced Technology. From the 24 existing colleges then receiving a 75 per cent grant for advanced work, the Government intended to develop as many as possible into such colleges, with self-governing bodies, research facilities and students of ability sustained on technical state scholarships or normal state scholarships. Further stimuli were applied to secure the release of technicians for part-time day release courses, and for the provision of facilities for women students of technology. Revealing statistics of Russian women in these fields were made available in *Cmd.* 9703. Following this the Willis Jackson Committee was appointed to investigate the training of teachers for technical colleges.

In three years (i.e. by 1959) eight colleges of advanced technology had been established with nearly 2,000 students reading for Diplomas in Technology. The number of students reading for advanced sandwich courses rose at the same time to 8,000, and a real dual system of higher technological education had been established. These eight, the Birmingham College of Technology, the Bradford Institute of Technology, the Battersea College of Technology, the Chelsea College of Technology, the Northampton College of Advanced Technology (London), Loughborough College. The Royal Technical College, Salford, and the Welsh College of Advanced Technology at Cardiff are really unique. They have claimed £10,000,000 of the £70,000,000 authorized in *Cmd* 9703 for building up to 1961. Together they are catering for 8,000 full-time and 'sandwich' students and 6,194 part-time day students. They have recently been joined by another, the Bristol College of Technology and almost certainly the Rutherford College of Technology, Newcastle as well.

Supplementing their work are 22 Regional Colleges which undertake advanced work, and sandwich courses; 160 Area Colleges which provide for courses up to and including the Higher National Certificate, and 270 Local Technical Colleges. A sequel to the establishment of the Diploma in Technology has been the institution in 1958 of an award to stimulate and recognize applied research programmes carried out in the colleges and industry: Membership of the College of Technologists (M.C.T.). This is open to university graduates working in industry. Conversely, universities are recognizing the Diplomates in Technology as equivalently eligible as University graduates for higher degrees.

Their increasingly specialized functions are a response to the growing specialized needs of industry.

THE HARVESTING OF TALENT

As Andrew Schonfield argued in 1958, 'The truth is that in one important sense the British economic problem today is the same as that of an under-developed country: it has to create the essential groundwork for an advance to a new and faster rhythm of industrial expansion.' It is this new and faster rhythm which British scientists and engineers will have to create, manipulate and master.

The flood of recruits to our secondary schools in 1958, and which will enter institutions of further education from 1963 onwards, will be the most valuable asset this country has acquired since the war. Graphically speaking, they represent a plateau of post-war live births on which gerontopia of the late twentieth century will rest. Their passage in the next ten years through schools, colleges, universities and industry will loosen still further our well-knit social system and continue the erosion of traditional curricula. Voluntarily or involuntarily, the configuration of education will be altered. Just how significantly is one of the guesses of today.

This providential accretion to our national resources could not arrive at a more opportune time. As an old, relatively slowly industrialized economy, we have not hitherto exhibited the desire for very rapid technological change. We are secretly rather proud of what Durkheim called *anomie*—our sense of planlessness, and there is a traditional belief that the word 'through' is a necessary concomitant of the process known as 'muddling'. But we are short of time, and the slowness of the elimination of the lag between our schools and our current needs may well be fatal. Such responses as it has exhibited to these current needs and necessities has been piecemeal. Experts in one field have been so far ahead of their counterparts in others that we are now beginning to realize how important is the science of communication. So the three traditional economic concepts—land, labour and capital—have been enriched by a fourth—time: time saved by automation.

This progress to *techtopia* poses several problems. First and most important is the sheer task of producing the requisite number of technically trained personnel: a task which the *Economist* in a special supplement on automation in July 1956 considered would take twenty years. Allied to this is the effect such training will have on the traditional hierarchies of craft and apprentice training. Not that hierarchies will disappear, for a third problem is the discovery and training of proper experts who will have an increasing role to play in such a society. To discover these, nooks and crannies of our social system will have to be thoroughly flushed. The three a's—age, ability and aptitude—are even more important in the age of the three a's—acceleration, automation and atomic power.

Acceleration is visible even in the national income, which has increased in the last ten years faster than at any other time in the century. The 5 per cent increase of annual manufacturing output and the $3\frac{1}{2}$ per cent increase in the annual productivity figure is not only continuing, but is likely to double the standard of living by 1984. This expanding economy needs energy: hence the power revolution, from fossil fuels to nuclear energy.

Nuclear energy involves a complex of educational and training programmes that lie outside the normal tertiary structure and is already hastening its transfiguration into a quaternary system. The M.C.T. the growth of post-graduate schools and research courses of all kinds are evidence of this. So, with an enlarged student body, an expanding economy, and a revolution in power production those responsible for the educational system of this country agree that the next decade offers as exhilarating and peremptory a challenge as any decade in our history.

The Sports Palace at Rome, designed by P. L. Nervi.

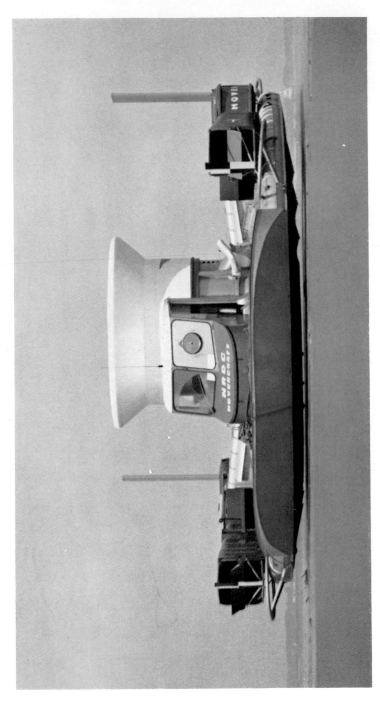

The S.R.N. 1 Hovercraft, built by Saunders-Roe for the National Research Development Corporation.

The Sinews of a New State, 1945–1960

THE CONTEMPORARY SOCIAL REVOLUTION

Engineering has effected and is effecting a major change in the standard of living and the social structure of modern England. The Barlow Report (*Cmd.* 6153) published in 1940 gave perhaps the most detailed evidence of the way in which industrial change has effected a re-distribution of the population. By 1948 a new standard industrial classification was adopted in preparation for the 1951 census. This census divided the occupations of Great Britain into twenty-four major groups and condensed these major groups into twelve broader groups. From this it appears that out of a total of 19,939,000 of the population who were gainfully employed, 7,526,000 (or 37·75 per cent) were engaged in the broad group 2—the manufacturing industries. These include no less than fourteen of the major groups, namely:

(3) Treatment of non-metalliferous mining products other than coal.
(4) Chemicals and allied trades.
(5) Metal manufacturing.
(6) Engineering, shipbuilding and electrical goods.
(7) Vehicles.
(8) Metal goods (e.g. tools and implements).
(9) Precision instruments.
(10) Textiles.
(11) Leather, leather goods and fur.
(12) Manufacture of clothing.
(13) Manufacture of food, drink and tobacco.
(14) Manufacture of wood and cork.
(15) Manufacture of paper and printing.
(16) Other manufacturing industries (e.g. tyres and tubes, other rubber goods, toys, pens and pencils).

When one adds to these the extractive industries which comprise the major groups 1—agriculture, forestry and fisheries, and 2—mining and quarrying, another 8·46 per cent are employed; and when one goes on to add the broad group transport and communications, a further 7·68 per cent of the population are embraced. So it is true to say that in all senses Britain is becoming an engineering state.

Put in other terms, the importance can be assessed not in the numbers employed in the engineering industry but in the percentage of the total industrial output and its value in millions of pounds. These figures, given below, tell their own story. They refer to the year 1955 and give a brief table of industrial production in Britain for that year.

	%	Millions of £'s
Shipbuilding, engineering and electrical goods	23·0	(£1,603·3)
Other manufacturing industries	19·8	(£1,382·9)
Textiles and clothing	13·6	(£ 946·9)
Vehicles	11·7	(£ 812·2)
Food, drink and tobacco	10·4	(£ 725·2)
Metal manufacture	8·8	(£ 620.6)
Chemicals	7·4	(£ 516·6)
Miscellaneous metal products	5·3	(£ 373·0)
		£6,980·7

In terms of the export trade and the battle for survival, the record of the engineering industry is even more impressive. For whereas textiles in 1958 only comprised 9·1 per cent of that trade (as compared to 23 per cent in 1938) engineering products comprised 40·7 per cent (as compared to 20 per cent in 1938). Indeed in 1958 Britain had the best balance of payments position since records had been kept. That this will be sustained depends largely on the imagination, vigour and industry of those who enter the engineering profession.

This sustention will also depend on research: a need which the British Association for the Advancement of Science recognized in 1952 when it appointed a committee to consider the problem. Its recommendations, presented by Professor C. F. Carter and B. R. Williams in *Industry and Technical Progress* (1957), *Investment in Innovation* (1958) and *Science in Industry* (1959) were for a larger investment in pure and applied research, greater numbers of scientifically literate people in industry, improved training, better communication of ideas and quicker apprehension of obsolescence. The first, represented by research teams in industry, also captured the attention of three other English economists who described them in 1958 as 'the most spectacular change in the activities of the industrial corporation'. These three, Professor Jewkes, Mr. Sawers and Mr. Stillerman, pointed out that of sixty important modern inventions, twenty-two had originated in industrial research laboratories.

CHAPTER 27

The Endless Frontier

THE ENGINEER'S AESTHETIC AND INTELLECTUAL CONTRIBUTIONS

There is a refreshing aesthetic in artefacts which Goethe (himself no mean scientist) described as 'frozen music'. Such an aesthetic is no less true of the ring formula for benzene than of a bridge. Both are ingenious, and both are exciting. The artistry of a bridge-builder is obvious to the naked eye, but the activities of the chemical engineer are not, until the products are bottled, batched or baled. Both profoundly affect the progress of mankind. One enables physical barriers to be crossed, the other helps biological ones to be circumvented.

Both are parts of a living development and are separate only in so far as orderly minds must separate in order to comprehend them. The visible achievements of Telford, Paxton, Labrouste, Contamin and Eiffel have been increasingly affected by the invisible work of mathematicians and scientists who have provided them with the formulae and materials they needed. And materials, scientifically tested, have become more important still in the hands of their successors—Hennibique, Freyssinet, Gropius, Nervi and Candela. Today's 'frozen music' is, more often than not, written, scored, and orchestrated, not in a boat on the Brenta, but at places like the Illinois Institute of Technology. There, since 1938, Mies van der Rohe has built and taught others to build the great structures of today: the General Motors Technical Centre at Detroit or the Kresge Auditorium at the Massachusetts Institute of Technology.

The engineer and technologist contribute something more than their structures to society. Without them science would probably degenerate into a series of casuistic exercises. Their strategic position between the endless frontiers of new knowledge on the one hand, and the equipment proved by success on the other, enable them to act as the most effective revolutionists of our time. So that misunderstood sect, the technocrats, might find that the prophecies of their messiah

Veblen are being fulfilled and that the engineers are really becoming the élite of our society.

The professional politicians and political theorists are already finding engineers among their ranks. Herbert Hoover was the first professional engineer to become president of the United States; while Georges Sorel, after twenty-five years as an engineer, elaborated his syndicalist doctrines which, as much as any others, have influenced political thinking in Europe in this century. Indeed in the field of pure political theory, Carl Schorlemmer was described by Engels as 'After Marx . . . undoubtedly the most eminent man in the European Socialist Party', whilst on the other hand Salvador de Madariaga, originally trained as an engineer, was one of the most articulate spokesmen for the Liberal tradition in the early half of the twentieth century.

Philosophy, the very core of a liberal education, has been consistently affected by engineers since the enthusiastic polytechnicians of nineteenth-century France elaborated positivism. Perhaps the most recent and notable example of the engineer philosopher at large in Britain was Ludwig Wittgenstein, who after graduating at the Technische Hochschule at Charlottenburg, Berlin, in 1908, came to Manchester to work as a research student in the engineering laboratory of the University on reaction jets at the tips of propeller blades. He became so interested in the design of propellers that he concencentrated on their mathematical theory. Eventually, on returning to Austria in 1914, he found himself in the army and concentrated on mathematics and philosophy. When he died in 1951 he had held the professorship of philosophy at Cambridge and was the author of the *Tractatus Logico-Philosophicus*. These 'highly syncopated pipings', as another philosopher called them, have proved to be most influential in the very institutions, like Oxford and Cambridge, which have exalted a training in words rather than things. For Wittgenstein taught that philosophy is an activity of making clear to people what they can and cannot say. It is perhaps significant that one of his notable disciples is Stephen Toulmin, whose *Philosophy of Science* (1953) might serve as a text for those wishing to ruminate about the meaning behind the social history of technology.

CRITICS OF TECHNOLOGY

Not all this change goes unchallenged. Today the optimism of the 'polytechnicians' has been increasingly flecked with despondency.

'The machine' has been blamed for many evils in modern society. It is seen as the symbol of hypertrophy, and of 'dominion of apparatus'. Its dehumanizing influence over human beings massed in large soulless cities has excited adverse comment. 'Proletarianized culture' (Arnold Toynbee) 'asphalt culture' (Emil Brunner), 'shiny barbarism' (Richard Hoggart) are phrases commonly bandied about. It was Aldous Huxley who gave us the best literary expression of this point of view in *Brave New World* (1932): a picture of an industrialized world state, where near-automata (instead of human beings) worshipped Ford and the Model T, and whose motto, Community, Identity and Stability reads strangely like the cult-language of some present-day psychologists. Yet it was also Aldous Huxley who in *Ape and Essence* sixteen years later, showed that such an outlook was based on a philosophy of the past that is really untenable. 'As I read History', says the Arch-Vicar to Dr. Poole, 'it's like this: for thousands of years, God and Belial fought an indecisive battle: then, three hundred years ago, almost overnight, the tide starts to run uninterruptedly in one direction.' The Arch-Vicar's opinion seems to be shared by many modern critics of technological progress, with the difference that Belial is given other names by other people: Anti-Christ (Berdyaev), The Prince of Lies (Auden) or The Antagonist (H. G. Wells). All agree that the tragedy is man's intoxication with the miracles of his own technology. All look at the Faust legend with their own spectacles and extrapolate on the stark facts of human development.

Yet there is more to be said for technological progress as the lightener of man's lot and load on earth. The energy-slaves at mankind's disposal have enabled more people not only to live but to live more abundantly than ever before. Electricity (unlike steam power) by no means involves that they live in large asphalt jungles. On the contrary it has made tolerable life in country places, and encouraged decentralization. It has (through television and cheap books) given media to the very prophets of woe enabling their misgivings to be communicated to millions.

SOME CURRENT SOCIAL IMPLICATIONS AND PROBLEMS

No, the problems of our society are not that technology has gone too far and too fast, but that its benefits are not shared equally throughout the world. It is true that it cannot be packaged and exported to backward areas, however gently, without disrupting their culture. Current indications of international responsibility for ensuring that all nations acquire better techniques of food growing, enhanced awareness of their national resources, an adequate supply of modern chemicals to combat disease, plagues and death, are being accompanied by a thoughtfulness and care by the World Federation of Mental Health to ensure that change does not detonate an explosion. Yet the problem of the backward areas is such that it may well be a race between donations and disaster. Some two-thirds of the 3,000,000,000 people of the world are increasing their population at such a rate that Sir Charles Darwin, in a particularly gloomy prognosis *The Next Million Years* (1952), foresees a densely crowded world with a population slowly approaching the margin of existence and compelled to work every day harder and harder. For, though the backward nations of the world are becoming industrialized, the rate of the industrialization does not seem to be closing the gap between themselves and the West. Measured in terms of net income per head, this gap is formidable, for the United States has a net income per head fifty times greater than that of China. That is why an energetic policy of mutual aid is so essential on the part of the advanced nations.

Conversely, in the highly developed areas, multiple problems present themselves. The first is the intensive face-to-face nature of society and the channelling of temperaments in impersonal organizations. These have given an ominous popularity to the term 'maladjustment' and led to a great outburst against 'organization' men by sociologists like David Riesman, William H. Whyte and Vance Packard. The second is the gearing of educational systems to state needs. The more science and technology impose qualifications on postulants for the professions, so the less freedom seems to be available to them to build up batteries of autonomy that will sustain the postulants in maturity. Not a little of the modern fiction is concerned with the dilemma of the scientist faced with moral problems, and readers of this book who have not discovered for themselves the novels of C. P. Snow are advised to do so, and not least because Sir Charles himself has climbed the scholarship ladder of science. Another engineer novelist, Nevil Shute, had also explored the moral hinterland of technological advance, notably in *On the Beach*.

A third problem is to prevent the new educational institutions now being so rapidly built all over the word, from succumbing to the sclerosis of organizational hardening: to ensure that they nourish, not sterilize, the flower of genius. The subtle and evanescent qualities impelling scientific advance need play (in the engineering sense) in any bureaucratic machine. Such institutions provide a stimulus to personality development hitherto afforded by others formerly catering for a different élite.

A fourth problem is to provide a moral substitute for war. As William James discovered, this is hard to find, but since his time we have learned a little more. Hitherto wars have summoned up the sinews of nations, and flexed their industrial strength. But today the flexing process is better understood, and senses of 'crisis' and 'emergency' are no longer needed to stimulate improvisations or innovations in the field of technology. Today we have substitutes for 'hot wars' in 'cold wars' against want and disease.

The last and deepest problem is to contain, or to come to terms, with the philosophy generated in the heyday of Britain's industrial hegemony by Karl Marx. This provides a seductive materialist interpretation of human amelioration that Arnold Toynbee has described as the last great Christian heresy. Its early fathers, Lenin and Mao Tse-Tung, have elevated the scientist into a priestly role. Though its contemporary neophytes make few, if any, hard and fast dogmas in the fields of mathematics, physics, engineering and chemistry, they are integrating biology, history and literature to serve the interests of a dynamic industrialism that captivates under-developed countries.

The Professional Spirit in Engineering

All these problems underline the changing function of the schools and colleges. Such diverse observers as Coleridge, Benjamin Disraeli and Lewis Mumford have noted that teachers are becoming (indeed have become) the new sacerdotal class of our society. More than any others they can humanise the capacity-catching machine, nourish the love of creation and adventure, cultivate the moral sensibilities and awaken the latent autonomy in each student. As Lewis Mumford has wisely written, 'We can no more continue to live in the world of the machine than we could live successfully on the barren surface of the moon. Man is at last in a position to transcend the machine and to create a new biological and social environment, in which the highest possibilities of human existence will be realized, not for the strong and lucky alive, but for all cooperating and understanding groups, associations and communities.' Of these groups, the Engineers are a growing body.

Compared to the close association of engineers and technologists

327

both in the U.S.S.R. and the U.S.A., Britain offers a sharp contrast. There is nothing obviously analogous in Britain to the American Society for Promoting Engineering Education (founded in 1893) or The American Engineering Council (founded in 1930—and abandoned in 1938), nor the Engineers' Council for Professional Development (founded in 1932) nor the National Society of Professional Engineers (founded in 1934). Yet there have been movements to form such organizations.

The British Science Guild for instance—during its existence fro 1905 to 1936—the Engineers' Joint Council (formed in 1922) and the Engineers' Guild (formed in 1938) have all in their way expressed the engineers' claim to exist as a professional group. The one attempt to found a General Engineering Council (analogous to the General Medical Council) never got past a first reading in the House of Commons on 29 April 1926. The General Strike, as much as anything, killed it.

The 'engineering point of view' in British public affairs is exhaled partly through the Parliamentary and Scientific Committee (a loosely knit body constituted in 1932), partly through the Royal Society and partly through the four senior institutions of the Civil, Mechanical, Electrical and Chemical Engineers. These, linked with other professional associations, institutions, and corporate groups in a loose but effective manner, and with local, regional, national and international associations and authorities, ventilate matters with the loose informality possible in these islands. Their multiple membership, informed by the regular issue of journals, and convened regularly are a guarantee that a healthy dialectic is sustained and that they can have their say if not their way. Sparking progress, ventilating problems and maintaining standards, they embody the vigorous group life of a modern state. They can act, at times, as pressure groups when and where a national issue demands an expression of opinion.

THE TRENDS OF HISTORY

It is difficult to offer any generalizing principles at the close of such a survey. Yet if principles cannot be brought forward propensities can. Six such propensities, according to W. W. Rostow in *The Process of Economic Growth* (1953) can be seen at work in the evolution of modern society. They are the propensity to develop fundamental science, the propensity to apply it to economic ends, the propensity to seek material advance, the propensity to accept innovations, the propensity to consume and the propensity to procreate. All are moderated or intensified by the institutions which they themselves generate.

The first two of these propensities embody the heart of the problem. Progressive improvement is wholly dependent on investment in education. Gone are the days of the brilliant empiric. The complex power-producing mechanisms of today need an ever-increasing cadre of professionals of high calibre, and a corps of technicians capable of understanding what they are about. These need selecting and training in colleges, universities and post-graduate schools. Investment in men is now at least as important as investment in plant because men react on plant. Conversely the hiatus between the progressive rationalization of the productive process and the apparent derationalization—or increased non-reflective automatic character—of those engaged in it needs resolving. Conditioning by the machine must, it is argued, be counteracted by the introduction of small units, co-operated activities and the investment of the job with a sense of drama and relevance. As A. R. Ubbelohde remarked, 'Technological integration is a necessity in many enterprises, but it sets a constant bad example of over-centralization in other matters.' Ubbelohde put his finger on one demoralizing facet of the use of energy slaves when remarking that they were 'constantly carving away and undermining the lower levels of the pyramid of human craftsmanship'.

Mumford has classified four phases of technological growth: the eotechnic, paleotechnic, neotechnic and biotechnic. As outlined in *Technics and Civilisation* (1934) and *The Culture of Cities* (1938), the eotechnic was based on wind, water and wood; the paleotechnic on coal and iron; the neotechnic on electricity, steel and light alloys, and the biotechnic on the application of science to life processes (insecticides, disinfectants, weed killers, chemotherapeutics, fertilizers and contraceptives) to control man's own evolutionary processes. He saw these periods as emergent or recessive in particular parts of the world, and also did his best to make his contemporaries aware of the environmental mesh through which they were evolving.

To Mumford's four phases we should now add a fifth: the spatio-technic, the emergent phase of our time. Today we live in an age when Moscow is less than four hours' plane journey from London and the earth itself is circled by an increasing number of man-made satellites which are going to enable us to predict the weather, investigate the solar system, and still further shrink the world to the significant dimensions of an early twentieth-century or a nineteenth-century nation. The organizations that now acquire significance are international ones. The International Power Conference (now twenty-five years old) and the International Atomic Energy Agency (the four-year-old world bank for fissile material, and scrutineer of health hazards) are but two of many international groups buttressing the work of the United Nations Organization. Aggregations of states based on planning independent of the United Nations have also been established in Western Europe (like the Coal-Steel Community, the Common Market and Euratom); in the Near East (like the Baghdad Pact); and the Far East (like the Colombo Plan). All three involve technical assistance from advanced Western nations in the production of power for the service of man. These three supra-national groups comprise in all some twenty-one countries: in the first Belgium, France, West Germany, Italy, Luxembourg and the Netherlands; in the second Iraq, Pakistan, Iran and Turkey; and India, Pakistan, Ceylon, Burma, Thailand, South Vietnam, Laos, Cambodia, the Philippines, Indonesia and Japan respectively.

ENERGY ON TAP

Energy on tap is the engineer's greatest contribution to civilization. One kilowatt hour of electricity can replace ten hours of human energy. By it fear can be abolished. Fear of slavery, hunger, cold, illness or ignorance. Because of it human senses have been strengthened, and human life has been lengthened. The countries consuming the most kilowatt hours of electricity naturally are the countries where we should expect to find the highest standards of living: Norway, Sweden, Switzerland and the U.S.A., followed by Great Britain, West Germany, Benelux, Austria and France. Death, within weeks or months, by starvation, disease or destruction would be the average human lot but for the collective ingenuity of technology. As this collective ingenuity improves, labour, as a heavy activity, is gradually being reduced. From a 70-hour week (as in the early stages of industrialization) we are reaching a 40-hour week. The built-in accelerator of our modern technological society is the increasing proportion of the national income set aside to train human beings to operate and improve it. And by improvement, the aesthetic as well as the utilitarian aspect of civilization is meant.

POWER: PAST AND FUTURE

One-seventh of all the people who have ever been alive are alive today. The expectation of life, thanks to technological developments, has increased from 30 years in the time of Augustus to 70. The rate of increase is such that, if it continues, today's population will double itself in 50 years. Populations need power in order to live, and Palmer C. Putnam has calculated that from the death of Christ to the beginning of the American Civil War, the total energy consumption of the world was from 6 to 9Q (Q = 10 B.T.U.),[18] i.e. ·45Q per century. From the beginning of the American Civil War to 1947 (i.e. 87 years) it was about 4Q, i.e. 4·5Q per century. Now it is equivalent to 20Q per century, and by the year 2000 is expected to be 100Q per century.

Of the energy reserves at mankind's disposal coal will (it is estimated) supply 36Q; oil 6Q, and uranium from 100 to 1,000Q (depending on the improvements in its use). Apart from these, and new man-made fuels which at present would need a great deal more research to be really practicable, there are five other sources of energy: wind, solar energy, tidal power, earth heat, sea temperature differences. The first of these has been explored by Palmer C. Putnam, who with Professor J. B. Wilbur of the Massachusetts Institute of Technology, constructed an experimental 1,250 kw. wind turbine at Grandpa's Knob in Vermont which operated with great success. Its performance, reported in *Power from the Wind* (1948), has inspired the construction of an experimental 100 kw. windmill, run in conjunction with a diesel electricity generating station, at Costa Hill in the Orkneys. There are several hundred other sites in Britain where wind can be similarly trapped to save some 4,000,000 tons of coal. An English writer, D. Brunt, has calculated in *Physical and Dynamic Meteorology* (1933) that the convertible total power in the atmosphere is 3×10^{17} kw. The Russians refer to it as 'blue' power and have a number of experiments in train.

The second, solar energy, has tempted man to tap it since the time of Archimedes. The energy received from the sun is equivalent to that of 2,000,000,000 nuclear power stations. The first serious discussion of it was a thesis written by John Ericsson in 1868. In 1870 Ericsson made the first sun motor by using a cylindrical mirror of parabaloid section with a cylindrical boiler in the focus line at New York. By 1875 he had made seven sun-motors. He obtained 1 h.p. per 100 square feet of mirror—and obtained about 10 per cent efficiency. To utilize solar energy he developed his caloric engine. Other engineers who have worked on solar energy include Carl Gunter of Austria, A. Mouchot and A. Pifre of France, and Henry Bessemer in England. In India William Adams constructed in 1876 a mirror-operated boiler with a steam train and pump at Bombay. In Chile, J. Harding built a water distiller that worked by solar power. In California, at Ostrich Farm, Pasadena, A. G. Eneas used 33 ft. 6 in. diameter mirrors to obtain heat. In Egypt Sir Charles Boys was a consultant to a large power plant built at Cairo in 1912. Solar water heaters have been built by the thousands in the southern part of the U.S.A. in the form of heat absorbers on the roof, whilst Dr Ghai, the Assistant Director of the Indian National Physical Laboratory, has utilized it for cooking. In 1940 the possibilities of utilising solar energy induced G. L. Cabot to bequeath an endowment to the Massachusetts Institute of Technology for work on photo-chemistry, thermo-electricity and other various applications of the principle. The then president of Harvard, J. B. Conant, was very interested in the possibilities. In 1953 the University of Wisconsin published *Solar Energy Research*, a result of a symposium organized on the subject. Solar energy is being more directly used in the United States. Photo-synthesis, the activation of silicon discs to produce electricity (one impregnated with arsenic as a negative, the other with boron as a positive) and the energizing of cadmium sulphide crystals are three present uses. Only 0·2 per cent of the available solar energy is being used, and to increase it, Dr. G. E. Fogg told the British Association in 1957 that it might be possible to grow chlorella and bacteria in association to break down the organic wastes of sewage. The subsequent harvesting of the microscopic algae (if a cheap method could be found) would provide sources not only of fuel but fodder and organic chemicals.

The third source of power, from the tides, suffers from two obvious disadvantages. There is no power at times of flood and ebb, and moreover these times change by forty minutes each day. J. B. S. Haldane has warned us of another: in his view, the development of tidal power would increase the braking action of the tides so that the rotation of the earth would slow down, the day would therefore become longer, and the moon would approach the earth. Yet, in spite of these warnings, engineers have let their imaginations and energies loose on tidal power ever since the century after the Norman Conquest when tidal mills were operating off the English coast. Peter Maurice's tidal wheels were operating on the system on the

Seine near Havre. Kaplan turbines, with blades of adjustable pitch, have suggested themselves in many quarters. Four main projects have received detailed study: the Severn Barrage (first proposed in 1849, studied in model form in 1947, and estimated to effect a saving of a million tons of coal); the Pebitcodiac and Meramcook Estuaries Scheme, in the Bay of Fundy in Canada; the Passamaquoddy Scheme in Maine, U.S.A., and the Rance and Mont St. Michel Schemes in Brittany. Britain has 66 favourable sites for tidal power stations.

The fourth source of power has perhaps received the least attention. In 1904 Parsons suggested to a large conference of engineers that a deep borehole should be made in the earth's crust to tap steam. He gave an estimate of its cost. Today, at Larderello in Italy, steam piped from subterranean volcanoes gives energy convertible to elecpower, whilst the Russian Academy of Science has one of its research teams working on earth heat in Kamchatka.

The last power source, utilizing temperature differences in the sea, is still being explored. At Abidjan on the French Ivory Coast there is a project which aims at using the difference between surface and sea-depth temperature as a power source.

INVENTIONS AND HUMAN DEVELOPMENT

Inventions have had an influence on history as decisive as any other factors. This was clearly seen by the National Resources Committee of the United States which in 1937 listed thirteen which would seem to have the most social influence of any in that country. As listed in *Technological Trends and National Policy* (Washington, 1937), they are the mechanical cotton-picker, air conditioning, plastics, the photo-electric cell, cellulosic fibres, synthetic rubber, prefab. houses, television, facsimile transmission, auto-trailers, gasoline from coal, steep-flight aircraft, tray agriculture (feeding plants in solution).

In England just after the first world war Sir Robert Hadfield was asked by the Institute of Patentees to nominate 'Things which required inventing' for their book, *What's Wanted*. As he turned the pages he noted such suggestions as—a cinema film that will speak; a method of utilizing atomic energy; a process for instantaneous colour photography; a method of conveying speech direct and readably to paper; a means of regulating the rain supply, or a means of inducing and preventing rain; the discovery of the mechanism which enables us to remember almost instantly in our brains without going through the mechanism of the card index and equivalent systems. Sir

Robert himself suggested four: an alloy, ferrous or non-ferrous, possessing 50 to 100 per cent higher tenacity than any known combination, whilst at the same time not being brittle; some form of lighting appliance capable of penetrating fog; refractory materials for the lining of steel melting furnaces and ladles that would be absolutely unattacked by erosion of the molten steel, or corrosion of the various metallic oxides formed during the steel-melting process; and a safe method of stopping steamers and ships from rolling, or at any rate not more than a few degrees, in the roughest weather. No more adequate comments on the speed with which science has been applied over the last forty years are needed than these.

For the gap between conception and creation has narrowed, and today's dreams become tomorrow's designs. In 1916 a Budapest professor, Sziland Zielinsky, conceived the idea of tapping the hot wells of Iceland. Today, three-quarters of the houses in Rekyavik are heated by them, and similar techniques are being explored in places as far apart as Alaska, Java and New Zealand. Other, more ambitious engineering projects like those of Pierre Gandrillon (1925) and W. C. Lowdermilk (1934) for a Jordan Valley Authority have now a very real relevance. So have the even more ambitious schemes of Herman Sörgel for dams at Gibraltar, Tunis, and Messina, and for inland seas in the Congo and Lake Chad. Perhaps most immediately significant is the Volta River project, which would make Ghana one of the world's leading aluminium producers, diversify its economy, and endow it with a 617,000 k.w. power station.

Listeners to Moscow Radio on 3 March, 1956 were astonished to hear A. Markin, a Soviet engineer, outline a scheme for linking Soviet Asia to Alaska by a dam that would both bar the Arctic from the Pacific and enable nuclear-powered pumps installed along it to transfer warm water from the latter to the former, thereby transforming the climate of continental Asia.

Man can radically alter his environment, and he can now alter his ways of thinking and his moods. The vast accumulation of chemotherapeutic inhibitor drugs that have been discovered since the war make even Aldous Huxley's 'Soma' appear an old-fashioned herbal medicament. That is why, as we operate on ever-narrowing segments of the endless frontier, we must keep our sense of direction. And to see where we want to go, we should, from time to time, reflect on the manner of our coming so far.

APPENDIX A

Bibliographies

(*Unless otherwise stated the place of publication is London*)

CHAPTER 1

R. J. C. Atkinson, *Stonehenge* (1958).
C. S. Coon, *The History of Man* (1955).
A. Lucas, *Ancient Egyptian Materials and Industries* (1948).
V. Gordon Childe, *The Dawn of European Civilisation* (6th edition, 1957).
George Sarton, *A History of Science* (Oxford, 1953).
A. J. B. Wace, *Mycenae* (Princeton, 1949).
Stuart Piggott, *Neolithic Cultures of Britain* (1954).
Albert Neuberger, *The Technical Arts and Sciences of the Ancients* (Methuen, 1930), trans. H. L. Brose, with 676 illustrations.

CHAPTER 2

Marshall Clagett, *Greek Science in Antiquity* (1957).
M. Cohen and E. I. Drabkin, *A Source Book in Greek Science* (New York, 1948).
B. Farrington, *Science in Antiquity*, H.U.L. (1936).
G. Sarton, *A History of Science: Ancient Science through the Golden Age* (1952).
S. Sambursky, *The Physical World of the Greeks* (1956).
W. W. Tarn, *Alexander the Great* i, (Cambridge, 1948), p. 145.
R. E. Wycherley, *How the Greeks built Cities* (1949).

CHAPTER 3

A. E. R. Boak, *Manpower Shortage and the Fall of the Roman Empire in the West* (Ann Arbor, 1955).
R. G. Collingwood and J. N. L. Myers, *Roman Britain and the English Settlements* (Oxford, 1937).
B. Farrington, *Science and Politics in the Ancient World* (1946).

335

T. Frank, *Economic Survey of Ancient Rome* (Baltimore, 1937). Volume III deals with Britain.

J. Lees-Milne, *Roman Mornings* (1956).

R. P. Oliver, 'A Note on the De Rebus Bellicis', *Classical Philology* (1955), pp. 113–15.

G. T. Rivoira, *Roman Architecture* (1925).

E. N. Stone, *Roman Surveying Instruments* (1928).

E. A. Thompson, *A Roman Reformer and Inventor* (Oxford, 1952).

Vitruvius (Loeb edition, 1931–4), *De architectura*.

F. W. Wallbank, *The Decline of the Roman Empire in the West* (1946).

CHAPTER 4

N. H. Baynes and H. St. L. B. Moss (Ed.), *Byzantium, An Introduction to East Roman Civilisation* (Oxford, 1948).

R. Bennett and John Elton, *History of Cornmilling* (1919).

Herbert Chatley, 'Far Eastern Engineering', *T.N.S.*, xxix (1958), 151–67.

Childe, V. G., 'Rotary Querns on the Continent and in the Mediterranean Basin', *Antiquity*, xvii, (1943), pp. 24–5.

E. C. Curwen, 'The Problem of Early Water Mills', *Antiquity* (1944), xviii, 130–46.

R. J. Forbes, *A Short History of the Art of Distillation* (Leiden, 1948).

S. C. Gilfillan, *Inventing The Ship* (Chicago, 1935).

Thomas Hodgkin, *Theodoric the Ostrogoth* (1891).

R. J. E. C. Lefebure des Noëttes, *L'Attelage* (Paris, 1931).

R. J. E. C. Lefebure des Noëttes, *De La Marine Antique à la Marine Moderne* (Paris, 1935).

De Lacy O'Leary, *How Greek Science Passed to the Arabs* (1949).

A. R. Lewis, *Naval Power and Trade in the Mediterranean*, A.D. 500–1100 (Princeton, 1951).

J. Needham, *Science and Civilisation in China* (Cambridge, 1954).

O. Neugebauer, *The Exact Sciences in Antiquity* (Copenhagen, 1951).

Steven Runciman, *Byzantine Civilisation* (1933).

C. Singer, *The Earliest Chemical Industry* (1948).

Lynn Thorndike, *History of Magic and Experimental Science* (8 vols., Columbia, 1951–8).

Lynn White, 'Technology and Invention during the Middle Ages', *Speculum*, xv (1940), 141–59) and 'Tibet, India, and Malaya as Sources of Western Medieval Technology' *American Historical Review*, lxv (1960), 515–26.

CHAPTER 5

S. J. F. Andreae, 'Embanking and Drainage Authorities in the Netherlands during the Middle Ages', *Speculum*, xxvii (1952), 158–67.

A. R. Bridbury, *England and the Salt Trade in the Later Middle Ages* (Oxford, 1955).

Martin S. Briggs, *The Architect in History* (O.U.P., 1927).

E. M. Carus-Wilson, 'An Industrial Revolution of the Thirteenth Century', *Economic History Review*, xi (1941), 39–60.

G. H. Cook, *The English Cathedral through the Centuries* (1957).

A. C. Crombie, *Robert Grosseteste and the Origins of Experimental Science*, 1100–1700 (1953).

F. H. Crossley, *English Church Craftsmanship* (1947).

J. G. Crowther, *The Social Relations of Science* (1942), 239.

John Harvey, *Early Medieval Architects* (1954).

G. P. Jones, 'Building in Stone in Medieval Europe', *Cambridge Economic History*, ii (1952), 494–518.

N. Pevsner, 'The Term "Architect" in the Middle Ages', *Speculum*, xvii (1942), 549–62.

S. Runciman, *A History of the Crusades*, iii (1954), 470.

Caroline Shillaber, 'Edward I, Builder of Towns', *Speculum*, xxii (1947), 298–309.

Lynn Thorndike, 'The Invention of the Mechanical Clock about A.D. 1271', *Speculum*, xvi (1941).

Lynn Thorndike, *A History of Magic and Experimental Science*, ii (Columbia, 1947).

T. F. Tout, *Medieval Town Planning* (Manchester, 1934).

T. F. Tout, *Studies in English Administrative History*, iv (1928), 473 ff.

CHAPTER 6

Ivor D. Hart, *The Mechanical Inventions of Leonardo da Vinci* (1925).

R. S. Kirby and P. G. Laurson, *Early Years of Modern Civil Engineering* (Yale, 1932).

F. C. Lane, *Venetian Ships and Shipbuilders of the Renaissance* (Baltimore, 1934).

E. McCurdy, *The Mind of Leonardo da Vinci* (1932).
The Notebooks of Leonardo da Vinci (1938).

Alfred Hooper, *Makers of Mathematics* (1948).

George Sarton, *Six Wings: Men of Science in the Renaissance* (1957).

CHAPTER 7

Georgius Agricola, *De Re Metallica*, translated H. C. and L. H. Hoover (1912).

W. Cunningham, *Alien Immigrants in England*.

M. B. Donald, *Elizabethan Copper* (1955).

P. Heyleyn, *Cosmographie* (1652).

Edward Taube, 'German Craftsmen in England during the Tudor Period', *Economic History*, iv (1939), 167–78.

J. U. Nef, 'Mining and Metallurgy in Medieval Civilisation', *Cambridge Economic History*, ii (1952), 430–93.

George Sarton, *Six Wings: Men of Science in the Renaissance* (1957).

CHAPTER 8

W. S. Abell, *The Shipwright's Trade* (Cambridge, 1948).

F. H. Anderson, *The Philosophy of Francis Bacon* (Chicago, 1948).

R. C. Cochrane, 'Francis Bacon and the Rise of the Mechanical Arts in Eighteenth-Century England', *A. of S.*, 12 (1957), 137–56.

H. C. Darby, *The Drainage of the Fens* (Cambridge, 1940).

M. B. Donald, 'Burchard Kranich' (*c.* 1515–1578), *A. of S.*, 6 (1950), 308–22.

F. N. Fisher, 'Sir Cornelius Vermuyden and the Dovegang Lead Mine', *J. of the Derbyshire Archaeological and Nat. Hist. Soc.*, lxxii (1952), 72–118.

F. W. Gibbs, 'The Furnaces and Thermometers of Cornelius Drebbel', *A. of S.*, 6 (1950), 32–43.

J. W. Gough, *The Superlative Prodigall* (Bristol, 1932), 115.

S. B. Hamilton, 'The French Civil Engineers of the 18th Century', *T.N.S.*, xii (1946), 149–59.

L. E. Harris, *Sir Cornelius Vermuyden* (1953).

T. R. Harris, 'Engineering in Cornwall before 1775', *T.N.S.*, xxv (1950), 111–22.

J. U. Nef, 'The Progress of Technology and the Growth of Large-scale Industry in Great Britain, 1540–1640', *Econ, H.R.*, v (1934), 3–24.

W. G. Perrin (ed.), *The Autobiography of Phineas Pett* (1918).

H. W. Robinson, 'Denis Papin, 1647–1712', *Notes and Rec. Roy, Soc.*, v (1948), 47–50.

W. W. Rostow, *British Economy in the Nineteenth Century* (Oxford, 1948), p. 17.

A. W. Skempton, 'The Engineers of the English River Navigations', *T.N.S.*, xxix (1958), 25–53.

E. G. R. Taylor, *Mathematical Practitioners of Tudor and Stuart Times* (Cambridge, 1954).

T. S. Willan, *River Navigation in England*, 1600–1750 (Oxford, 1936).

E. Zilzel, 'The Genesis of the Concept of Scientific Progress', *J.H.I.*, vi (1945), 325–49.

CHAPTER 9

W. H. G. Armytage, 'Thomas Paine and the Walkers', *Pennsylvania History*, xviii (1951), 16–30.

H. W. Dickinson, *Robert Fulton: Engineer and Artist* (1913).

H. W. Dickinson, *A Short History of the Steam Engine* (Cambridge, 1939).

R. S. Fulton and A. P. Wadsworth, *The Strutts and the Arkwrights* (Manchester, 1958).

G. C. Hopkinson, 'The Development of the South Yorkshire and North Derbyshire Coalfield', *T. of the Hunter Arch. Soc.*, vii (1957), 295–319.

Rhys Jenkins, 'Savery, Newcomen and the Early History of the Steam Engine', *T.N.S.*, iv (123–4), 113–30.

A. Raistrick, 'The Steam Engine on Tyneside', *T.N.S.*, xvii (1938), 131–63.

John Rowe, *Cornwall in the Age of the Industrial Revolution* (Liverpool, 1953), pp. 126 ff.

A Catalogue of the Civil and Mechanical Engineering Designs 1741–1792 of John Smeaton, F.R.S., in The Royal Society (1950).

R. E. Schofield, 'Membership of the Lunar Society of Birmingham', *A. of S.*, xii (1956), 119–36.

R. Stuart, *History and Descriptive Anecdotes of Steam Engines* (1829).

R. H. Thurston, *A History of the Growth of the Steam Engine* (1878).

O. A. Westworth, 'The Albion Steam Flour Mill', *Economic History*, ii (1932), 380–95.

CHAPTER 10

W. H. G. Armytage, 'Sir Godfrey Copley and his Friends', *N. and R.R.S.*, x (1954), 22–37.

G. R. de Beer, 'The Relations between Fellows of the Royal Society and Men of Science when France and Britain were at War', *N. and R.R.S.*, x (1952), 244–99.

Sir R. Blomfield, *Vauban* (1938).

H. Brown, *Scientific Organisations in Seventeenth-Century France, 1620–1680* (Baltimore, 1934).

C. C. Gillispie, 'The Natural History of Industry', *Isis*, xlviii (1948), 398–407.

D. Hudson and K. W. Luckhurst, *The Royal Society of Arts, 1754–1954* (1954).

Shelby T. McCloy, *French Inventions of the Eighteenth Century* (Kentucky, 1952).

A. E. Richardson, *Robert Mylne, Architect and Engineer* (1955).

L. T. C. Rolt, *Thomas Telford* (1958).

R. E. Schofield, 'Josiah Wedgwood and a proposed Eighteenth-Century Industrial Research Organisation', *Isis*, xlvii (1956), 16–19.

Samuel Smiles, 'The Grand Canal of Languedoc and its constructor Pierre Paul Riquet de Bonnepas', in *Lives of the Engineers Myddleton, Brindley and etc.* (1904), pp. 361–72.

'The Society of Civil Engineers', *T.N.S.*, xvii (1938), 51–71.

E. C. Wright, 'The Early Smeatonians', *T.N.S.*, xvii (1937), 57, xviii (1938), 101–10.

CHAPTER 11

John P. Addis, *The Crawshay Dynasty, 1765–1867* (Cardiff, 1957).

G. C. Allen, *The Industrial Development of Birmingham and the Black Country, 1860–1927* (1929).

E. V. Armstrong and H. S. Lukens, 'Jean Antoine Chaptal, Comte de Chanteloupe', *Journal of Chemical Education*, xiii (1936), 257–62.

T. S. Ashton, *The Industrial Revolution* (H.U.L.).

W. H. B. Court, *The Rise of Midland Industries* (Oxford, 1938).

H. W. Dickinson, 'Joseph Bramah and his Inventions', *T.N.S.*, xxvi, (1946), 169–86.

C. C. Gillispie, 'The Discovery of the Leblanc Process', *Isis*, xlviii (1957), 152–70.

P. Gooding and P. E. Halstead, 'The Early History of Cement in England', *Proceedings of the Third International Symposium on the Chemistry of Cement* (1954), pp. 1–26.

G. S. Graham, 'The Transition from Paddle Wheel to Screw Propeller', *Mariners' Mirror*, xliv (1958), 35–48.

H. Heaton, 'Benjamin Gott and the Industrial Revolution in Yorkshire', *Econ.Hist. Rev.*, iii (1931), 61.

Historical Tinned Foods (1939).

S. Lilley, 'Nicholson's Journal' (1797–1813), *Ann. Sci.*, vi (1950), 78–101.

D. McKie, *Antoine Laurent Lavoisier* (1952).

J. Mirsky and Allan Nevins, *The World of Eli Whitney* (New York, 1952).

M.P.I.C.E., xxvii (1868), 181–3.

L. T. C. Rolt, *Thomas Telford* (1958).

'The Society of Civil Engineers', *T.N.S.*, xvii (1938), 51–71.

'William Smith, Civil Engineer, Geologist, 1769–1839', *T.N.S.*, xxiii (1948), 93–8.

R. H. Spiro, 'John Loudon McAdam and the Metropolitan Turnpike Trust', *J.T.H.*, ii (1955), 207–13.

R. Taton, 'The French Revolution and the Progress of Science', *Centaurus*, iii, 73–9.

CHAPTER 12

Asa Briggs, *The Age of Improvement* (1959).

James Dugan, *The Great Iron Ship* (1953).

W. O. Henderson, *Britain and Industrial Europe, 1750–1870* (Liverpool, 1954).

B. C. Hunt, *The Development of the Business Corporation in England, 1800–1867* (Harvard, 1936).

J. B. Jefferys, *The Story of the Engineers, 1800–1945* (1945).

C. L. V. Meeks, *The Railroad Station: An Architectural History* (New Haven, 1956).

S. Miall, *A History of the British Chemical Industry* (1931).

J. F. Petrie, 'Maudslay Sons and Field as General Engineers', *T.N.S.*, xv (1934), 39–61.

L. T. C. Rolt, *Isambard Kingdom Brunel* (1957).

CHAPTER 13

H. W. Dickinson, 'Richard Roberts, his Life and Inventions', *T.N.S.*, xxv (1950), 123–37.

J. E. Hodgson (ed.), *Notebook of Sir George Cayley* (1933).

Minutes of the Institution of Civil Engineers, xxviii, 1868, 573 ff.

Thomas Kelly, *George Birkbeck: Pioneer of Adult Education* (Liverpool, 1957).

A. E. Musson, 'James Nasmyth and the Early Growth of Mechanical Engineering', *Econ. Hist. Rev.*, x (1957), 121–7.

J. L. Pritchard, 'Sir George Cayley (1773–1857): A Pioneer of Science and Engineering', *J. of the Roy. Soc. Arts*, cvi (1958), 197–210.

J. W. Roe, *English and American Toolbuilders* (New Haven, 1916).

D. Thompson, 'Queenwood College, Hampshire', *A. of S.*, xi (1956), 246–54.

Mabel Tylecote, *The Mechanics' Institutes of Lancashire and Yorkshire before 1851* (Manchester, 1957).

CHAPTER 14

R. F. Colvile, 'The Navy and the Crimean War', *J.R.U.S.I.*, lxxxv (1940), 73–8.

'The Baltic as a Theatre of War', *J.R.U.S.I.* lxxxvi (1941), 72–80.

S. Pollard, 'Laissez Faire and Shipbuilding', *Economic History Review*, v (1952), 98–115.

M. Robbins, 'The Balaclava Railroad', *J.T.H.*, i (1953), 28–43, and ii (1955), 51–2.

W. H. Chaloner, *The Social and Economic Development of Crewe, 1780–1923* (Manchester, 1950).

The Perkin Centenary Volume: 100 Years of Synthetic Dyestuffs (Pergammon Press, 1958).

CHAPTER 15

Sir Percival Griffiths, *The British Impact on India* (1952).

D. G. Harris, *Irrigation in India* (1923).

Henry C. Hart, *New India's Rivers* (Orient Longmans, Bombay, 1956).

L. S. S. O'Malley (ed.), *Modern India and the West* (Oxford, 1941).

H. G. Rawlinson, *The British Achievement in India* (1948).

Brigadier-General H. A. Young, *The East India Company's Arsenals and Manufactories* (Oxford, 1937).

CHAPTER 16

H. Barger, *The Transportation Industries* (1889–1946), (New York, 1951).

Ralph S. Bates, *Scientific Societies in the United States* (New York, 1945).

I. B. Cohen, 'Science and the Civil War', *Technology Review*, xlviii (1946), 167–70, 192–3.

J. C. van Dyke (ed.), *A. Carnegie: Autobiography* (1933).

Charlotte Erickson. *American Industry and the European Immigrant, 1860–1885* (Harvard, 1957).

H. U. Faulkner, *American Economic History* (1949).

A. P. van Gelder and H. Schlatter, *History of the Explosives Industry in America* (New York, 1927).

F. S. Haydon, *Aeronautics in the Union and Confederate Armies* (Baltimore, 1941).

Louis C. Hunter, *Steamboats on the Western Rivers: An Economic and Technological History* (Cambridge, 1949).

L. H. Jenks, 'Railroads as an Economic Force in American Development', *J. of Econ. History* iv (1944).

Walter Millis, *Armies and Men. A Study in American Military History* (1958).

Nathan Reingold, 'Science in the Civil War', *Isis*, xlix (1958), 307–318.

Earle D. Ross, *Democracy's College: The Land Grant Movement in the Formation Stage* (Ames, Ohio, 1942).

B. Thomas, *Migration and Economic Growth: A Study of Great Britain and the Atlantic Economy* (1954).

Thomas Weber, *The American Railroads in the Civil War, 1861–1865* (New York, 1952).

E. V. Wills, *The Growth of American Higher Education* (Philadelphia, 1936).

E. Younger (ed.), *Inside the Confederate Government: The Diary of Robert Garlick Hill Kean* (New York, 1957), p. 13.

A. J. Youngson Brown, *The American Economy, 1860-1940* (1951), p. 76.

CHAPTER 17

W. F. Bruck, *Social and Economic History of Germany* (Cardiff, 1938).

J. H. Clapham, *The Economic Development of France and Germany* (Cambridge, 1936).

G. Haines, 'German Influence on English Education', *Victorian Studies*, i (Bloomington, 1958), 213–44.

W. O. Henderson, 'Peter Beuth and the Rise of Russian Industry, 1810–1845', *Econ. Hist. Rev.*, viii (1956), 222–3.

Bernard Menne, *Krupp* (1937).

Richard Sasuly, *I. G. Farben* (New York, 1947).

Special Reports on Educational Subjects, ix (H.M.S.O., 1902).

Thorstein Veblen, *Imperial Germany and the Industrial Revolution* (New York, 1915).

James Kendall, F.R.S., *Michael Faraday* (1955).
A. and N. Clow, *The Chemical Revolution* (1952).

CHAPTER 18

Rollo Appleyard, *Charles Parsons, His Life and Work* (1933), and *The History of the Institution of Electrical Engineers* (1939).
F. G. C. Baldwin, *The History of the Telephone in the United Kingdom* (1925).
A. A. Bright, *The Electric Lamp Industry* (1949).
John Dummilow, *1899–1949* (Manchester, 1949).
J. H. Dunning, *American Investment in British Manufacturing Industry* (1958).
R. H. Parsons, *The Development of the Parsons Steam Turbine* (1936).
W. L. Randell, *S. Z. de Ferranti and his Influence upon Electrical Development* (1946).
J. Scott, *Siemens Brothers, 1858–1958* (1958).
G. Siemens, *History of the House of Siemens* (trans. A. F. Rodger) (Munich, 1957).
W. C. Stevens, *The Story of the E.T.U.* (Manchester, 1952).
Proc. Roc. Soc., xv, (J. G. Appold).
      lxxxviii A 1912–13 (Osborne Reynolds).
      lxxxiv A 1911 (Sir Benjamin Baker).

CHAPTER 19

H. Barron, *Modern Synthetic Rubbers* (1949).
Department of Explosives Supply, *Second Report on Costs and Efficiencies for H.M. Factories controlled by Factories Branch* (H.M.S.O., 1919).
L. F. Haber, *The Chemical Industry in the Nineteenth Century* (Oxford, 1957).
D C Hague, *The Economics of Man-made Fibres* (1957).
E. A. Hauser, *Latex* (New York, 1930).
A. J. Ihde, 'Chemical Industry, 1780–1900', *Journal of World History* iv (Paris, 1958), 957–84.
D. B. Keyes, 'History and Philosophy of Chemical Engineering Education', *Chemical Engineering Progress* xlix (1935), 635.
W. Lewis, *The Light Metals Industry* (1949).
W. J. S. Naunton, *Synthetic Rubber* (1937).
D. M. Newitt, 'The Origins of Chemical Engineering', in H. W. Cremer (ed.) *Chemical Engineering Practice*, i (1956), 1–52.

Appendix A: Bibliographies

W. J. Roff, *Fibres, Plastics and Rubber* (1956).

N. Swindin, *Transactions of the Institution of Chemical Engineers*, xxxi (1953), 187.

Transactions of the Institution of Chemical Engineers, i (1923), vii–x.

William Woodruff, *The Rise of the British Rubber Industry during the Nineteenth Century* (Liverpool, 1958).

A. and N. Clow, *The Chemical Revolution* (1952).

J. Vargas Eyre, *Henry Edward Armstrong* (1958).

Chapter 20

T. C. Carpenter and H. Diederichs, *The Development of the Internal Combustion Engine to 1860* (1908).

C. F. Caunter, *The History and Development of Light Cars* (H.M.S.O. 1957).

E. A. Cowper, *Description of Cugnot's Locomotive* (1853).

W. Fletcher, *Steam on Common Roads* (1891).

A. Gordon, *Elemental Locomotion by means of Steam Carriages on Common Roads* (1832).

W. Hancock, *Narrative of Twelve Years of Experiments, 1824–36 (with Steam Carriages)* (1839).

John Head, *Rise and Progress of Steam Locomotion on Roads* (1873).

R. W. Kidner, *The First Hundred Road Motors* (1958).

G. Maxcy and A. Gilberston, *The Motor Industry* (1959).

J. B. Rae, *American Automobile Manufacturers* (New York, 1959).

L. T. C. Rolt, *Horseless Carriage: The Motor Car in England* (1956).

C. W. Scott-Giles, *The Road Goes On* (1946).

G. I. Savage, *The Economic History of Transport* (1959).

The Times, 16 February 1940, p. 11.

S. and B. Webb, *The Story of the King's Highway* (1926).

W. Worby Beaumont, *Motor Vehicles and Motors* (1900).

C. H. Gibbs-Smith, *A History of Flying* (Batsford, 1953).

Chapter 21

Aims and Objects, Sixty Years of Progress: The Institution of Public Health Engineers (1955).

Cyril Bibby, *T. H. Huxley: Scientist, Humanist and Educator* (1959).

J. Vargas Eyre, *H. E. Armstrong: Pioneer of Technical Education* (1958).

S. E. Finer, *The Life and Times of Sir Edwin Chadwick* (1952).

Journal of the Institute of Water Engineers, x (1956), 509–14.

Sir Eric Ashby, *Technology and the Academics* (1958).

D. L. Burn, *Economic History of Steelmaking* (Cambridge, 1940).

D. S. L. Cardwell, *The Organisation of Science in England* (1957).

G. C. Allen, and A. G. Donnithorne, *Western Enterprise in Far Eastern Economic Development: China and Japan* (1954).

CHAPTER 22

P. W. S. Andrews and E. Brunner, *Life of Lord Nuffield* (Oxford, 1955).

British Association for the Advancement of Science.
> *Britain in Depression* (1935).
> *Britain in Recovery* (1938).

C. F. Carter and B. R. Williams, *Industry and Technical Progress* (O.U.P., 1957).

Colonel R. E. Crompton, *Reminiscences* (1925).

Rees Jeffreys, *The King's Highway* (1949).

H. S. Hele-Shaw, *Obituary Notices of Fellows of the Royal Society*, iii, 791 ff.

W. Hoffman, 'The Growth of Industrial Production in Britain: a Quantitative Survey', *Economic History Review*, ii (1949), 165 ff.

A. E. Kahn, *Great Britain in the World Economy* (1946).

C. L. Mowat, *Britain Between the Wars, 1918–1940* (1956).

P.E.P. Engineering Reports, *Motor Vehicles* (1950).

A. C. Pigou, *Aspects of Economic British History. 1918–1925* (1948).

A. Plummer, *New British Industries in the Twentieth Century* (1937).

R. S. Sayers, 'Springs of Technical Progress in Britain, 1919–1939', *Economic Journal*, lx (1950), 275–90.

Charles Wilson, *History of Unilever* (1954).

A. J. Youngson, *The British Economy 1920–1957* (1960).

CHAPTER 23

O. E. Anderson, *Refrigeration in America: A History of a new Technology and its Impact* (Princeton, 1957).

C. A. and Mary Beard, *The Rise of American Civilisation* (1927).

E. F. L. Brech, *Organisation: The Framework of Management* (1957).

Wallace Clark, *The Gantt Chart: A Working Tool of Management* (New York, 1922).

David L. Cohn, *Combustion on Wheels* (Boston, 1944).

F. B. Copley, *F. W. Taylor, Father of Scientific Management* (1923).

H. B. Drury, *Scientific Management: A History and Criticism* (Columbia, 1922).

Appendix A: Bibliographies

G. Everson, *The Story of Television*: *The Life of Philo T. Farnsworth* (New York, 1948).

H. U. Faulkner, *American Political and Social History* (6th edn., 1952)

J. W. Frey and H. C. Ide, *A History of Petroleum Administration for War* (Washington, 1946).

John Gunther, *Inside U.S.A.* (1947) contains a good bibliography.

L. A. Hawkins, *Adventure into the Unknown. The First Fifty Years of the General Electric Research Laboratory* (New York, 1950).

Jerome C. Hunsaker, *Aeronautics at Mid-Century* (New York, 1952).

Gerselte Mack, *The Land Divided: A History of the Panama Canal and other Isthmian Canal Projects* (New York, 1944).

A. Nevins and F. E. Hall, *Ford, the Times, the Man, the Company* (New York, 1953).

B. W. Niebel, *Motion and Time Study* (Irwin, Hornewood, Illinois, 1955).

Charles A. Siebman, *Radio, Television and Society* (New York, 1950).

W. R. Spriegel and Clark E. Myers, *The Writings of the Gilbreths* (R. D. Irwin, Hornewood, Illinois, 1953).

Willson Whitman, *David Lilienthal* (1948).

CHAPTER 24

A. Baykov, *The Development of the Russian Economic System* (Cambridge, 1946).

Abram Bergson (ed.), *Soviet Economic Growth* (New York, 1953).

C. A. G. Bridge (ed.), 'The Russian Fleet under Peter the Great', *Navy Records Society*, xv (1899).

Maurice Dobb, *Soviet Economic Development since 1917* (1948).

'The Ural Metal Industry in the Eighteenth Century', *Econ. H.R.*, iv (1951), 252–5.

A. Gershenkron, 'The Rate of Industrial Growth in Russia since 1885', *Journal of Econ. Hist.*, Supp. vii (1947), No. 2, 1952.

P. Grierson, *Books on Soviet Russia, 1917–1942* (1943).

S. B. Hamilton, 'Captain John Perry, 1670–1732', *T.N.S.*, xxvii (1951), 241–53.

Walther Kirchner, 'Samuel Bentham and Siberia', *Slavonic and East European Review*, xxxvi (1958), 471–80.

A. G. Korol, *Soviet Education for Science and Technology* (1957).

L. R. Lewitter, 'Peter the Great, Poland, and the Westernisation of Russia', *J.H.I.*, xix (1958), 493–506.

M. Mitchell, *The Maritime History of Russia, 848–1948* (1949).

Appendix A: Bibliographies

George A. Modelski, *Atomic Energy in the Communist Bloc* (Melbourne, 1959), pp. 81–195.

Lowell R. Tillet, 'Soviet Historians and the World's First Airplane', *South Atlantic Quarterly*, lviii (1959), 409–20.

Demetri Shimkin, 'Scientific Personnel in the U.S.S.R.', *Science*, cxvi (1952), 512–13.

Alexander Vucinich, *The Soviet Academy of Sciences* (Stanford, 1956).

W. W. Rostow, *The Dynamics of Soviet Society* (1953).

P. P. Zabarinsky, 'The Earliest Steam Engine in Russia', *T.N.S.*, xvi (1936), 57–67.

CHAPTER 25

J. G. Crowther and R. Whiddington, *Science at War* (1948).

W. Dornberger, V_2 (translated by James Cleugh and Geoffrey Halliday) (1954).

R. S. Edwards, *Co-operative Industrial Research* (1950).

R. H. Goddard, *Rockets* (New York, 1946).

William Hornby, *Factories and Plant* (H.M.S.O., 1958).

K. A. H. Murray, *Agriculture* (H.M.S.O., 1955).

G. Edward Pendray, *The Coming Age of Rocket Power* (1947).

P.E.P. Engineering Reports (i) *Agricultural Machinery* (1949), (ii) *Motor Vehicles* (1950).

M. M. Postan, *British War Production* (H.M.S.O., 1952).

J. D. Scott and R. Hughes, *The Administration of War Production* (H.M.S.O., 1955).

CHAPTER 26

L. Landon Goodman, *Man and Automation* (Penguin Books, 1957).

P.E.P. Broadsheet 380 (1955), *Towards the Atomic Factory*.

D. C. Marsh, *The Changing Social Structure of England and Wales* (1958).

R. Holroyd, 'The Development of the Petroleum Chemical Industry in Britain', *Chemistry and Industry*, No. 29 (1958), 900–9.

Institute of Production Engineers, 'The Automatic Factory—What does it Mean?' Report of the Conference held at Margate 16–19 June 1955.

G. Walker, *Economic Planning by Programme and Control in Great Britain* (1957).

A. Schonfield, *British Economic Policy since the War* (Penguin Books, 1958).

J. Jewkes, D. Sawers and R. Stillerman, *The Sources of Invention* (1958), p. 127

Appendix A: Bibliographies

CHAPTER 27

F. R. Allen, H. Hart and Others, *Technology and Social Change* (New York, 1957).

R. N. Anshen, *Science and Man* (New York, 1942).

Rollo Appleyard, *Charles Parsons, His Life and Work* (1933).

M. Margaret Ball. *N.A.T.O. and the European Union Movement* (1959).

J. Blair, 'Technology and Size', *American Econ. Review*, xxx, ciii (1948), 121.

F. Brown (ed.), *World Power Conference: Statistical Year Book* (1957).

Jose de Castro, *Geography of Hunger* (1957).

Norman Davy, *Studies in Tidal Power* (1922).

H. Finer, *The T.V.A.: Lessons for International Application* (1944).

Denis Gabor, 'Inventing the Future,' *Encounter* xiv (1960), 12.

J. K. Galbraith, *The Affluent Society* (1958).

Lord Dudley Gordon, *Engineering*, clxxvi (1953), 381.

Sir Robert Hadfield, *Metallurgy and its Influence on Modern Progress* (1925).

Ervin Hesner, *The International Steel Cartel* (North Carolina, 1943).

H. Heywood, 'Solar Energy, Past, Present and Future Application', *Engineering*, *Section D, British Association*, clxxvi (1953), 377, 409.

W. Hoffman, *British Industry, 1700–1950*, trans. W. D. Henderson and W. H. Chaloner (Oxford, 1955).

Seymour Lipset and Richard Bendix, *Social Mobility in Industrial Society* (California, 1959).

Margaret Mead, *Cultural Patterns and Technical Change* (Mentor Books, 1957).

S. Milman, *Dynamic Factors in Industrial Productivity* (1956).

N. J. G. Pounds and W. N. Parker, *Coal and Steel in Western Europe* (1957).

P. C. Putnam, *Energy in the Future* (1954).

B. D. Richards, 'Tidal Power: Its Development and Utilisation', *Journal of the Institution of Civil Engineers*, xxx (1948), 104–49.

W. W. Rostow, *The Process of Economic Growth* (1943).

R. Steel (ed.), *Biochemical Engineering* (1958).

E. C. Wassink, 'Problems in the Mass Cultivation of Photo-autotrophic Micro-organisms' in *Autotrophic Micro-organisms: Fourth Symposium of the Society for General Microbiology held at the Institution of Electrical Engineers, London, April 1954* (Cambridge, 1954), pp. 247–70.

349

E. S. and W. S. Woytinsky, *World Population and Production* (New York, 1957).

A. R. Ubbelohde, 'Towards Tektopia', *The Twentieth Century*, September 1955, pp. 226–35.

General Bibliography

An analytical bibliography of the history of engineering and applied science has been appearing in the *Transactions of the Newcomen Society for the Study of the History of Science and Technology*. From vol. v, this includes lists of monographs and theses. A similar bibliography, covering wider aspects of science, including medicine, appears in *Isis*.

Three other general bibliographies are:

A. G. S. Josephson, *A List of Books on the History of Industry and the Industrial Arts* (Chicago, 1915–16).

A. D. Roberts, *A Guide to Technical Literature* (1939).

Bernard Barber, *Sociology of Science: A Trend Report and a Bibliography* (U.N.E.S.C.O., Paris, 1956).

General works which will be found useful are:

J. Beckman, *A History of Inventions, Discoveries and Origins*, trans. by W. Johnston, 4th edition enlarged by W. Francis and A. W. Griffith (1846).

S. Smiles, *Lives of the Engineers* (1874), 5 volumes.
    Men of Invention and Industry (1884).
    James Nasmyth's Autobiography (1883).

A. P. M. Fleming and H. J. Brocklehurst, *A History of Engineering* (A. and C. Black, 1925).

George Sarton, *Introduction to the History of Science* (Baltimore, 1927–47), supplemented, if possible, by the bibliographies in *Isis*, the journal he founded.

H. P. Vowles, *The Quest for Power* (1932).

A. Wolf, *A History of Science, Technology and Philosophy in the 16th and 17th Centuries* (London and New York, 1935).

Lewis Mumford, *Technics and Civilisation* (1934).

Appendix A: Bibliographies

E. C. Smith, *A Short History of Marine Engineering* (1937).

E. Cressy, *A Hundred Years of Mechanical Engineering* (1937).

C. Matchoss, *Great Engineers* (trans. by H. S. Hatfield) (G. Bell, 1939).

A. Wolf, *History of Science, Technology and Philosophy in the 18th Century* (1939).

Lynn Thorndike, *A History of Magic and Experimental Science during the first Sixteen Centuries of our Era* (New York, 1923–41) (6 vols.).

J. G. Crowther, *The Social Relations of Science* (Macmillan, 1942).

Since the war a large number of studies in the history of engineering have been published, together with a number of histories of particular aspects of engineering history. Amongst the general works, the readability or reliability of the following are commended:

H. F. Heath and A. L. Hetherington, *Industrial Research and Development in the United Kingdom* (Faber, 1946).

F. D. Klingender, *Art and the Industrial Revolution* (Noel Carrington, 1947).

Siegfried Giedion, *Mechanisation takes Command. A Contribution to Anonymous History* (New York, 1948).

S. Lilley, *Men, Machines and History* (Cobbett Press, 1948).

H. S. Hatfield, *The Inventor and his World* (Penguin Books, 1948).

J. Gloag and D. Bridgwater, *A History of Cast Iron in Architecture* (George Allen and Unwin, 1948).

J. U. Nef, *War and Human Progress. An Essay on the Rise of Industrial Civilisation* (Harvard, 1950).

R. J. Forbes, *Man the Maker* (Constable, 1950).

J. K. Finch, *Engineering and Western Civilisation* (McGraw-Hill Book Company, 1951).

Hans Staub, *A History of Civil Engineering* (L. Hill, 1952).

A. and N. Clow, *The Chemical Revolution* (1952).

S. P. Timoshenko, *History of Strength of Materials* (McGraw-Hill, 1953).

A. P. Usher, *A History of Mechanical Invention* (2nd edn., Harvard U.P., 1954).

J. D. Bernal, *Science in History* (Watts, 1954).

Mitchell Wilson, *American Science and Invention* (Simon and Schuster, New York, 1954).

R. S. Kirby, S. Withington, A. B. Darling and F. G. Kilgour, *Engineering in History* (McGraw-Hill Book Company, 1956).

John W. Oliver, *History of American Technology* (Ronald Press, New York, 1956).

J. Needham, *Science and Civilisation in China* (Cambridge, vol. i, 1954; vol. ii, 1956).

C. Singer, E. J. Holmyard and A. R. Hall, *A History of Technology* (Oxford, vol. i, 1954; vol. ii, 1956; vol. iii, 1957; vols. iv and v, 1958).

J. G. Crowther, *Discoveries and Inventions of the 20th Century* (4th edn., Routledge, 1955).

P. F. R. Venables, *Technical Education* (Bell, 1955).

A. Hunter Dupree, *Science in the Federal Government* (Harvard, 1957).

J. Jewkes, D. Sawers and Richard Stillerman, *The Sources of Invention* (1958).

For those who would like to supplement their reading by dipping into foreign works, the following are recommended:

Theodor Beck, *Beiträge zur Geschichte des Maschinenbaues* (Springer, Berlin, 1900).

Ludwig Darmstaedter, *Handbuch zur Geschichte der Naturwissenschaften und der Technik* (Berlin, 1906). (An invaluable work of reference.)

René Taton, *Histoire Générale des Sciences* (Paris, 1957).

Arturo Ucelli, *Enciclopedia Storica Delle Scienze E Delle Loro Applicazioni* (Ulrico Hoepli, Milan, 1941, 42, 43) is superbly illustrated and easy to read, even for those with little knowledge of Italian.

There are several Russian works. One, edited by I. M. Rabinovitch, deals with Building Mechanics, Transport and Mining, and is published by Jostekhizdat, Moscow, in 1957. Another, edited by L. D. Belkind, deals with the history of power engineering, and two volumes of this were published by Josenrjoizat, Moscow, in 1957. At the time of writing, neither of these have been translated.

The Science Museum has published an excellent series of booklets, of which the following can be heartily recommended:

C. F. Caunter, *The History and Development of Cycles* (1955).

C. St. C. B. Davidson, *History of Steam Road Vehicles* (1953).

M. J. B. Davy, *Interpretive History of Flight* (1948).

H. T. Pledge, *Science since 1500* (1946).

G. R. M. Garratt, *One Hundred Years of Submarine Cables* (1950).

Appendix A: Bibliographies

For the reader who would like a classified list of historical events, that compiled by G. F. Westcott (H.M.S.O., 1955) is invaluable. It covers the whole field of mechanical engineering including energy conversion; transmission and storage; atomic energy; pumping, blowing and compressing machinery; explosives and ordnance.

Biographical information can be found in the obituary notices in the *Proceedings*, *Minutes* or *Transactions* of most of the professional societies mentioned in the text. Those in the *Proceedings* (1854–1901; 1904–32), *Obituary Notices* (1934–54) and *Biographical Memoirs* of the Royal Society, the *Minutes of Proceedings* (1837 to date) and *Journal* of the Institute of Civil Engineers (1935 to date) will be found especially useful and informative. It is for this reason that dates of birth and death have, in many cases, been attached to the significant figures in the body of the text.

Local studies abound. To supplement the many monographs on firms, trades, districts and towns the reader is recommended to examine the studies published under the auspices of the British Association for the Advancement of Science to mark its annual meetings. Since the war, these generally take the form of separate volumes, but before the war, can often be found in its Proceedings.

APPENDIX B

List of Professional Institutions

Professional organizations are independent institutions devoted to furthering relevant knowledge and skills, securing high standards of competence and conduct in their members. They set appropriate entry standards for those wishing to obtain membership and help to secure professional status and remuneration. Some have attained such eminence and influence in the community that they are granted a Royal Charter.

Now a particular institution may not devote its main efforts to all of these aims, and may scarcely undertake one or other of them. Whether it does or no depends on its history and on its relationship to other organizations. Thus the Chemical Society undertakes to further knowledge and skills by publishing research papers, whereas the Royal Institute of Chemistry publishes more generalized yet equally valuable surveys of particular modern developments. Similarly the Engineers' Guild concentrates on professional status and remuneration, yet the Royal Institute of Chemistry is showing renewed interest and concern with it, partly because of the growing importance of this problem in the nationalized industries.

It is natural that, in maintaining and advancing professional standards, these professional organizations should exert a decisive influence on education for industry and commerce. For not only do they work formally with the technical colleges, but also provide, through their members, a two-way channel for ideas and information on theory and practice in engineering, science and professional preparation for it.

Such professional preparation involves standards of knowledge and skills, and these involve different kinds of membership. These standards are determined by the Councils of the respective professional institutions, usually on the advice of their appropriately constituted Education and Examinations Committees. Such membership is usually of two kinds.

354

Appendix B: List of Professional Institutions

(*a*) *Corporate Membership*, with full voting and other rights and responsibilities. There are normally two main grades, e.g. *Associate Membership*, gained after passing the final examinations of the institution, or of other approved examinations which exempt therefrom, *and* after satisfactorily completing a prescribed period of practical, industrial, commercial or other professional training, and attaining a mature age, e.g., 30 years. Examples are designated by A.M.I.Mech.E., A.M.I E E., and A.R.I.C.

Membership or *Fellowship*, which is a senior grade gained after many years of responsible professional experience, e.g., M.I.Mech.E., M.I.E.E., and F.R.I.C.

(*b*) *Non-Corporate Grade*, i.e. not full membership and without its appropriate privileges and responsibilities. Usually two grades, i.e. *Student Membership* into which newcomers are admitted. They remain in this grade until they have passed the final examinations, which admits them to the second non-corporate grade of *Graduateship*. In some cases this may be designated by letters, e.g. G.I.Mech.E.

Where the tertiary and quaternary levels of education in the country during this century, embodied in the university technological departments and technical institutions of all kinds, provide satisfactory courses, students are allowed to pass, or be exempted from, the final examinations of professional institutions. In the technical colleges, by far the greatest concern has been with devising courses which would be approved for exemptions, wholly or in part, from the examinations of the professional institution. This has been particularly true of the National Certificate and Diploma courses, and a new phase has been entered upon with courses leading to the Diploma in Technology.

Reports, journals, proceedings and transactions of these professional institutions will be found invaluable, and those interested in particular subjects are recommended to consult them. The historical surveys, appreciations, obituaries and centennial tributes they publish are often of a very high quality and the reader could do no better than refer to them.

(*a*) *The Aristocratic Period*

Date of
Foundation

1660 The Royal Society
1754 The Royal Society for the Encouragement of Arts
1771 The Royal Physical Society of Edinburgh

355

1783   The Royal Society of Edinburgh
1799   The Royal Institution of Great Britain
1804   The Royal Horticultural Society
1831   The British Association for the Advancement of Science

(b) The General-Professional Period

Date of
Foundation

1818   The Institution of Civil Engineers
1819   The Royal Microscopical Society
1834   The Institute of Builders
1834   The Royal Institute of British Architects
1838   The Royal Agricultural Society of England
1841   The Chemical Society
1847   The Institution of Mechanical Engineers
1860   The Institution of Naval Architects
1863   The Institution of Gas Engineers
1866   The Royal Aeronautical Society
1869   The Iron and Steel Institute
1871   The Institution of Electrical Engineers
1873   The Institution of Municipal Engineers
1874   The Society for Analytical Chemistry
1874   The Physical Society
1876   The Royal Society of Health
1877   The Institute of Chemistry
1881   The Society of Chemical Industry
1886   The Institute of Brewing
1889   The Institute of Marine Engineers
1889   The Institution of Mining Engineers
1892   The Institution of Mining and Metallurgy
1896   The Institution of Water Engineers
1897   The Institution of Heating and Ventilating Engineers

(c) The Specialist-Professional Period

Date of
Foundation

1900   The Ceramic Society
1900   The Institute of Refrigeration
1901   The British Standards Institution
1903   The Faraday Society

1904 The Institute of British Foundrymen
1906 The Institution of Automobile Engineers
1908 The Institute of Metals
1908 The Institution of Structural Engineers
1911 The Institution of Locomotive Engineers
1913 The Institute of Petroleum
1916 The Society of British Aircraft Constructors
1917 The Institute of Quarrying
1918 The Institute of Physics
1919 The Institute of Transport
1920 The Institute of the Motor Industry
1921 The Institution of Production Engineers
1921 The Institution of the Rubber Industry
1922 The Institution of Chemical Engineers
1923 The Institute of Welding
1925 The British Institute of Radio Engineers
1925 The Textile Institute.
1927 The Institute of Fuel
1927 The British Boot and Shoe Institution
1930 The Institution of Highway Engineers
1931 The Institute of Housing
1931 The Plastics Institute
1950 The Institute of Biology

INDEX OF SUBJECTS AND PLACES

Index

Index

Moslems, 37, 39, 40, 41, 52, 54, 56, 60, 161
Motor Manufacturing Industry (National Advisory Council), 304
Motor Union, 224
Motor Van and Wagon Users Association, 225
Motorized equipment, 296
Mulberry, 296
Munich, 262
Municipal engineering, 139, 244
Musselburgh, 104
Muswell Hill, 235
Muzzle loaders, 67
Mystery plays, 51

Nagasaki, 301
Naples, 52, 58
Napoleonic Wars, 106–24, 125, 127, 136
National Academy of Sciences, 177; Advisory Council on Education for Industry and Commerce, 317; Aerostatic School, Meudon, 110; Association for Promotion of Technical Education, 240; Association of Colliery Managers, 239; Certificate System, 257–8; Colleges, 303, 317; Institute of Agricultural Engineering, 303; Institute of Industrial Psychology, 315; Physical Laboratory, 135, 226, 295, 311; Research Development Corporation, 309, 310, 312; Resources Committee of United States, 333; Society of Professional Engineers, 328; Telephone Company, 233
Nature, 221, 230, 234
Navigation, 54, 56
Navy Department's Permanent Commission, 177
'Neptune's staircase', 120
Neptunist, 68
Netherlands, 74
New Atlantis, 76, 185
Newcastle, 69, 70, 112, 113, 159, 199, 202, 231, 214
New River, 72
New York, 107, 108, 114, 117, 136
Niagara, 137, 166, 196, 202
Nickel carbonyl, 209
Nitre and Mining Bureau, 172, 173
Nitric acid, 299
Non-ferrous metals, 19, 202–4, 238–9, 246, 252, 274–5
North British Rubber Company, 212
North German Lloyd Shipping Company, 192
North-Western Provinces, 163
Norwich, 75
Nottingham University, 232

Nova Scotia, 134
Nürnberg, 54, 65
Nylon, 305 ff.

Oleomargarine, 210
Okinawa, 300
Ordnance Department, 114, 160
Ostia, 29, 30
Ostrich farm, Pasadena, 332
Oxford, 62, 76

Paddington, 136
Padua, 54
Pantheon, 31
Papacy, The, 52
Paper, 54, 211
Parachutes, 55
Parallelograms, 68
Paris, 48, 55, 98, 110, 120, 123, 127, 190, 230, 239
Parliamentary and Scientific Committee, 328
Parliamentary Committee on Small Arms, The, 160
Partington, 308
Passy, 106, 107
Patricroft, 126
Pearl Harbour, 295
Pedal Velocipedes, 220, 221
Penicillin, 297
Peoples' Colleges, 231
Percy Committee, 316, 317
Perivale, 252
Persians, 21–3, 39
Petrochemical industry, 159 ff.
Petrol engine, 166
Petroleum, 159 ff., 190, 207, 227, 306, 307
Peru, 63, 112
Philosophical Magazine, 123
Philosophical Transactions, 78
Phoenicians, 22
Photography, 210
Photo-synthesis, 332
Physical and Dynamic Meteorology, 331
Physical Society, 186, 206
Physicians, College of, 238
Physicist Engineers, 300
Physics, 57
Physikalische Technische Reichsanstalt, 193
Piston, 57, 62, 189 ff.
Pittsburg, 199, 202
Plastics, 212, 304
'Pluto', 296
Political and Economic Planning, 294
Polymerization, 304
Pompeii, 30
Pontoon bridges, 103
Poplar, 127

364

INDEX OF NAMES

Index